D0576244

The African Food Crisis

Lessons from the Asian Green Revolution

The African Food Crisis

Lessons from the Asian Green Revolution

Edited by

Göran Djurfeldt

Department of Sociology,
University of Lund, Sweden

Hans Holmén

Department of Geography,
Linköping University, Sweden

Magnus Jirström

Department of Social and Economic Geography,
University of Lund, Sweden

and

Rolf Larsson

Department of Sociology,
University of Lund, Sweden

CABI Publishing

CABI Publishing is a division of CAB International

CABI Publishing
CAB International
Wallingford
Oxon OX10 8DE
UK

Tel: +44 (0)1491 832111
Fax: +44 (0)1491 833508
E-mail: cabi@cabi.org
Website: www.cabi-publishing.org

CABI Publishing
875 Massachusetts Avenue
7th Floor
Cambridge, MA 02139
USA

Tel: +1 617 395 4056
Fax: +1 617 354 6875
E-mail: cabi-nao@cabi.org

A catalogue record for this book is available from the British Library, London, UK.

Library of Congress Cataloging-in-Publication Data

The African food crisis : lessons from the Asian Green Revolution / edited by Göran Djurfeldt . . . [et al.]

p. cm.

Includes bibliographical references and index.

ISBN 0-85199-998-0 (alk. paper)

1. Agriculture--Asia. 2. Green revolution--Asia. 3. Food supply--Africa. I. Djurfeldt, Göran, 1945– II. Title.

S470.A1A35 2004
630'.96--dc22

2004016109

ISBN 0 85199 998 0

Typeset by AMA DataSet Ltd, UK.
Printed and bound in the UK by Biddles Ltd, King's Lynn.

Contents

Contributors

Tunji Akande, *Agriculture and Rural Development Department, Nigerian Institute of Social and Economic Research (NISER), PMB 5 UIPO, Ibadan, Nigeria.*

Gasper C. Ashimogo, *Department of Agricultural Economics and Agribusiness, Sokoine University of Agriculture, PO Box 3007, Morogoro, Tanzania.*

Göran Djurfeldt, *Department of Sociology, Lund University, PO Box 114, SE-221 00 Lund, Sweden.*

Steven Haggblade, *International Food Policy Research Institute, PO 32481, Lusaka, Zambia.*

Hans Holmén, *Department of Geography, Linköping University, SE-581 83 Linköping, Sweden.*

Aida C. Isinika, *Institute of Continuing Education, Sokoine University of Agriculture, PO Box 3007, Morogoro, Tanzania.*

Magnus Jirström, *Department of Social and Economic Geography, Lund University, Sölveg. 10, SE-223 62 Lund, Sweden.*

Joseph T. Karugia, *Department of Agricultural Economics, University of Nairobi, PO Box 29053, Nairobi, Kenya.*

Rolf Larsson, *Department of Sociology, Lund University, PO Box 114, SE-221 00 Lund, Sweden.* (deceased).

James E.D. Mlangwa, *Faculty of Veterinary Medicine, Sokoine University of Agriculture, PO Box 3007, Morogoro, Tanzania.*

V. Kwame Nyanteng, *Insitute of Statistical, Social and Economic Research, ISSER, University of Ghana, PO Box LG74 Legon, Ghana.*

Willis Oluoch-Kosura, *Department of Agricultural Economics, University of Nairobi, PO Box 29053, Nairobi, Kenya.*

Keijiro Otsuka, *Foundation for Advanced Studies on International Development (FASID), GRIPS/FASID Joint Graduate Program, 2-2 Wakamatsu-cho, Shinjuku-ku, Tokyo 162-8677, Japan.*

A. Wayo Seini, *Institute of Statistical, Social and Economic Research, ISSER, University of Ghana, PO Box LG74, Legon, Ghana.*

Takashi Yamano, *Foundation for Advanced Studies on International Development (FASID), GRIPS/FASID Joint Graduate Program, 2-2 Wakamatsu-cho, Shinjuku-ku, Tokyo 162-8677, Japan.*

Rolf Larsson in memoriam

One of the editors of this book, Dr. Rolf Larsson met with a fatal accident in Tanzania on October 19, 2004. The book is dedicated to his memory.

Preface

This volume is the result of a 3-year project coordinated by a group of Swedish researchers and with collaborating scholars from all over the world.[1] The project aimed to systematically probe a common comparison, namely between Asian agricultural development during the so-called Green Revolution and the current problematic agricultural situation in sub-Saharan Africa. Coordinated by the Swedish group, scholars from eight African countries made case studies of their own countries. These studies involved two kinds of data. First, macro level data about national agricultural trends were collected in an attempt to trace developments before, during and after Structural Adjustment Programmes (SAPs). These data were drawn from secondary sources mainly. Secondly, based on a common survey format, the participating teams conducted interviews with more than 3000 farm households in more than 100 villages. The resulting country studies fed into a comparative and continental analysis made by the Swedish team, who further made analyses of seven Asian countries, focusing on the early period of the Green Revolution. This leg of the project was based mainly on secondary sources but also included study trips and interviews with key persons.

The project took off with a joint methodology workshop in Lund in May 2001 and concluded with a summarizing workshop in Nairobi in January 2004, during which the draft material for this book was reviewed by a number of internationally well-reputed scholars in this field.

Throughout, the project has benefited from generous inputs from its group of advisors and other scholars, colleagues and institutions. We acknowledge our indebtedness to all of them. First of all our partners in Africa: Mulat Demeke and Teketel Abebe, Addis Ababa University; Willis Oluoch-Kosura and Joseph T. Karugia, Nairobi University; Frank Muhereza and the late Bazaara Nyangabyaki, Centre for Basic Research, Kampala; Gasper C. Ashimogo, Aida C. Isinika and James E.D. Mlangwa, Sokoine Agricultural University, Morogoro; James Milner and John M. Kadzandira, University of Malawi, Zomba; Oliver Saasa and Mukata K.W. Wamulume, Institute of Economic and Social Research, University of Zambia; Patrick Kormawa, formerly with the International Institute of Tropical Agriculture (IITA), and R. Okechukwu with the same Institute, Ibadan; Tunji Akande, Nigerian Institute of Social and Economic Research (NISER), Ibadan; Wayo Seini, and V. Kwame Nyanteng, Institute of Statistical, Social and Economic Research, (ISSER) Accra. The list of enthusiastic field workers

[1] The basic finance for the project was generously extended by the Bank of Sweden Tercentenary Foundation and from Sida (Swedish International Development Cooperation Agency), who financed the participation of African researchers. The views expressed in this book are those of the authors and not necessarily of sponsoring institutions or of the organizations to which they are affiliated.

in all the eight countries is too long to reproduce, but we want to extend our thanks to all who contributed, not least to the survey respondents who did their best to answer all our queries.

A great contribution has been made by our advisors. We owe gratitude to Göran Hydén for encouragement and constructive and creative criticism of the research design and of the draft chapters. Ruth Oniang'o was a great source of inspiration throughout. Similarly, the critical comments of Deborah Bryceson and Kjell Havnevik were crucial. We also owe special thanks to Michael Lipton, whose critical comments at an early stage radically improved the project design. Similarly, Mike Mortimore at an early stage gave a crucial input. Deeply felt thanks to you all!

At Lund we particularly want to thank Mikael Hammarskjöld, who acted as a documentalist and generously contributed with erudition to this project. Similarly Ditte Mårtensson contributed with resourcefulness, competence and humour, and Olle Frödin helped in editing the manuscript. We would also like to thank Anders Danielsson, Franz-Michael Rundquist, Staffan Lindberg, Stig Toft Madsen and Christer Gunnarsson. As reviewers, the latter, together with Karl-Erik Knutsson and Hans-Dieter Evers, contributed with inspiring critiques and thus to the end product. Eidi Genfors and Marija Brdarski at Sida contributed not only with generous time, but also with sharp and incisive comments! Thanks also to colleagues in the Development Studies seminar at the Department of Sociology in Lund.

We furthermore acknowledge the contributions of Essie Blay, Charlotte Wonyango, Lucy Binauli, Yeraswork Admassie, Steven Haggblade, Marco Quinones, Paul Mosley and Astrig Tasgian.

In the Philippines, our programme was competently and generously organized by Mercedita A. Sombilla and Mahabub Hossain. We met and gained from the knowledge and experience of Cristina C. David, Mario Lamberte, D.F. Panganiban, Segfredo R. Serrano, Bruce J. Tolentino, Eliseo R. Ponce, Victoriano B. Guiam, Ramon L. Clarete, Gelia Castillo, Leo Gonzales, Agnes C. Rola, William G. Padolina, Leocadio S. Sebastian, Leah J. Buendia, Santiago R. Obien, Mahar Mangahas, Orlando J. Sacay and Ric Reyes.

In Indonesia, our programme was organized by Dwi Astuti and her colleagues, whom we remember with much warmth; they were introduced to us by Olle Törnquist. We met a number of resourceful and knowledgeable persons, among them Mely G. Tan, Siswono Yudo Husodo, Effendi Pasandaran, M. Ali Iqbal, Rusli Marzuki, Thee Kian Wie, Sediono Tjondronegoro, Mubyarto, Gunawan Wiradi, Hans Antlöv, Mochammad Maksum and Francis Wahono.

In India, we were much inspired by interviews with two of the most prominent names in the Indian agricultural debate, M.S. Swaminathan and G.S. Bhalla. Thanks to Partha N. Mukherji, Rahul Mukherji, Venkatesh Athreya, Lawrence and Pushpa Surencra for friendship and support.

In Bangladesh, finally, Mahabub Hossain introduced us to parts of his vast network. We acknowledge the contributions of Kari M. Badruddoza, Showkat Ali, Hamid Miah, Noel B. Magor, Matia Chowdhury, Hassanuzzaman, Syeduzzaman, A.M.A. Muhit, Syed A. Samad and Masikur Rachman.

The usual riders apply. We alone are responsible for possible errors and mistakes. If some food for thought comes out of this project, it is because we have been riding on the shoulders of all these committed people!

Göran Djurfeldt
Hans Holmén
Magnus Jirström
Rolf Larsson

Lund, June 2004

1 African Food Crisis – the Relevance of Asian Experiences

Göran Djurfeldt,[1] Hans Holmén,[2] Magnus Jirström[3] and Rolf Larsson[1]
[1]Department of Sociology, Lund University, Lund, Sweden; [2]Department of Geography, Linköping University, Linköping, Sweden; [3]Department of Social and Economic Geography, Lund University, Lund, Sweden

This book looks at the African food crisis against the background of the Asian experience. The enquiry starts out from one remarkable fact, *viz.* that 30 to 40 years ago the Asian food situation was described much in the same apocalyptic terms as those that tend to be reserved for Africa today.

Although we are beginning to get used to it, just as remarkable is the fact that the threat of famine did not materialize in Asia at large. With exceptions like China in 1959–1961 and North Korea more recently, post-colonial Asia has been largely successful at famine prevention, and a number of then food-deficit countries in Asia are now food exporters.

The ghost of Thomas Malthus, revived by demographers and others in the West from the 1940s onwards, added to the perception of crisis that then prevailed. The 'population bomb' was ticking for Paul Ehrlich in the late 1960s (Ehrlich,1968). Roughly at the same time Georg Borgström metaphorically described the effect of population growth on the environment as a nuclear bomb.[1] In fact, ecological disaster for China had already been forecast by Buck (1937) in the 1930s. The crisis mindset was reinforced by the notion that poverty and hunger would make the Asian masses an easy prey for communist agitation.

High population growth rates, widespread poverty, hunger and malnutrition gave credibility to the messages of doom. So did the apparent ineffectiveness on the part of the newly independent states in taking over their national polities and in imputing dynamism to their development policies. It is significant that the concept of a 'soft state' originally referred to Asia and its alleged notoriously corrupt governments, lacking the 'social discipline' to carry out policies that they paid lip service to. Today, even more significantly, the same term is usually reserved for Africa. Neither is it a coincidence that the concept of a 'soft state' was coined by a Western, i.e. not Asian, scholar, by another Swede, Gunnar Myrdal (Myrdal, 1968). The allegedly soft states epitomized the Western pessimism about Asia's development at the eve of its Green Revolution. This pessimism is today reserved for Africa south of the Sahara.

The familiar catalogue of threat of famine, chronic food shortages, rampant poverty, rapid population growth, soft states and corrupt governments belong to the *standard narrative* (Toft Madsen, 1999) about sub-Saharan Africa. The HIV/AIDS pandemic seems to be the only new element, and adds to the credibility of the apocalyptic discourse. The narrative belongs to the paraphernalia of Western pessimism, obviously nurtured by more than three decades of apparent stagnation in the subcontinent. The fact that the

 The African Food Crisis
(eds G. Djurfeldt, H. Holmén, M. Jirström and R. Larsson)

same narrative until fairly recently was standard in Asian studies is commonly forgotten. For us it is the starting point.

Obviously it would be naïve to assert that just because the standard narrative proved to be a poor prognosis for Asia, it will prove equally poor in Africa. We are not claiming that, but rather that the narrative loses some of its credibility, given how poorly it has fared in Asian studies. Studies of African development must start from another platform than that of apocalypses and professional pessimism. It is evidently no solution to stand pessimism on its head, for example by trying to reassert the principled optimism of modernization theories, which have been thoroughly discredited. Just as there is no law-like Malthusian descent into doom, there is no law taking the world to a consumerist Utopia, although the marketing industry apparently would like us to believe so.

This book starts out from the simple question: *If Asia could do it, why not Africa?* The question does not preclude a pessimistic answer, but, unlike many other approaches, it does not presuppose one. As it stands, however, it is too broad and we will shortly reformulate and narrow it down.

The African Food Crisis

At the time of independence, most of sub-Saharan Africa was self-sufficient in food. In less than 40 years, the subcontinent went from being a net-exporter of basic food staples to reliance on imports and food aid. In 1966–1970, for example, net exports averaged 1.3 million tons/year, three quarters of which were non-cereals. By the late 1970s, sub-Saharan Africa imported 4.4 million tons of staple food per year, a figure that had risen to 10 million tons per year by the mid-1980s (Paulino, 1987). Cereal imports increased from 2.5 million tons per year in the mid-1960s to more than 15 million tons in 2000 and 2001 (FAOSTAT data, 2004). Since independence, agricultural output per capita remained stagnant and, in many places, declined. Africa is the only continent where cereal production per capita was less in 2001 than in 1961 (Fig. 1.1). Notwithstanding the seriousness of the situation, it should also be noted that after independence sub-Saharan Africa faced the highest rate of population growth ever recorded. Growth actually took place over the last decades, but it has not been rapid enough.

The stagnating or falling per capita production of cereals in Africa over the last 40 years is in great contrast to the development in East and South-east Asia, where per capita production increased during this period (Fig. 1.1). Comparing the first and last 5-year annual averages during the entire period, 1961–2001, per capita output in Asia grew by 24% while it decreased by 13% in sub-Saharan Africa.

Are there lessons to be learned from the Asian experience that could benefit national food security in sub-Saharan Africa? We believe there are.

A Model of the Green Revolution

The Green Revolution is a much misunderstood and maligned process (e.g. Shiva, 1991; Madeley, 2002; deGrassi and Rosset, 2003), so much slandered that the term itself may have grown largely worthless. We will try to resuscitate the term while claiming that the

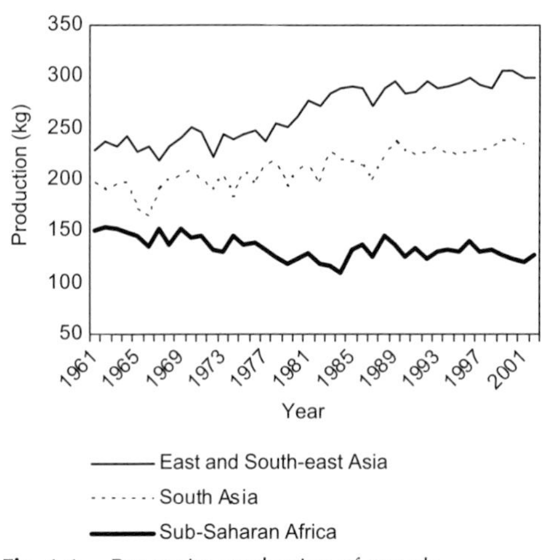

Fig. 1.1. Per capita production of cereals, sub-Saharan Africa and Asian regions. (Source: FAO (FAOSTAT data 2004) on http://apps.fao.org)

Green Revolution should be given much of the credit for relegating the threat of famine to Asian history. Our rescue operation for the term is to redefine it. We claim that *the Green Revolution is too narrowly defined when seen as a package of technology.*

Misplaced assumptions have marred the discussion of an African Green Revolution. Assuming a narrow technological definition of the Green Revolution, the discussion easily turns into debate about *transferability*. Given the radically different agroecological conditions, the answer is given: Asian technologies, on the whole, are not transferable. *Inter alia,* this is because the scope of irrigation in sub-Saharan Africa is much below that in Asia, making rice much less of a dominant crop than in Asia. African Green Revolutions must build on another crop-mix and therefore also on other technologies.

The perspective in this book is less centred on technology than is the conventional account. We regard the Green Revolution in Asia as a *state-driven, market-mediated and small-farmer based strategy to increase the national self-sufficiency in food grains* in a string of Asian countries, from the mid-1960s onwards. Technology was an important precondition for the results attained, and the development of agricultural technology was both an important part and a result of the process.

Our understanding of the Green Revolution is graphically rendered in Fig. 1.2.

The model stresses the following:

- The Green Revolution was state-driven, i.e. states or governments were *driving* the development of the food-grain commodity chains (see Djurfeldt, Chapter 2 this volume, for an elaboration of this argument).
- Green Revolutions were driven by states towards the goal of self-sufficiency in food grains, a goal that was motivated not only by the threat of famine, but also by the volatile world markets for grain, which made vulnerable those countries that depended on import.
- Asian Green Revolutions were market-mediated, i.e. markets played a fundamental role in different parts of the chain, with regard to both farm inputs and the trade and processing of grains. In other words, we are not talking about socialist models like those followed by China and Vietnam until the late 1970s and by North Korea even today.
- The Green Revolutions were small-farmer based, i.e. they were not based on large-scale mechanized farming. Asian rice farming was and remains dominated by small-sized family farms.
- Finally, we point to the crucial geopolitical as well as domestic political dimensions of the Asian Green Revolutions, which have to be kept in mind when discussing the African ones. One of these dimensions is obviously industrialization, which has been running parallel to agricultural development in Asia. This subject is not dealt with in this book, although the subject is briefly touched upon in Chapter 4 (Djurfeldt and Jirström, this volume).

We want to stress that the model is used not as a normative precept, but as a *causal and explanatory model.* We contend that this model

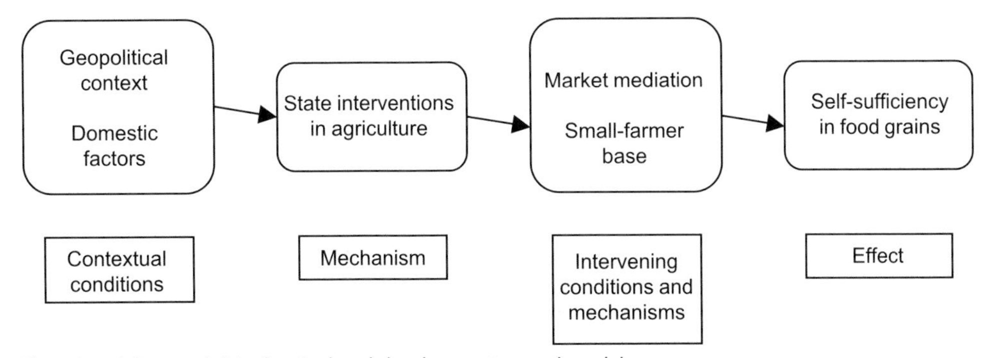

Fig. 1.2. Asian model(s) of agricultural development, causal model.

is useful in trying to *explain* the Asian Green Revolution. A further elaboration of the model is given in Chapter 2 (Djurfeldt), while in Chapters 3 (Jirström) and 4 (Djurfeldt and Jirström) in this volume we proceed to use the model in further understanding agricultural development in seven countries, from Japan in the north, to India in the south. In the African leg of the study, we use the model as a heuristic device in trying to understand what has happened and, equally interesting, what has not happened in sub-Saharan Africa, from the 1960s onwards.

We do not go into contested issues like the social or ecological effects of the Asian Green Revolution. Freebairn (1995) has demonstrated that the methodological basis of many of the early studies of the social consequences of the Asian Green Revolution was weak. Especially in the popular literature, many authors contend that the Green Revolution has increased poverty, alienated peasants from their land and promoted large-scale agriculture. We believe they are wrong and that methodologically more stringent studies corroborate that.[2]

Similarly, the ecological effects of the Green Revolution are often described in apocalyptic terms. Without going into detail, we think this is misleading and we regard the ecological problems of Asian rice farming as comparable to those in the West. This is bad enough, but does not foreshadow an *apocalypse*.

A commonly spread 'truth' is that a green revolution unavoidably leads to a loss of biodiversity (Shiva, 1991; Madeley, 2002). These critics usually envisage large tracts planted with only one crop, as for example in the irrigated Indian river-plains or the vast rice tracts in Malaysia or the Philippines. However, it should be noted that rice was associated with monoculture long before the Green Revolution (deGregori, 2004). The alleged loss of biodiversity is moreover not uncontested. Borlaugh (2002) makes the point that 'the high yields of the green revolution . . . had a dramatic conservation effect: saving millions of acres of wildlands all over the Third World from being cleared for more low-yield crops'. Thereby, from a biodiversity point of view, valuable rainforests have been saved thanks to the Green Revolution. Moreover, Dawe (2003, quoted in deGregori, 2004:33) found that the share of rice in the total harvested area in Asia has declined in almost all of Asia since 1970. Dawe concludes that overall cropping diversity seems to have increased since the beginning of the Green Revolution and that Asian farmers plant a wider variety of crops today than they did in 1970 (deGregori, 2004:33). The Green Revolution is better than its reputation.

With these brief remarks, we will leave the social and ecological dimensions and the normative domain and continue our quest for explanations – of success in Asia and of failures in Africa. First, however, a few words about the policy implications of our study.

Policy Implications

To the extent that our model proves well-grounded, it casts doubts on current policies, as pursued by many African governments, and as recommended by both donors and non-governmental organizations (NGOs).

Ever since the Structural Adjustment Programmes were thrust on more or less willing governments, the mainstream approach has been that free markets and the institutions necessary for their functioning are going to energize agricultural development. Our model implies that, although well-functioning markets are essential, they are not sufficient. We emphasize the state-drivenness of agricultural development, and we claim that there is an anti-state bias in almost the whole development community. To the extent that we succeed in corroborating our hypothesis about state-drivenness, it throws doubt on donor policies against African governments. Our approach implies that governments need to establish ownership over their agricultural policies and that donors need to assist them in achieving that.

As will be shown in the following, agricultural policies in Africa have seldom been small-farmer based, as our model requires. We claim that the pervasive bias against the small farm sector is a major hindrance to increased food security in the subcontinent.

Furthermore, in Africa small-farmer based agricultural growth is, we claim, an efficient means of poverty reduction and will remain so as long as a majority of the population directly or indirectly subsists from agriculture and lives in rural areas. Again, our results will, to the extent that they are tenable, have fundamental consequences for the policies pursued by governments and donors. More specifically, we contend that the Poverty Reduction Strategy Papers (PRSPs) worked out by most African governments poorly reflect this insight.

Another myth, underlying many policy interventions by donors and NGOs, is that Green Revolution technologies are not applicable in sub-Saharan Africa. One consequence has been under-investment and misdirected policy directives in crop breeding and agricultural research in general, as well as counter-productive dismantling of extension services. In the following we aim to show that this has had deleterious consequences for food security and contributed to the African food crisis.

Finally, many governments have been led to rely on imports of food grains, both by economic incentives (i.e. low prices) and by misdirected advice. If our model proves well-founded, it throws doubt on the heavy reliance on imports. African governments had better protect their farmers against import of low-price grains and utilize the room that the WTO gives for such protection on the part of the so-called HIPC (heavily indebted poor countries).

Thus, our approach has several implications for policy which may prove inconvenient both to donors and governments. This much said about policy, we will stick to our explanatory framework and return to policy implications only in the concluding chapter.

Methodology

In the Asian leg of the study, we have made a historical and comparative study of agricultural development in seven Asian countries. We start with Japan and continue with Taiwan and South Korea in East Asia and we conclude that there are important continuities between the agricultural policies pursued by these early starters and those followed in the more classical Green Revolution cases of Indonesia, Philippines and India. We have added Bangladesh to the set, because its Green Revolution took off only in the 1980s and during macro-economic conditions which resemble those in Africa today. Since most of these processes are well documented, our case studies primarily build on secondary sources. We have complemented these with interviews with key persons in Indonesia, Philippines, India and Bangladesh. The interviews have given us important insights into agricultural policy making.

Sub-Saharan Africa is still full of white spots on the social scientist's map. Basic statistics are missing or are of low quality. Existing research is illuminating spot-wise, but it cannot throw light on all the issues raised in this project. Therefore the African leg of the study had another strategy compared with the Asian one. We have made case studies in eight countries: Ethiopia, Ghana, Kenya, Malawi, Nigeria, Tanzania, Uganda and Zambia.

In all these countries, contracts were made with local scholars. They conducted two types of study for the project: on the one hand macro studies wherein they were commissioned to look at secondary data and conduct interviews with key persons. Following an analytical framework elaborated from the model of Asian Green Revolution presented above, our partners attempted to document agricultural development and, by means of the model, to explain what had and had not happened in their respective countries.[3]

Also, our partners conducted surveys in their respective countries with a questionnaire[4] developed in collaboration with all partners in the project. The project and survey design presupposes that the potential for intensification in food crop production is more likely to be found in more well-endowed areas, with better than average rainfall and access to markets. Thus we excluded the Sahelian countries from the country sampling frame, limiting the selection to the group of countries located in what may be labelled the 'maize and cassava belt'. Despite a clear potential for an agriculture-led development, these countries all face problems with low agricultural performance, rural poverty and recurrent food shortages.

The household sample consists of more than 3000 households in more than 100 villages (Table 1.1). Also in this case, the sampling design reflects the agricultural potential of the regions in which the households reside. This is illustrated by Fig. 1.3, showing 'agricultural dynamism' as a continuum, where 'low' depicts low productivity potential due to aridity or remoteness to markets. At the other extreme, 'high' refers to areas where ecological endowments and marketing infrastructure have combined to create some of the most dynamic and productive environments in Africa (examples are Mount Kilimanjaro in Tanzania, parts of the Kenyan highlands, areas surrounding the main cities).

Our intention has been to capture the dynamism in regions that are 'above average' in terms of ecological and market endowments but exclude the most extreme cases in this regard. While the households sampled are not representative of farmers in rural Africa as a whole, the encircled area can nevertheless be said to be typical of the environment in which a majority of the smallholder population in sub-Saharan Africa reside. This area seems sufficiently diverse to throw light on crucial conditions for farmer performance.

The sampling was thus a multistage one:

Stage 1. Countries (purposive sample) – Ethiopia, Ghana, Kenya, Malawi, Nigeria, Tanzania, Uganda and Zambia.
Stage 2. Agroecological regions (purposive sample) – total 20.
Stage 3. Villages (purposive sample) – total 103.
Stage 4. Farmer households (random sample) – total 3097.

Apart from the survey targeting the 3000 farm households, informal interviews were conducted with village leaders and farmer groups with the purpose of gaining additional information about conditions above household level (e.g. population densities, market access, land-use pattern, land availability, rainfall, state and donor activities).

Overview of the Book

In Chapter 2, Djurfeldt elaborates on the causal model of the Green Revolution and reviews theories of agricultural development as well as existing research. Central concepts like state-driven and market-mediated development are defined and discussed. The chapter concludes with a review of alternative and competing explanations of the Green Revolution.

Chapter 3, by Jirström, is a study of Asian precursors to the Green Revolution, namely Japan, Taiwan and South Korea. The author shows that there are important continuities between these pioneers and the later starters in South-east and South Asia. State-drivenness, small-farmer base and market-mediation were all there, as well as geopolitical conditions rewarding attempts to increase national self-sufficiency in rice. The technological basis of these early 'Green Revolutions' was different, however, although there are substantial similarities in the way in which technology was diffused

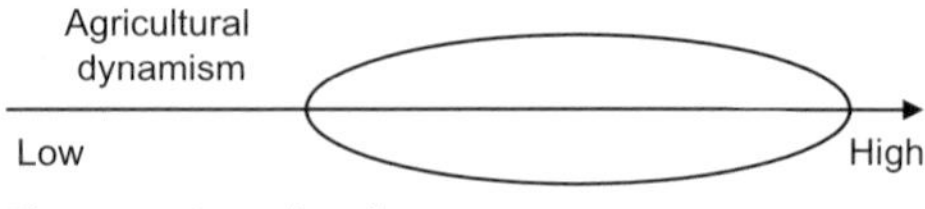

Fig. 1.3. Sampling frame.

Table 1.1. Countries, number of regions, villages and farm households.

	Country								
	Ethiopia	Ghana	Kenya	Malawi	Nigeria	Tanzania	Uganda	Zambia	Total
Regions	4	2	2	4	2	2	2	2	20
Villages	4	8	10	8	49	10	5	9	103
Households	322	416	298	400	495	403	320	443	3097
% Female-headed	5	17	43	40	12	20	14	24	22

among smallholders; the participatory element is especially noteworthy. Finally, foreign aid was instrumental in propelling processes which were essentially driven by nationalist motivations.

In the subsequent chapter (4), Djurfeldt and Jirström try to resolve the puzzle of why, almost simultaneously, Indonesia, Philippines and India made a U-turn in their agricultural policies, going against the mainstream prescriptions on agricultural price policies.

With the model in hand, we leave Asia in Chapter 5 where Holmén gives an overview of agricultural development in the eight case-study countries and discusses prevalent theories about agricultural intensification, the role of the African state in agricultural development, market institutions and the position of African smallholders.

Holmén continues the comparative exercise in Chapter 6 by making a comparative study of the eight case-study countries based on the macro reports already mentioned. He shows that whereas there has been no shortage of efforts to introduce 'Green Revolution technologies' in sub-Saharan Africa, diffusion has been limited and improvements short-lived. In contrast to Asia, food crop intensification has neither been driven by the States, nor have they been market-mediated or smallholder inclusive. Moreover, the Structural Adjustment Programmes implemented during the last two decades have had a largely negative impact on food security.

In Chapter 7 Larsson gives an account of the agricultural crisis by highlighting some of the factors that determine farmers' options regarding the production of food staples. The analysis is based on data deriving from the *Afrint* survey of more than 3000 farming households. It argues that farmers' access to 'Green Revolution technologies' and to viable and stable markets is among the most important effective means for raising agricultural productivity. It concludes that the crisis is policy related in that the majority of the farm population are trapped in a situation of financial and institutional insecurity in which inadequate on-farm resources result in low labour and area productivity.

In Chapter 8, Haggblade presents some results from a study of success in African agriculture. His analysis of these cases turns out conclusions which fit very well into those drawn from the *Afrint* case studies, so that the findings enrich and complement each other.

After these comparative analyses, four case studies follow. In Chapter 9 Akande gives an overview of the Green Revolution in Nigeria – if there is one! – while in the subsequent chapter Olouch-Kosura and Karugia tell the story of the stalled maize revolution in Kenya. In Chapter 11, Isinika *et al.* review the experience of agricultural intensification in Tanzania under the heading 'From Ujamaa to Structural Adjustment'. The final case study is from Ghana, where Seini and Nyanteng focus on the role of the smallholders in Ghana's agricultural development after Structural Adjustment.

In the penultimate chapter Otsuka and Yamano give a different perspective on Asian Green Revolutions, bridging Asia and Africa by discussing both continents. The authors summarize the important research tradition in agriculture economics and apply their Asian perspective to the African Green Revolution with a case study of what they call an 'Organic Green Revolution' based on 'crop–livestock–agroforestry tree interactions'.

In the ultimate chapter, Akande, Djurfeldt, Holmén and Isinika, four of the collaborating researchers, draw the overall conclusions and policy implications of this exercise in a comparative study of agricultural development.

Notes

[1] Borgström's bibliography includes several English titles dealing with his apocalyptic vision of a food crisis: 1965, 1969, 1973a, b, c, d.

[2] There is a long list of studies to be referred to here. See for example, Lipton (1989), Hazell and Ramasamy (1991), David and Otsuka (1994), Pingali *et al.* (1997), Datt (1998) and Palmer-Jones and Sen (2001).

[3] These 'macro reports' are available at the *Afrint* Home Page: www.soc.lu.se/afrint

[4] The questionnaire is available online at www.soc.lu.se/afrint/publics.htm

References

Borgström, G. (1965) *The Hungry Planet: the Modern World at the Edge of Famine*. Macmillan; Collier–Macmillan, New York and London.

Borgström, G. (1969) *Too Many: a Study of Earth's Biological Limitations*. Macmillan, London.

Borgström, G. (1973a) *Focal Points: a Global Food Strategy*. Macmillan; Collier–Macmillan, New York and London.

Borgström, G. (1973b) *The Food and People Dilemma*. Duxbury Press, North Scituate, Massachusetts.

Borgström, G. (1973c) *Harvesting the Earth*. Abelard-Schuman, New York.

Borgström, G. (1973d) *World Food Resources*. Intext Educational Publishers, New York.

Borlaugh, N. (2002) We can feed the world. Here's how. *The Wall Street Journal*, http://www.ifpri.org/media/innews/2002/051302.htm (13 May, 2003).

Buck, J.L. (1937) Nan-ching ta hsèueh. College of Agriculture and Forestry. Department of Agricultural Economics and Institute of Pacific Relations. *Land Utilization in China, a Study of 16,786 Farms in 168 Localities, and 38,256 Farm Families in Twenty-two Provinces in China, 1929–1933*. The Commercial Press Ltd, Shanghai, China.

Datt, G. (1998) *Poverty in India and Indian States: an Update*. International Food Policy Research Institute, Washington DC.

David, C. and Otsuka, K. (eds) (1994) *Modern Rice Technology and Income Distribution in Asia*. Lynne Rienner, Boulder, Colorado.

deGrassi, A. and Rosset, P. (2003) A new Green Revolution for Africa? Myths and realities of agriculture, technology and development. Open Computing Facility (23 September, 2003).

deGregori, T.R. (2004) Green Myth vs. the Green Revolution. *The Daily Nation*, http://www.butterfliesandwheels.com/printer_friendly.php?num=50 (7 March, 2004).

Ehrlich, P.R. (1968) *The Population Bomb*. Ballantine Books, New York.

FAOSTAT data (2004) Food and Agriculture Organisation of United Nations, http://www.fao.org/waicent/portal/statistics_en.asp

Freebairn, D.K. (1995) Did the green revolution concentrate incomes? A quantitative study of research reports. *World Development* 23, 265–279.

Hazell, P.B. and Ramasamy, C. (1991) *The Green Revolution Reconsidered: the Impact of High-yielding Varieties in South India*. The Johns Hopkins University Press, Baltimore, Maryland.

Lipton, M. (1989) *New Seeds and Poor People*. Johns Hopkins University Press, Baltimore, Maryland.

Madeley, J. (2002) *Food For All: the Need for a New Agriculture*. Zed Books, London.

Myrdal, G. (1968) *Asian Drama: an Inquiry into the Poverty of Nations*. Penguin Books, Harmondsworth, Middlesex, UK.

Palmer-Jones, R.W. and Sen, K. (2001) *What Has Luck Got To Do With It. A Regional Analysis of Poverty and Agricultural Growth in Rural India*. Understanding Socio-economic Change through National Surveys.

Paulino, L.A. (1987) The evolving food situation. In: Mellor, J.W., Delgado, C.L. and Blackie, M.J. (eds) *Accelerating Food Production in Sub-Saharan Africa*. John Hopkins University Press, Baltimore, Maryland.

Pingali, P.L., Hossain, M. and Gerpacio, R.V. (1997) *Asian Rice Bowls: the Returning Crisis?* CAB International, Wallingford, UK.

Shiva, V. (1991) *The Violence of the Green Revolution: Third World Agriculture, Ecology and Politics*. Zed Books, London.

Toft Madsen, S. (1999) *State, Society and the Environment in South Asia*. Curzon Press, Richmond, UK.

2 Global Perspectives on Agricultural Development

Göran Djurfeldt
Department of Sociology, Lund University, Lund, Sweden

The research reported on in this volume aims to systematically probe a comparison often made between Asian agricultural development during the so-called Green Revolution and current agricultural development in sub-Saharan Africa.

This chapter will elaborate on the theoretical perspective informing this exercise in comparative analysis of agricultural development.

The project started with a conception of the Asian Green Revolution as state-initiated and partly planned processes aimed at achieving national self-sufficiency in food grains. In an attempt to formulate a generalized model, we stress four aspects of these processes, namely that they were:

- State-driven, i.e. that states or governments were *driving* the development of the food-grain commodity chains (see further below).
- Market-mediated, i.e. that markets played a fundamental role in different parts of the chain, both with regard to farm inputs and in the trade and processing of grains.
- Small-farmer based, i.e. not based on large-scale mechanized farming.
- Finally, we point to the crucial geopolitical dimensions of the Asian Green Revolutions, which have to be kept in mind when discussing African ones.

Please refer to Fig. 1.2 in Chapter 1, where the model is depicted.

A Causal and Explanatory Model

Let us elaborate on the elements of the model and their interrelation. We conceive of a model with four links in a causal chain containing: (i) contextual conditions or parameters; (ii) a driving causal mechanism; (iii) intervening conditions and mechanisms; and (iv) an effect or outcome.

For the actors driving the process, its effect was an intended goal: an increased self-sufficiency in food grains. As we will see in Chapters 3 and 4, the urgency of national self-sufficiency in food grains was brought home to the national leaderships in Asian countries at various stages in their history and by both domestic factors and international circumstances. Among these were, of course, famines or threatening famines, but there were other factors as well, like instability and insufficiency of world markets, exposing the vulnerability of import dependence, or vulnerability to political pressure from grain exporting countries.

Setting a goal of increasing production of basic food grains is obviously not a matter belonging to the routine affairs of a government and the linkage between agricultural growth and governance is worth looking at in more detail. It is a central theme in this book that the role of the state in agricultural development is an insufficiently researched and theorized issue. Before continuing to discuss the role of the state, we will first discuss agricultural growth, as such.

Intensifying Production

Viewed from a purely agronomic point of view, increasing production of food grains may be brought about by two means, either by expanding the area under these crops, or by intensifying production. The first option implies either growing grains at the expense of other crops or bringing new land under the plough, i.e. extensive growth.

Going from extensive to intensive agricultural growth is no easy matter and it is no easier to achieve in a situation where land reserves are still available. As Boserup was one of the first to stress (Boserup, 1965), in a subsistence peasant economy, extensive growth is often favoured for the simple reason that the output per working day tends to be higher.

If the peasant knows that the grain needed to feed the family can be produced with less input of labour by extending the farm by a hectare or two, than by slogging to increase production on a constant area, he or she would, other things being equal, prefer the extensive option to the intensive one.

The conditions for intensification have therefore always been associated with a closed land frontier and with pressure from a growing population. Highlighting this connection, Boserup used it to formulate her polemic against Malthus: an increasing population does not necessarily lead to a food crisis, it may equally well stimulate intensified production and higher yields, permitting the growing population to be fed from a constant area.

Boserup inspired much research on the interconnection between population and production all over the developing world. Summarizing these results, it has been demonstrated over and again that Boserup's contention holds water. As she insisted, there is no automatic connection between a growing population and agricultural intensification.

In a situation of land scarcity there is a possibility that a population-induced downward trend in per capita production releases a process of intensification, but there is no guarantee for this. Moreover, researchers have shown that intensification is by no means always driven by population growth and closed land frontiers. Commercial forces may be more important. One example is from Africa, where the case studies edited by Turner II *et al.* showed that in heavily populated areas, agricultural growth tended to be associated more with demand in neighbouring urban markets than with population pressure as such (Turner II *et al.*, 1993). One can draw the same conclusion from Tiffen's *et al.* famous Machakos study: dynamic links to an urban economy means more for the management of land resources than the demographic factors as such (Tiffen *et al.*, 1994).

The Asian experience is a forceful corrective to the Malthusian perspective in which population growth inadvertently leads to a deepening food crisis. However, Boserup's anti-Malthusian thesis also has to be modified by pointing to the loose causal connection between demographic factors and agricultural intensification and by stressing the importance of markets and commercial opportunities as driving forces of intensification. Their potency depends on the development of markets, infrastructure and institutions, as we will discuss later in this chapter.

Pre-industrial Methods of Intensification

Boserup's work is best read as an agro-historical one. The connections she pointed to between population growth and agricultural development seem to be well established historically and over a longer term. There is a definite historical trend towards increased intensity of land use, not in a strict sense determined by, but at least correlated with, population growth (Evans, 1998).

The development of agricultural technology associated with this long-term trend involves methods such as fallowing, crop rotation, use of nitrogen-fixing crops, manuring, composting, integration between land and animal husbandry, irrigation, especially small-scale systems, etc. We term these technologies *pre-industrial methods of intensification*, because they do not require industrial inputs, but only resources which are available locally, or in a smaller area.

There is ample evidence that pre-industrial methods of intensification, and the knowledge associated with them, have developed in several locations and independently of each other, for example in China, Europe, Central America and West Africa (e.g. see Netting, 1993). Such methods of intensification depend on farmers' own ingenuity and innovativeness. They may spread over wider areas as farmers travel and learn from observing others. They draw mainly on local resources and farmers use them when required in order to increase production, either for reasons of subsistence or when stimulated by demand in local and international markets.

Increasing intensification drawing on pre-industrial methods, conditioned by an increasing population and stimulated by demand in local markets, accounts for much growth, not only historically and over the longer run, but also contemporaneously and in recent history. As we will show in detail later in this volume, there is much to Wiggins' contention that agricultural development in sub-Saharan Africa since the mid-1980s can be described in such terms. Production of food staples in Africa has kept pace with the growing agricultural and rural population, but it has not been able or been allowed to meet the demand from a growing urban population[1] (Wiggins, 2000). Much of this growth is, however, extensive rather than intensive, although it certainly includes some growth of the latter type, but based mainly on pre-industrial methods.

To take an example from Asia, pre-Green Revolution agricultural growth in India had a similar character, based mainly on extension of the area cultivated and just about keeping pace with the growing agrarian and rural population (Bhalla and Singh, 1997). Such a pre-industrial agrarian system is sensitive to disturbances from drought and flood, or from war and conflicts. It can hardly be the basis for the food supply in an industrialized and urbanized society.

Taken as an approach to agrarian history, Boserup's theory is firmly grounded. Taken as a theory of contemporary development, or of development during the last century, however, her perspective needs to be complemented, as she herself strived to do in the works published after *The Conditions of Agricultural Growth* in 1965. As we see it, it is necessary to bring in two other factors: (i) industrialization and the spur to urbanization which it creates; and (ii) the development of state power into a driving force of economic and agricultural development. We will shortly return to this theme, but before doing so we have to conclude the discussion of intensification.

Scientific and Industrial Inputs

The Green Revolution is obviously a process of intensification, although based not on pre-industrial methods, but on scientific and industrial inputs. While it is often reduced to a matter of technology, a reductionism to which we are critical, one of the defining traits of the Green Revolution is obviously the technologies it built on. They would have to be treated as part of the contextual factors making the revolution possible.

As is well-known, the Asian Green Revolution was based on breakthroughs in crop breeding, achieved, first for wheat in Mexico by the team led by Borlaug, and later for rice by the International Rice Research Institute (IRRI) in the Philippines, and even earlier in China (Barker *et al.*, 1985; Conway, 1997). The Rockefeller Foundations' Mexican Agricultural Program (MAP), conducted in cooperation with the Ministry of Agriculture, started in 1943 and focused on wheat and maize. In the mid-1950s the incorporation of plant dwarfness through the introduction of the *Norin 10* genes led to the development of a number of Mexican semi-dwarf wheat varieties.[2] The relatively short stem of the new varieties allowed them to respond to higher levels of nitrogen fertilizer and yet not lodge (fall over). The Mexican miracle varieties where soon to be released to farmers in Mexico and then rapidly diffused internationally – India and Pakistan were the first to profit substantially from the breakthrough. The Mexican success triggered cooperation between the Ford and Rockefeller Foundations and the government of the Philippines in establishing IRRI in 1962 – the first in the series of international agricultural research institutes. Like Borlaug and his team in Mexico before them, rice breeders at IRRI searched to incorporate dwarfness into rice plants that would thrive under tropical conditions. The breakthrough came with IR-8 – a crossing of a Taiwanese and an Indonesian variety – released in 1966.

Dwarf and semi-dwarf wheat and rice varieties had been known long before the initiation of the mentioned breeding programmes (see Jirström, Chapter 3 this volume). In the 19th century they were, however, more of a curiosity than of commercial value. It was not until the advent of chemical fertilizer and the subsequent development of the industrial capacity in this field that the dwarfing characteristic became significant (Dalrymple, 1978; Perkins, 1997). The fertilizer-responsive and non-lodging varieties came in demand as fertilizers became affordable and accessible. Some countries, especially Japan and its two colonies Taiwan and Korea, were pioneers on this technology path but for most countries the dramatic yield improvements following the development of semi-dwarf wheat and rice varieties started only after 1945 in the developed countries and in the 1960s and 1970s in the developing countries.

The introduction of these technologies makes the farmers dependent on scientific knowledge developed outside the farm and outside the farming community. Making agriculture dependent on scientific–industrial inputs has a deep significance, since it implies a much deeper incorporation of the sector into a national and international division of labour. With the exception of large-scale irrigation works, pre-industrial methods of intensification, as we already noted, depend on local knowledge and localized circuits of reproduction. The Green Revolution, on the other hand, constitutes a widening of the geographical scope of agricultural circuits of reproduction, from localized loops to nationalized and globalized chains.

Thus we are arguing that a demographic theory of agricultural development, whether Malthusian or Boserupian, is incomplete. When doing so we are not, however, arguing for an economic theory to replace the demographic ones. Our argument is broader than that and we are more inspired by political economy than by economics.

It is a key contention of this research that food grain commodity chains during the late 20th century have been state-driven and that this contributes heavily to explaining the Green Revolution. The notion of state-drivenness is inspired by commodity chain theory and requires some explanation. First, a few words about commodity chains before we come to the notion of state-drivenness.

Commodity Chains

Commodity chain theory focuses on the whole chain from raw material producers to retailing of end products. It incorporates a notion of *actors driving a chain*. The latter may vary from the raw material producers, like in the oil and mining industries, to retailers, for instance in the fast food or ready garments industries (Gereffi, 1999; Raikes *et al.*, 2000; Gibbon, 2001).

Many associate the concept of commodity chains with dependency and with world system theories. It is correct insofar as Wallerstein, the founding father of world system theory, together with Hopkins, was the first one to use the concept of *global commodity chains* in a systematic fashion (Hopkins and Wallerstein, 1994; Rammohan and Sundaresan, 2003:903–904). Gereffi, who has done much to popularize the concept, is himself a product of the world system school, but his results are notable in putting doubts on some of the assumptions often made by adherents of the school, for example that countries are bound to remain in a dependent position and that further integration into the world system is bound to deepen dependence rather than the reverse. In his studies of the apparel industry, Gereffi found that Asian actors are moving out of dependence and into more central positions in the chain (Gereffi and Kaplinsky, 2001).

Commodity chain theory has focused much on the global dimension, for example in the works by Nolan *et al.* (1999), but there are a number of authors who have used the commodity chain concept while dropping the 'global' prefix (e.g. Long and Villarreal, 1998; Ribot, 1998). When we apply the concept to the food grain chain, the global prefix similarly recedes somewhat into the background, both because food grains are not heavily traded internationally compared with many other commodities, and because food grain markets are heavily regulated and protected by national governments. They remain so even under the current WTO regime, and they have been so at least since World War II.

We use commodity chain theory mainly because we find the notion of actors driving a chain fruitful. More specifically, authors like Gereffi argue that several characteristics of the chain and the commodity determine at which node in the chain the power to control it will reside. We would add that contextual factors are important as well. Thus, if we want to explain why the multinational oil companies control the oil industry and the downstream links from the industry, e.g. petrol bunks, we have to look into the characteristics of the industry, its history and its relation to state power, where the strategic importance of oil becomes an important factor.

Talking of agriculture, we can easily work out a catalogue, for example, showing that roasters control the international coffee chain (Ponte, 2002), supermarkets control the global commodity chain for fresh vegetables (Nolan *et al.*, 1999), etc. While it is true that international trade in grains, especially wheat and maize, is dominated by a few multinationals (Kloppenburg, 1988), this does not imply that the food grain chain on the whole is driven by the multinationals. We contend that food grain chains are driven by states or regional states, like the European Union. More specifically, we argue that the Asian Green Revolutions were driven by states and governments taking the leading position in the food grain chains. We proceed to demonstrate this in a later chapter, continuing the discussion here by pointing to the contextual factors making for central positions of the states in food grain chains.

State-drivenness and the International System of States

How has it happened that the state is driving the food grain chain, but not any and every other commodity chain? The question needs to be historically situated to be properly answered: how did it come about that in the mid-20th century, the state came to be driving the food grain chains practically all over the world?

We can approach an answer to this question by first noting that there are no important exceptions to the rule. In the West, state regulation of grain production and trade was pervasive at least since the 1930s. One of the important features of Roosevelt's New Deal, and part of the package that helped overcome the Great Depression, was obviously the farm price support and the restrictions on foreign trade in grains (Skocpol and Finegold, 1982). Similarly, in Western Europe, major countries had their own varieties of this policy before we got the Common Agricultural Policy, at the heart of which is a similar regulation (Tracy, 1989). The Soviet Union and China were obviously no exceptions either, since their experiments with socialism implied an attempt to entirely do away with the market, and replace it with state control.

The era of de-colonization which started in the late 1940s implied a prominent role for the state in the economy and in development policies, also in countries outside the Soviet bloc. This role has been much discussed, for example in the now classical debate on the Developmental State, but the debate was mainly normative and raised questions like, what ought the state to do, and how should it proceed to guarantee results? Little scholarly effort has been spent in trying to explain the very phenomenon of the Developmental State. Why did the states during the second half of the 20th century get or take this prominent role in agricultural development (earlier in Germany and Japan)?[3] This is obviously not the place to rectify this bias in scholarly attention. We will, however, suggest that one reason for the universality of state-drivenness in the food grain chain has to do with *the international system of states*, as it functioned during this epoch.

In order to move away somewhat from a normatively loaded discourse on the state, it is sobering to study the writings of the so-called realist school in political science. They make it their task to explain the role of the state, rather than legitimating or criticizing it. Two important texts are Jackson and Rosberg's mini-classical paper (Jackson and Rosberg, 1982) and a more recent volume by Herbst (Herbst, 2000). In their perspective, a Weberian understanding of the state is fundamental. For Weber, the state is defined by its monopoly on legitimate use of violence within its territory.

If monopoly on force is fundamental to the state, the paradox is obvious in the case of one of the continents we are dealing with: *African states have no uncontested such monopoly*. The list of challenges to monopoly is long, and a series of African states are in constant armed conflict with insurrectionary groups within their territory: Angola, Congo, Rwanda, Burundi, Uganda, Sudan, Ethiopia, Sierra Leone, etc.[4] The contrast is striking, both with Europe during the time of its state formation from the 16th century and onwards, and with Asia. In Europe, the state monopoly was built up in confrontation between states contesting the territorial control of their adversaries, as Herbst notes, quoting Tilly (Tilly, 1985; Herbst, 2000). In Asia, similar confrontations promoted the development of state monopolies in countries like China, Japan, Taiwan, India, Pakistan and Bangladesh.

Weber's definition of the ideal-typical state was formulated in a European setting and with the European historical background taken as given (Weber, 1971). The history of European states like England and France can be written as one of state formation, which in its turn is equivalent to the gradual establishment of the monopoly on force. In Jackson and Rosberg's terminology, this is the *empirical conception* of statehood.

Another aspect of the history of the European state, which Weber gave less attention to and which has come to the fore in later scholarship, is the emergence of *the international state system*. The establishment of fully fledged European states, like England and France, led to the emergence of a system of inter-state relations. In the ideal-typical definition of this system, the principle of *non-interference* in the internal affairs of other states is a core one. For Jackson and Rosberg, this is the juridical or *international law conception* of the state. A neo-Weberian conception of the state is thus double-sided and refers to both the empirical and the juridical conceptions of state.

The interesting fact is that, while in Europe empirical statehood preceded the juridical one, in Africa it is the other way round. The realist perspective on the African states is guided by the insight that these states did not develop from within and in conflict with each other, as did the European states. They are the *products of the international system of states*. All of them, with a possible exception of Liberia,[5] have a colonial background, which led to scholars dubbing them as post-colonial states. The implicit focus in this label is on the colonial heritage and its importance for the functioning of the African states. This is certainly important, but the focus of the realists lies elsewhere: when the African colonies became independent, they became fully fledged members of the international system of states – in spite of the fact that they did not command a monopoly on force within their territories.

This history of the African states illustrates the importance of a state system perspective. The emergence of a global system of states reverses the causality in the formation of states, from endogenous causes to exogenous ones. Therefore no universal theory of state-formation is possible, except one which incorporates a reference to the global system and which is moreover contextually and historically grounded.

African states

There are wide-ranging implications of the fact that the newly independent African states did not control their territories. As part of the colonial heritage they had artificial borders, dividing big ethnic groups and merging small ones into political entities, to which they had little allegiance. Against this background, it is remarkable that the African state system has remained as stable as in fact it has. Of all the secessionist struggles, only the Eritrean one has so far succeeded. Despite being brittle, the sub-Saharan states have all survived, except the disintegrating Somalian one. Notwithstanding their being stamped as 'rogue', 'criminal', 'corrupt' and 'soft' (softened by the 'economy of affection'), the sub-Saharan African states continue to function as *members of the international state system*. They remain members of the UN, maintain diplomatic missions, fulfil their obligations to the Bretton Woods institutions and pay the heavy instalments and interest on their international debts. The African capitals remain better connected to the international system of states than to their own hinterland. Although tenuously connected to their own territories and societies, these states remain stable.

According to the realist school, there are two basic reasons for the paradoxical stability of the African state system. First, after independence African political leaders quickly realized that, if they wanted to remain uncontested by their neighbours and be left to solve their domestic law-and-order problems, they had to respect the right of their colleagues in other states to do so. This is the principle of non-interference, basic to the international system of states, applied to the African case. According to Jackson and Rosberg this also explains the sustainability of the fragile Organisation for African Unity (OAU), located in Addis Ababa, now being replaced by the African Union.

The other basic reason for the sustainability of the African state system is the support received from the international system of states, including its supra-statal members, like the World Bank, the IMF and the United Nations. This support, however, has definite limits:

> the survival of Africa's existing states is largely an international achievement . . . But there are definite limits to what international support can contribute to the further development of the capabilities of African states. A society of states that exists chiefly in order to maintain the existing state system and the independence and survival of its members cannot regulate the internal affairs of its members without the consent of their governments. It is therefore limited in its ability to determine that the resources transferred to the new states are effectively and properly used.
>
> (Jackson and Rosberg, 1982:22)

The last sentence in this quotation deserves being stressed: the international system has few possibilities to enforce a domestic order, for example in the system of administration or revenue. Or, is this where time passed this mini-classic? The Structural Adjustment Programmes are exactly such attempts to enforce domestic orders, with the internal debt of the client countries as levers, a point we shall come back to. The fact that implementation of these programmes is faltering (see, e.g. Jayne *et al.*, 2002, Chapter 6) has a given background exactly in the mechanisms identified by Jackson and Rosberg, why it would be premature to argue that time has passed their work in this respect.

Realism is a geopolitical perspective on the African state which is fundamental to an understanding of development potentials in Africa. The gist of the argument could be said to be that, while the international system of states gives a solid, if not perfect, guarantee for the territorial integrity of the African states, it does not guarantee either the establishment of a monopoly on force within the territories of its members or, of course, the establishment of the other capabilities of a *developmental state*.

With more than 10 years' perspective on the end of the Cold War it is evident that the post-communist world order did not bring any fundamental change in the international system of states. Admittedly, the end of superpower rivalry implied lower volumes of military and development aid. This is turn led to an escalation of civil wars and insurgency in sub-Saharan Africa. It is probably also a basic reason for lower volumes of international development assistance received by the subcontinent in the 1990s.

Despite these epochal changes in the international system, the basic traits of the African state system remain intact and the principle of non-interference between states still holds, despite obvious strains, for example in the Congo.

Asian states

From a realist point of view, there are striking similarities as well as differences between Asia and Africa. In both cases, the end of colonialism had a similar effect, leading to the establishment of a regional state system with remarkable stability.

De-colonization implied a trauma for British India with the division into India and Pakistan in 1947 and the delayed secession of Bangladesh in 1971. The ensuing regional stability of South Asia, including the inability of politicians in the regions to solve endemic conflicts, like those in Kashmir and Sri Lanka, can perhaps be attributed to the stabilizing influence of the international state system – after the initial shocks suffered in the 1940s. The regional conflicts in South Asia probably facilitated the 'broadcasting of power', to use a term from Herbst, within the borders of each state. However, India or Pakistan have still got their unruly provinces, while the state in Bangladesh seems close to having a monopoly on violence within its territory.

State formations like those in Indonesia, the Philippines and Laos have striking similarities to several states in Africa. We are referring to multi-ethnic states with both pre-colonial state formations and 'stateless' societies within the same borders.

Even India could be brought into the comparison with Africa because, as is well-known, British India consisted of both territories under direct administration and the subjugated 'princely states', which were the surviving pre-colonial state formations. The forested areas in south, central and north-western India, which are usually classified as 'tribal', have obvious similarities with parts of Africa, with a predominance of forest-dwellers, subsisting on swidden agriculture, hunting and gathering. Against the prognoses of many pessimists, and against the life-long struggles of secessionists in various corners of the country, the Indian Union has survived and today seems stronger than ever – the Kashmir conflict and the secessionist movements in Punjab and in the north-east notwithstanding.

Coming to the differences, the balance between pre-existing state formations and 'stateless' societies is certainly another one in Asia and Africa, and a larger share of Asia's pre-colonial state formations survived colonialism as political entities than was the case in Africa, where only Ethiopia could be placed in that category. China, Japan, Korea, Annam (Vietnam), Siam (Thailand), Burma and Ceylon (Sri Lanka) are important examples of state formation pre-dating and surviving colonialism.

The difference in state forms between Asia and Africa has an obvious background in agrarian ecology. Settled agriculture was much more widespread in Asia than in Africa. In sub-Saharan Africa, plough-agriculture was only widespread in Ethiopia (McCann, 1995). In the rest of the subcontinent, intensive agriculture was typically concentrated in small pockets, where ethnic minorities were locked into small territories, like the Kofyar on the Jos plateau in Nigeria (Netting, 1993; Stone, 1996).

The regional post-colonial state system in Asia thus had a different composition from the one in Africa, with more states pre-dating colonialism in Asia. Pre-colonial state formation is an obvious advantage to a post-colonial state, since state capacity is higher and the tax-base not only more resourceful, but easier to exploit. Inter-state rivalries were similarly much more entrenched in Asia than in Africa. These may be factors facilitating the emergence of the *developmental state* in Asia and its absence or impotence in Africa.

The Developmental State

In the 1950s and 1960s, when most former colonies gained their independence, mainstream development theory, multilateral organizations and donor practice emphasized the importance of the state in the development process. With only a small and undercapitalized private sector, it seemed natural that the state should be given prominence in any development strategy. The concept of the 'developmental state' met few objections in these early years of independence.

More recently Castells has made an attempt to resuscitate the concept of developmental state and to use the concept to explain the NIC-phenomenon, i.e. the industrialization that began with the 'four tigers' and is now spreading all over Asia, including China. Along with other scholars, but in disagreement with neo-liberal analyses of the NIC-phenomenon, Castells contends that industrialization in Asia cannot be understood without analysing the role of the state.

We would like to add to this that neither can Western industrialization be understood without reference to the role of the state. Although this is not the place to demonstrate this in detail, we contend that not only industrialization, but the phenomenal human development that has occurred in the West over the last one and a-half or two centuries, cannot be explained by an economistic theory. In other words, like industrialization, human development is not driven solely by economic growth. Development presupposes an active role on the part of the state in driving the process (Ranis *et al.*, 2000).

To define a state as 'developmental' does not only refer to the state's assuming leadership in the development process. More important is that the state's development policy is inclusive, namely that it has the effect of involving all or at least the majority of people and the whole economy in the transformation process. The opposite, a 'non-developmental' state, is exclusive, thus preserving 'modernization' for a small segment of the population. The latter case seems to have been more pronounced in Africa where a small urban-based elite and large cash crop producers excluded the majority of the population from development, leaving them more or less to eke out a living in subsistence agriculture and informal sectors.

In both Asia and Africa, nation building was part of the post-independence imperative. It can be argued that, in the 1960s, the situation facing the Asian governments was both easier and more difficult than that facing the African governments at the time. In Asia, recurring famines, dependence on food imports and threats of war were difficulties that urgently needed to be resolved – if the governments were to survive. In contrast to many Asian governments, an urgent task for the new leaders of African countries was to create nations out of superficially integrated territories, some of which were very large, rendering transport, communication and the exercise of power difficult.

Moreover, in Asia, there was a much more elaborate division of labour, and a higher degree of urbanization and industrialization than in Africa, where national governments faced the structural constraints of a colonial economic legacy oriented towards export crop production and mineral extraction. Bearing this in mind, and taking account of the fact that the food situation in the 1960s was relatively good, the development path pursued by African governments made sense – at the time. In this retrospective light, it appears that in Africa the nation-building project took precedence over development in general and over food self-sufficiency in particular. In both Asia and Africa, the state assumed the role of 'primus motor' of development. As has already been mentioned, governments in Asia were more successful than their African counterparts in using nationalist ideologies to motivate political interventions directed against sectoral vested interests. They also more successfully managed to pursue flexible development strategies and at the same time encourage the participation of small farmers, entrepreneurs and market forces in the process.

The development project in Africa took a rather different direction with the state assuming a dominating, and often exclusive, role as development agent (Hydén, 1983; Bell, 1986; Scott, 1998). Control, it appears, was given a higher priority than development. As a consequence few if any African governments have managed to establish themselves as 'developmental states' despite their willingness to present themselves as such. Moreover, while Asian governments in most cases have gained legitimacy carrying out their development project, African leaders and governments appear to have lost much of their initial legitimacy because they failed to fulfil a similar role.

We argue that the Green Revolution is a state-driven process. In fact, it can be seen as an epitomized case of a Developmental State project. In this perspective, then, the discussion about agricultural intensification in Africa boils down to a question of the capacity of the African states to drive such a development. Alternatively, of course, one may ask if in the last few decades the world has changed so much that state-driven development of the food grain chain is no longer an option or a possibility. We will return to this question in the last chapter of this volume.

A halfway summary of the argument so far is that the self-sufficiency in food grains which has been the effect of the Green Revolutions in Asia, but not so far in Africa, was brought about by state interventions in agriculture. Important mechanisms for these interventions were the existence of developmental states. Their way of functioning was conditioned by contextual factors, both domestic and geopolitical, some of which we have discussed above.

Market Mediation

It is curious that we need a new term to express what is very commonplace, namely that really existing markets are not like the ideal-typical ones which economists build their theories around. *Really existing markets are not free ones,* they show tendencies to monopoly and monopsony and, most important of all, they are institutionalized and regulated by the surrounding society and in particular by the state. Too much of the contemporary discussion has centred on the antinomies of free-market and market-free societies while real markets and real societies always fall somewhere in between the extremes.

Even in socialist Soviet Union and China, where there was a political programme to liberate society from markets, they continued to exist, both formally and informally. With the failure of cooperative and state farming and food chains controlled by the state, the food problem was eased by the small-scale gardening sector, and the mainly informal systems of exchange built around it (Wädekin, 1990).

With the exception of China before 1978 and Vietnam, Asian Green Revolutions were state-driven and *market-mediated,* in the sense that governments relied on private business to handle at least parts of the provision of inputs (seed, fertiliser, irrigation etc.) and to distribute the food grain produced from the farm gate to the consumer. These markets were not free and in important respects they were controlled by the state. In many countries state agencies operated in farm markets, both on the input and the output side, alongside private dealers. Details about these 'mixed economies' will be given in later chapters.

In the historical epoch we are talking about, especially the late 1960s and early 1970s, traditional socialist ideas still had a high credibility, both among intellectuals and among others. There were many proposals, for example, to nationalize grain trade or manufacturing and trade of fertilizers, but nowhere were they carried through. Indonesia under Soeharto may have gone farthest on this road, with its BULOG, a state trading agency with a pervasive role in the grain trade (Timmer, 1997).

Private traders had a role in all Asian Green Revolutions, outside the socialist bloc. The reason why they were not done away with was not only or mainly ideological, given the hostility which many politically influential intellectuals showed to them. It would fit the facts better to explain the role of the private sector by reference to *Realpolitik* rather than ideology. The business communities were well entrenched all over Asia. Politicians were not powerful enough to manage what, for example, the Ugandan government achieved with its Asian community, i.e. to throw them out, even if Indonesia tried to do the same with its Chinese minority.

Often more by expedience than by principle, then, the agricultural development strategies of Asian governments came to rely on private merchants for handling many of the transactions of the food grain chain. That is why we describe these strategies as state-driven *and* market-mediated. We will proceed to show how this came about in the seven Asian countries dealt with later.

Although this is impossible to substantiate, it might be that the market mediation of the Asian Green Revolutions was a precondition for their successes. The fact that China and Vietnam did not take off in their attempts to solve their food problems before they gave up their attempt to do away with the market mechanisms at least gives some support to this speculation. So does, of course, the food crisis in North Korea around the turn of the millennium.

Small-farmer base

A final characteristic of the Asian Green Revolutions was that they were *small-farmer based*. This contention requires a discussion of terminology and definition. One way to start is with the distinction that Johnston and Kilby introduced between unimodal and bimodal agrarian structures (Johnston and Kilby, 1975, 1982). Bimodal agrarian structures have normally been polarized between an estate or landlordist sector on the one hand, and a large number of smallholdings on the other. Unimodal structures are dominated by small and medium-sized holdings.

It is no easy matter to define small and big holdings when trying to operationalize the character of agrarian structure. The most common method is, of course, to define 'small' and 'big' in area terms, but it is not a very satisfactory method, not mainly because any borderline is bound to be artificial, but because the significance of area varies with type of crop and intensity of cultivation and it changes with increasing labour productivity and mechanization. What is 'small' today might have been 'big' yesterday.

In other publications, Djurfeldt *et al.* have tried to operationalize small and big in other, more theoretically meaningful, terms. 'Small' in this discourse ideal-typically refers to holdings which are worked and managed by a family or household and the production of which goes to meet the subsistence needs of the same family. Formulating this ideal type in more precise terms and operationalizing it is not a trivial matter.[6] In low-wage, labour-surplus economies like the Asian ones, the matter is made more complicated by the fact that cheap labour is used to substitute family labour even on very small farms. Thus, labour-hiring, which has frequently been used as an indicator of large-scale holdings, is not very useful either.

The easiest way out may be to study the land distribution and use the Gini index or some other indicator of inequality, together with some measure of landlessness. Increasing inequality in the distribution of land and increasing landlessness or proletarianization would be indictors of increasing bimodality, while the reverse movement would indicate increasing unimodality.

In intellectual discussions of agricultural development there has always been a strong bias against smallholders, stressing the importance of scale. Big farms have been seen as necessary for modernized agriculture. Alongside this dominating bias, there has been another tradition stressing that 'small is beautiful', for example among the *narodniks* in the Russian agrarian debate. In the academic community, there was an important break in the 1960s, when the works of Chayanov became available to the English reading public (Chayanov, 1966) and when Schultz published his 'Transforming traditional agriculture' (Schultz, 1964, 1983). The former was more influential in Europe and the latter in the USA and among economists.

A common theme for both Chayanov and Schultz is that smallholders are efficient producers, and by now this has been accepted by social scientists working with agriculture. It has not had the same impact in the scholarly world as a whole, and it has had even less influence among intellectuals in general or among the educated public, where 'big is beautiful' is still a hit.

The left, both in academia and more generally, has always had an ambivalent position when it comes to the small versus big issue. On the one hand, leftists have been taken in as much as everybody else by the presumed superiority of big units and their necessity for modernization. On the other hand, leftists are rightly worried about increasing inequality and proletarianization, and political activists have tried to use it as a means for mobilizing landless labour and poor peasants. When in political office, leftists have tended to go for big is beautiful, while in opposition, they have fought against policies promoting large-scale agriculture.

One might argue that the fascination with large-scale agriculture was one of the main reasons for the complete fiasco of Soviet agrarian policies, and those of China before 1978, as Djurfeldt and others have done (see Djurfeldt, 2001, Chapter 7; and, for example, Hedlund, 1989; Shanin, 1989; Riskin, 1995; and Selden, 1995). As pointed out in Chapter 1 of this volume, the apprehension that the Green Revolution would be promoting polarization and proletarianization has on the whole proven wrong.[7]

Widely seen by the left and by others as a strategy betting on the strong and promoting increased inequality and landlessness, the Asian Green Revolutions on the whole seem to have had the opposite effect. The movement has been towards unimodality rather than in the other direction. That's why we are claiming that the Green Revolution was small-farmer based. This was an effect both of the technology with its scale-neutrality, and of the major shift in pricing policies which were part of the policy packet associated with the Green Revolution, both in South and South-east Asia. More about this in a later chapter.

One might think that the shift in policies was due to the influence of scholars like Chayanov and Schultz, but as we will show, it is difficult to prove that influence. It is more likely that the shifts had to do with shifting power alliances and with shifting interests making their influence felt on state policies. As always, Minerva's owl flies after the fact: when scholars of agriculture began widely to realize that an agricultural development strategy promoting unimodality is indeed an effective, although not necessarily efficient, way to solve the world's food problem, this had been proven by Asian (and Western) governments going less by intellectual analysis and more by pragmatic concerns about electoral support and legitimacy.[8]

Other Hypotheses

The above is an attempt theoretically to underpin the model of the Asian Green Revolutions summarized in Fig. 1.2 in Chapter 1 and the starting point and inspiration of the research reported on in this volume. To repeat, *the Asian Green Revolutions were state-led projects aimed at securing national self-sufficiency in food grains, conditioned by geopolitical and domestic factors and relying on markets and small farmers in order to reach their goals*. In a later chapter we will try to demonstrate this, after which we will proceed to investigate the relevance of this model to Africa.

Before concluding, it might be asked, if the above model is read as a hypothesis about agricultural development in Asia, what in that case are the zero or alternative hypotheses?

One alternative hypothesis would of course be the effectiveness of pre-Green Revolution policies in bringing about the same effect of national self-sufficiency in food grains. In South and South-east Asia, governments were pursuing industrialization aiming at import substitution, policies which, in line with the orthodoxy of the day, stressed a low-price policy for food grain, in order to channel the resources of society into the big drive to industrialize. A zero hypothesis could be that these policies would equally well have solved the food problem.

It is obviously not possible to conduct a strict test of this alternative hypothesis, since it is counter-factual. The fact remains, however, that after the policy shifts the agricultural growth reached new levels. With pre-Green Revolution rates of growth, the goal of national self-sufficiency would presumably have taken much longer to be reached – if ever it had been reached.

We have already touched upon yet another alternative interpretation, namely that of the left, which tends to see the Green Revolution as a capitalist project, leading to development of capitalism in agriculture. We have no space to go into the definitional issues of what characterizes a capitalist project and a capitalist development in agriculture. Our hypothesis of the Green Revolution being state-led obviously contradicts the hypothesis of an economically determined process, which the capitalist hypothesis necessarily implies. Similarly, the prognoses made by many left-leaning scholars, of the Green Revolution as leading to increased polarization, proletarianization and pauperization, have on the whole not proven true.

What other alternative hypotheses are there? What would a neo-liberal standpoint on the Asian Green Revolution imply? From a normative or policy point of view, the answer is obvious. A neo-liberal would argue that a pure market solution would have been the most efficient way to modernize Asian agriculture. One could even imagine that a neo-liberal would grant that our model adequately describes what actually happened in a number of Asian countries and that the state-led projects have in fact been effective. They have reached their goals, *but not in an efficient manner*, would be the argument. Much more could have been achieved with less state-intervention and freer markets.

There is a curious overlapping here, between the leftist and the neo-liberal critique. Both critics claim that they know more efficient and/or equitable ways to do it. Both ground their argument in a theoretical discourse, one dealing with the wonder of free markets, the other with the marvels of co-operation and equitable solutions. The standpoint in this book is that theoretical speculation may be interesting but not very fruitful as a strategy of empirical research. Looking at what really happened in Asia, it is evidently not particularly well described, neither by the leftists nor by the liberals. We think there is a lesson in this, and we will try to spell it out more explicitly in the concluding chapter.

With reference to Africa, our standpoint is similar. It is easy to spell out the implications of a liberal or leftist programme for African agriculture, or for that matter a green one, but it is not the most interesting task! It is more interesting to find out what is really happening on the ground and to confront this – not first-hand with a normative and programmatic model, but with an empirical one. That is why we will try to use the very model developed from the study of Asian Green Revolutions in understanding what is happening in Africa. Hopefully, there will be a lesson to draw also from this exercise.

Already at this point we may note a similarity between *pre-Green Revolution* policies in South-east and South Asia and current policies in Africa, namely their reliance on a low-price policy for food grains brought about by low tariffs to import. As in contemporary Africa, there were many voices in Asia of the 1950s and 1960s arguing that the productivity problems in agriculture could be solved with pre-industrial methods of intensification, and by institutional reforms. It sounds so familiar, given the contemporary African debate! Compare the following description of the Indian policy debate in the late 1950s:

> At the same time, the planners informed the National Development Council that domestic output of fertilizer was expected to reach only one-half the projected level, and 'on account of the foreign exchange shortage, supplies of chemical fertilizers were bound to fall short of the demand which was itself increasing.' The gap had to be filled through 'greater efforts in the direction of green manures and other manurial resources.' In fact, given the stringent financial situation, it was more important than ever that 'local participation and community effort . . . be enlisted on the largest scale possible in support of agricultural programs.'[9]

The difference as it appears from this point of view is that African states today are not firmly seated at the driver's wheel in agricultural development. They still have to start driving and here there is of course a great contrast with Asia in the epoch of import substitution. In the 1950s and 1960s, a prominent role for the state in economic development was taken for granted. There was no question in Asia of the state getting into a driving position, but rather about the methods to be followed in achieving growth in agriculture. The paradigmatic shift, as we will see, was in allowing the market and the small farmers a seat and a ticket, while remaining driving.

Notes

1 Tiffen argues that African agriculture does in fact provide for both the rural population and to an increasing extent for the urban one as well (Tiffen, 2003).

[2] The *Norin 10* variety was developed in Japan (*Norin* is an abbreviation of the Ministry of Agriculture and Forestry) and was in 1946 brought to the USA by Dr S.C. Salmon, a USDA scientist acting as agricultural advisor to the occupation army (Dalrymple, 1978).

[3] This has been discussed somewhat regarding primarily industrial development in the last mentioned cases, but it has evoked scant attention in the cases we are interested in here.

[4] The implicit universe here and in the following is Africa south of the Sahara, with Ethiopia and South Africa as exceptional cases, the former with a longer history of state formation and a shorter one of colonial occupation, the latter with a strong, well-established state.

[5] Jackson and Rosberg mention only Liberia, but Ethiopia could have been added here.

[6] See the following references: Athreya (1990); Djurfeldt (1996); Errington (1996).

[7] For an interesting aside, see the study by Freebairn of the social science literature on the Asian Green Revolutions. He shows that conclusions on the social effects of agricultural development depended both on the ethnicity of the scholar and on the methods used in the study (Freebairn, 1995).

[8] This can be taken as support of Hayami and Ruttan's theory of induced agricultural growth, which stipulates that new agricultural technologies develop as the demand for them from the farmers grows. Note, however, that in this case demand seems to have been channelled through politics, rather than through the market. We still feel Hayami and Ruttan's formulation is too economistic and that it takes too little account of the political element, which is prominent in our treatment of the subject (Hayami and Rutton, 1971).

[9] Frankel (1978) quoting from *India, Planning Commission, Appraisal and Prospects of the Second Five Year Plan*, New Delhi, 1958, p. 40 och p. ii.

References

Athreya, V.B., Djurfeldt, G. and Staffan, L. (1990) *Barriers Broken: Production Relations and Agrarian Change in Tamil Nadu*. Sage, New Delhi, India.

Barker, R., Herdt, R.W. and Rose, B. (1985) *The Rice Economy of Asia*. Resources for the Future, Washington, DC.

Bell, M. (1986) *Contemporary Africa: Development, Culture and the State*. Longman Scientific & Technical, Essex, UK.

Bhalla, G.S. and Singh, G. (1997) Recent development in Indian agriculture: a state level analysis. *Economic and Political Weekly* (Review of Agriculture, March 29), A1–A14.

Boserup, E. (1965) *The Conditions of Agricultural Growth*. George Allen & Unwin, London.

Chayanov, A.V. (1966) *A.V. Chayanov and the Theory of Peasant Economy*. Richard D. Irwin, Homewood, Illinois.

Conway, G. (1997) *The Doubly Green Revolution: Food for All in the Twenty-first Century*. Penguin Books, London.

Dalrymple, D.G. (1978) *Development and Spread of High-yielding Varieties of Wheat and Rice in the Less Developed Nations*. Washington, DC.

Djurfeldt, G. (1996) Defining and operationalising family farming from a sociological perspective. *Sociologia Ruralis* 36, 340–351.

Djurfeldt, G. (2001) *Mera Mat: Att Brödföda en Växande Befolkning*. Arkiv förlag, Lund, Sweden.

Errington, A. (1996) A comment on Djurfeldt's definition of family farming. *Sociologia Ruralis* 36, 352–355.

Evans, L.T. (1998) *Feeding the Ten Billion: Plants and Population Growth*. Cambridge University Press, Cambridge, UK.

Frankel, F.R. (1978) *India's Political Economy, 1947–1977: the Gradual Revolution*. Princeton University Press, Princeton, New Jersey.

Freebairn, D.K. (1995) Did the green revolution concentrate incomes? A quantitative study of research reports. *World Development* 23, 265–279.

Gereffi, G. (1999) International trade and industrial upgrading in the apparel commodity chain. *Journal of International Economics* 48, 37–70.

Gereffi, G. and Kaplinsky, R. (2001) *The Value of Value Chains: Spreading the Gains from Globalisation*. Institute of Development Studies, Brighton, UK.

Gibbon, P. (2001) Upgrading primary production: a global commodity chain approach. *World Development* 29, 345–363.

Hayami, Y. and Ruttan, V.W. (1971) *Agricultural Development: an International Perspective*. Johns Hopkins University Press, Baltimore, Maryland.

Hedlund, S. (1989) *Private Agriculture in the Soviet Union*. Routledge, London.

Herbst, J. (2000) *States and Power in Africa: Comparative Lessons in Authority and Control*. Princeton University Press, Princeton, New Jersey.

Hopkins, T.K. and Wallerstein, I. (1994) Commodity chain in the capitalist world economy prior to 1800. In: Gereffi, G. and Korzniewicz, M. (eds) *Commodity Chain and Global Capitalism*. Praeger, New York, pp. 17–50.

Hydén, G. (1983) *No Shortcuts to Progress. African Development Management Perspectives*. University of California Press, Berkeley, California.

Jackson, R.H. and Rosberg, C.G. (1982) Why Africa's weak states persist: the empirical and the juridical in statehood. *World Politics* 35, 1–24.

Jayne, T.S., Mwanaumo, A., Nyoro, J.K. and Chapoto, A. (2002) False promise or false premise? The experience of food and input market reform in eastern and southern Africa. *World Development* 30, 1967–1985.

Johnston, B.F. and Kilby, P. (1975) *Agriculture and Structural Transformation: Economic Strategies in Late-developing Countries*. Oxford University Press, New York.

Johnston, B.F. and Kilby, P. (1982) 'Unimodal' and 'bimodal' strategies of agrarian change. In: Harriss, J. (ed.) *Rural Development, Theories of Peasant Economy and Agrarian Change*. Routledge, London and New York, pp. 50–65.

Kloppenburg, J.R., Jr (1988) *First the Seed: the Political Economy of Plant Biotechnology, 1492–2000*. Cambridge University Press, Cambridge, UK.

Long, N. and Villarreal, M. (1998) Small product, big issues: value contestations and cultural identities in cross-border commodity networks. *Development and Change* 29, 725–750.

McCann, J. (1995) *People of the Plow: an Agricultural History of Ethiopia, 1800–1990*. Wisconsin University Press, Madison, Wisconsin.

Netting, R.M. (1993) *Smallholders, Householders: Farm Families and the Ecology of Intensive, Sustainable Agriculture*. Stanford University Press, Stanford, California.

Nolan, C., Humphrey, J. and Harris-Pascal, C. (1999) *Horticulture Commodity Chains: the Impact of the UK Market on the African Fresh Vegetable Industry*. School of Development Studies, University of East Anglia, Norwich, UK.

Perkins, J.H. (1997) *Geopolitics and the Green Revolution: Wheat, Genes and the Cold War*. Oxford University Press, New York.

Ponte, S. (2002) The 'latte revolution'. Regulation, markets and consumption in the global coffee chain. *World Development* 30, 1099–1122.

Raikes, P., Jensen, M.F. and Ponte, S. (2000) Global commodity chain analysis and the French *filière* approach: comparison and critique. *Economy and Society* 29, 390–417.

Rammohan, K.T. and Sundaresan, R. (2003) Socially embedding the commodity chain: an exercise in relation to coir yarn spinning in southern India. *World Development* 31, 903–923.

Ranis, G., Stewart, F. and Ramirez, A. (2000) Economic growth and human development. *World Development* 28, 197–219.

Ribot, J.C. (1998) Theorizing access: forest profits along Senegal's charcoal commodity chain. *Development and Change* 29, 307–341.

Riskin, C. (1995) Feeding China: the experience since 1949. In: Drèze, J., Sen, A. and Hussain, A. (eds) *The Political Economy of Hunger. Selected Essays*. Clarendon Press, Oxford, UK, pp. 401–444.

Schultz, T.W. (1964, 1983) *Transforming Traditional Agriculture*. Repr., University of Chicago Press, Chicago, Illinois.

Scott, J.C. (1998) *Seeing Like a State. How Certain Schemes to Improve the Human Condition Have Failed*. Yale University Press, New Haven, Connecticut.

Selden, M. (1995) Yan'an communism reconsidered. *Modern China* 21, 8–45.

Shanin, T. (1989) Soviet agriculture and perestroika: four models: the most urgent task and the furthest shore. *Sociologia Ruralis* 29, 7–22.

Skocpol, T. and Finegold, K. (1982) State capacity and economic intervention in the early New Deal. *Political Science Quarterly Author* 97, 255–278.

Stone, G.D. (1996) *Settlement Ecology: the Social and Spatial Organization of Kofyar Agriculture*. The University of Arizona Press, Tucson, Arizona.

Tiffen, M. (2003) Transition in sub-Saharan Africa: agriculture, urbanization and income growth. *World Development* 31, 1343–1366.

Tiffen, M., Mortimer, M. and Gichuki, F. (1994) *More People, Less Erosion. Environmental Recovery in Kenya*. John Wiley & Sons, New York.

Tilly, C. (1985) War making and state making as organized crime. In: Evans, P.B., Rueschemeyer, D. and Skocpol, T. (eds) *Bringing the State Back In*. Cambridge University Press, New York.

Timmer, C.P. (1997) Building efficiency in agricultural marketing: the long-run role of BULOG in the Indonesian food economy. *Journal of International Development* 9, 133–145.

Tracy, M. (1989) *Government and Agriculture in Western Europe 1880–1988*, 3rd edn. Harvester Wheatsheaf, London.

Turner, B.L., II, Hydén, G. and Kates, R. (eds) (1993) *Population Growth and Agricultural Change in Africa*. University of Florida Press, Florida.

Wädekin, K.-E. (ed.) (1990) *Communist Agriculture*, 2 vols. Routledge, London.

Weber, M. (1971) *Max Weber: Makt og Byråkrati*. Gyldendal Norsk Forlag, Oslo, Norway.

Wiggins, S. (2000) Interpreting changes from the 1970s to the 1990s in African agriculture through village studies. *World Development* 28, 631–662.

3 The State and Green Revolutions in East Asia

Magnus Jirström
Department of Social and Economic Geography, Lund University, Lund, Sweden

Based on the model presented in Chapter 1 (Djurfeldt, this volume) this chapter analyses the agricultural intensification processes of three East Asian countries – Japan, Taiwan and South Korea. The objective of this and the following chapter is to show how a number of Asian states during critical periods of structural change of their economies and in the face of domestic food insufficiency, decided to support and promote a science-based agricultural intensification process in the small-scale, family farm sector. Based on high-yielding technologies, the three countries were able to increase their staple-food production to the extent that they reached their national goals of food self-sufficiency. Our main focus will be on the role of the state and its different forms of intervention in agriculture. However, as the private sector played a significant role in the process and small-scale family farms were the actual production units, these actors will also receive attention.

From Drama to Miracle

In 1977, the Asian Development Bank's Second Agricultural Survey, *Rural Asia: Challenge and Opportunity*, projected a 'substantial supply–demand deficit in rice, wheat and maize by 1985, possibly ranging from 24 to 30 million tons' (Asian Development Bank, 1977). The survey was headed and monitored by some of the leading experts on Asian rural and agricultural development and their projection was not exceptional at the time. Into the late 1970s, Asia was given little hope of ever being able to meet its rapidly growing food demand and the old notion of an Asian dilemma of 'too little land and too many people' was still enduring. The ongoing production success of the Green Revolution was hard to grasp in the midst of the process, although the spread of the high-yielding technology was in full swing.

Less than a decade later, starting in 1982 and peaking in 1986, Asian rice farmers experienced the so called 'crisis of success' as the world market price of rice plummeted (Timmer, 1992). All of a sudden there was too much rice as production outran demand. Many traditional rice-importing nations such as India, Indonesia, the Philippines and Bangladesh had more or less achieved self-sufficiency in production and some had even begun exporting rice. East, South-east and South Asia's share of world market imports shrank from approximately 68% in 1961 to 16% in 1992 (Pingali *et al.*, 1997). In less than three decades, following the launch of IR-8 in 1966, production of rice – the main staple crop of Asia – had almost doubled. In the wheat growing regions, mainly in India and Pakistan, high-yielding wheat varieties had resulted in quantum leaps in production. Per capita food crop production in Asia had

 The African Food Crisis
(eds G. Djurfeldt, H. Holmén, M. Jirström and R. Larsson)

increased by more than 25% during the period and Asia had moved from a situation broadly characterized as a food crisis to high levels of national food security in most countries. It certainly was an unexpected development for those involved, including farmers, policymakers, academic experts and the media. Today few people would think of Asia in terms of a food crisis.

The Green Revolution in tropical Asia was, however, not unique. Although less spectacular, dramatic increases in production had occurred in East Asia prior to the development in South-east and South Asia. Starting in Japan in the second half of the 19th century and continuing in its former colonies Taiwan and Korea in the 1920s and 1930s, processes sharing several features with those of the Green Revolution in the tropics contributed to the transformation of these societies and their economies.

Countries Selected and Omitted in the Survey

We have selected countries which have all gone through agricultural intensification processes in which the combination of government interventions, private sector and small-scale, family farms has played a central role. China is not included in the analysis. Arguably, it could have been. After the agricultural crisis in 1959–1961, during which 30 million people are estimated to have died of starvation and malnutrition, the Chinese government gave greater emphasis to modern agricultural technologies and inputs (Lin, 1998). As in the rest of East Asia agricultural intensification as such did not represent a new path of development. China's historical emphasis on cropping intensities – increasing the annual number of crops on existing farmland – through water control and early maturing varieties continued after 1949. Although the new government initiated highly successful breeding programmes,[1] the agricultural development strategy that was followed in the 1950s was one of collectivization of agriculture and of mass mobilization of rural labour into various projects and practices. On the one hand these included labour-intensive investments in irrigation, flood control and land reclamation projects. On the other, they implied an intensified application of traditional methods, such as more careful weeding, closer planting and manufacturing of compost (Barker *et al.*, 1985; Lin, 1998).

According to Hayami and Ruttan (1985), this strategy represents the most serious effort in recent history to develop agriculture within the so called conservation model framework. The growth rates that could be achieved were, however, not compatible with the demands for agricultural output. A stronger emphasis on modern inputs as well as a partial retreat in 1962 from the Soviet-style collectivization of 1958 did increase somewhat the growth rate in yields and production, but not sufficiently and the overall performance of agriculture continued to lag behind (Lin, 1998).

The picture of a stagnating agricultural sector changed only in 1978 when China started a series of fundamental reforms in the rural sector.[2] The new policies dominated by the household responsibility system gave greater incentives and more flexibility in decision making to producers and output growth accelerated to a rate several times the long-term average in the period 1949–1977 (Lin, 1998:528).

Post-revolution agricultural development in China illustrates that the introduction of new high-yielding technology in itself does not guarantee high growth rates in agriculture. The same goes for state intervention in the agricultural sector. In China state intervention was (and remains) indeed pervasive, but as a centrally planned economy relying on rural collectivization and communalization and a development strategy heavily biased towards industry and urban consumers, the character of state intervention in China differs substantially from that of the countries selected for analysis in this study. In relation to our tentative model, it can be argued that not until 1978, when all three of the main factors were interacting – government interventions, market mediation and small-scale, private family farms – did China experience the fast growth rates normally associated with the Green Revolution.

East Asia: the Pre-Green Revolution Case

Although the overall pattern of agricultural change in the East Asian region is complex and varied, it is still possible to identify a sufficient set of commonalities to argue that a particular path of agriculture development took place in all the three countries. This path has emerged as a consequence of a number of common environmental, economic and institutional characteristics conditioning agriculture and, partly, as a result of certain similarities in the agricultural development policies implemented in the three countries. These policies were influenced by specific geopolitical conditions.

Our focus is on both the pre-war and the post-war development processes. Although post-war development carried with it dramatic changes in all three countries, important foundations for rapid agricultural growth based on the application of high-yielding technologies were already laid prior to the miracle years. In terms of rapid agricultural growth all three countries experienced pre-war periods of historically very high growth rates. We begin in Japan in the Meiji period, 1868–1911.

The Japanese origin of the East Asian agricultural mode

Because of a very real danger of colonization by the Western powers in the late 19th century, Japan's new leadership, the Meiji government, embarked on a fast road towards modernization. Progress in education, science and technology was viewed as essential. Throughout the Meiji period the question of how best to combine the necessary Western knowledge with Japanese culture and traditions was debated. The relatively liberal model of countries like the USA was found less appropriate than the more authoritarian model of countries like Prussia. The Japanese turned increasingly to the ideas of the German historical school, which recognized the need for the state to play an active role in driving the economic development of late-industrializing nations. By avoiding *laissez faire* ideals, which indirectly posed a threat to Meiji leadership's own power, and instead selecting an approach which seemed more compatible to Japan's circumstances, the Meiji government to some extent tried to follow the idea of a combination of 'eastern ethics and western science' (Morris-Suzuki, 1998).

Japan started its transition to modernity as a predominantly agrarian economy. For the Meiji leadership, growth in the agricultural sector was imperative as it constituted the only realistic way of financing industrialization and modernization. At the same time, the new regime was forced to consolidate the nation and improve the situation of the vast majority of the people – the farmers. Population growth and urbanization meant that more food had to be grown on existing farm land and surpluses at the farm level had to reach the growing non-agricultural share of the population at reasonable costs. With a very limited scope for expanding the cultivated area, the required growth had to come from a more intensive use of the land.

There was, of course, no blueprint for agricultural development that the Meiji leaders could have followed and we should avoid interpreting policies and their implementation as if they had followed a master plan. The role of the state in agricultural development only gradually emerged and, as forcefully shown by Francks (1984), the political recognition grew step by step that agriculture could not be squeezed unless simultaneously stimulated.

An early learning experience, which also illustrates the urgency with which Meiji government acted, was the failed attempt to develop agriculture based on Western large-scale farm technology. During a 10-year period, the Japanese leadership, impressed by the superiority of Western industrial technology, organized the import and subsequent domestic production of large-scale farm machinery. As it turned out, demonstrations at agricultural stations and training led by British and American instructors at the growing number of agricultural universities and colleges did not, on the whole, give the expected results. Unlike the case in industry, the borrowing of mechanical technology from the Western world was unsuccessful due to its incompatibility with the factor endowments in Japanese agriculture. Efforts to introduce exotic plants and animals were equally unsuccessful (Hayami and Ruttan, 1985).

The government quickly realized its failure, abandoned the Anglo-American model and instead invited German agricultural chemists and soil scientists. This formed part of the search for land-saving, bio-chemical technologies that would raise yields. As it developed, the so-called 'fertilizer consuming rice culture' proved compatible with the resource endowment situation at the time. The biochemical path of technological development in agriculture characterized the Meiji period and dominated Japanese agriculture up until the 1960s, when the overall level of mechanization increased rapidly. During the Meiji period Japan was able to develop a unique and highly productive system of agricultural technology often referred to as the *Meiji Noho* (Meiji agricultural methods) (Francks, 1984; Hayami and Ruttan, 1985). The introduction of these technologies signalled the start of a period (1880–1910) during which Japan experienced rapid increases in agricultural production and productivity.

The Meiji Noho

The *Meiji Noho* bears striking similarities to the modern Green Revolution. Central to the package of technologies and inputs was the use of high-yielding seed varieties of rice. These varieties, like those of the modern Green Revolution, had genetic characteristics that allowed them to absorb large quantities of fertilizer, resulting in larger yields of grain. Most important was the shorter and sturdier stem which could resist lodging and carry the heavier panicles. The absorption of additional nutrients required a well-controlled supply of water and, as in the case of the modern Green Revolution, it was in the areas with well-developed irrigation systems that the *Meiji Noho* package made its first success.

The high-yielding varieties were, however, not created through breeding programmes in the way they were in the modern Green Revolution. Instead the increasing use of the high-yielding seeds was the result of diffusion processes in which farmers seeking for higher-payoff techniques adopted varieties that had been in use for some time in certain locations. Japanese farmers had been experimenting with high-yielding dwarf varieties of rice and wheat prior to the opening of the country in 1868.[3] As shown by Francks (1984), high-yielding varieties were selected and grown in locations with superior irrigation and drainage.

These seed varieties are known as *rono* (veteran farmers) varieties. In 1885, the Ministry of Agriculture and Commerce (founded in 1881) established an itinerant instructor system, in which a combination of veteran farmers and graduates from the agricultural schools were employed as instructors. Travelling throughout the country holding agricultural extension meetings, the instructors contributed to the diffusion of both the best practical farming experience and new scientific knowledge. In some cases adoption was anything but smooth and had to be enforced by the sabres of the police (Hayami and Ruttan, 1985). The most famous of the *rono* varieties was named *Shinriki* (power of the gods). It was selected in 1877 by a veteran farmer surprised by its high yield and soon became widely diffused first in western and then in eastern Japan. Through the late 1920s it was Japan's leading improved variety.

The *rono* varieties performed well only under high levels of fertilizer application. As self-supplied sources were inadequate and also too expensive (extremely labour consuming), farmers had to rely on commercial fertilizers. Fertilizer supply firms exploited the opportunity. Lower transportation costs reduced the cost for herring meal from Hokkaido. After the Russo-Japanese war in 1904–1905, the enormous inflow of a cheaper alternative – the Manchurian soybean cake – followed along with the spread of *Shinriki*. The switch from organic to inorganic fertilizer started in the 1930s with the establishment of the domestic chemical fertilizer industry (Barker *et al.*, 1985).

While gradual in nature the agricultural changes during the Meiji period were indeed dramatic to large segments of farm households. Smaller-scale owners who failed to cope with the new economic and technical conditions lost land and fell into the category of small-scale tenants or part-tenants. Different types of non-farm part-time employment became important for large groups of farm households. However, despite a polarization of the agricultural structure, the great majority of farmers gradually learned to utilize the labour-intensive technology to their advantage. In combination with the expansion of non-farm income opportunities, this contributed to a relatively low degree of inequality in incomes and life-styles.

Cultivating landlords, family-sized farms and farmers' societies

The cultivating landlords, typically two or three households in a 19th century Japanese village, may have owned above average size landholdings, but they remained small-scale operators who personally farmed part of their land. An indication of the difference between this group of landlords and the rural elites in many other societies, then and now, is the fact that the vast majority of Japanese cultivating landlords owned less than 5 ha (Francks *et al.*, 1999). Accordingly, it was their initial commercial advantages rather than technical economies of scale that made them early adopters of technologies. In their role as intermediaries engaging in marketing and the provision of credit as well as rural-based manufacture and small-scale industry, they were able to obtain a number of initial economic and political advantages.

However, as later proved to be the case in the modern Green Revolution, the smaller-scale farm households gradually overcame institutional obstacles. After the turn of the century it seems that Japanese farmers became even more efficient in utilizing the new commercial technology. As it turned out, it was the medium-scale (0.5–2.0 ha) farms which came to form the backbone of Japanese agriculture after 1900. These households, often both owning and renting land, could combine the intensive use of family labour and high-yielding technologies with part-time employment in the growing industrial sector.

Central to the farm household was its integration into the village community and its participation in numerous forms of cooperation at village level. The dependence on irrigation systems encouraged group solidarity and led to the gradual development of communal groups such as the water use associations operating at different hierarchical levels in the irrigation system. This greatly facilitated the spread and further development and refinement of agricultural technologies. In later years, when the development of new, more science-intensive technologies was to replace the *Meiji Noho* techniques, the strong links between farmers and their organizations on the one hand and government research institutions on the other were of great importance. In a sense, the Japanese example of farmer–scientist cooperation predates by decades the participatory approaches presently promoted in agricultural development project in Third World countries.

The increasing role of the state

Apart from a technological backlog in the form of high-yielding varieties, Meiji Japan also inherited a relatively well-developed agricultural infrastructure from the Tokugawa period. Already, at the beginning of the Meiji period almost 100% of the rice fields in Japan were irrigated, although sufficient water supply and drainage was a problem in many regions. As a result, paddy yields in Japan in the early Meiji period of 2.3 tons/ha were a ton above the post World War II level of most South and South-east Asian countries (Barker *et al.*, 1985).

Although starting from a high level of land productivity, Japanese farmers were successful in further intensifying their cultivation and thereby sustaining a steady average annual growth rate of the sector of approximately 1.6–2.0% during the entire Meiji period (Hayami and Ruttan, 1985; Francks *et al.*, 1999). This stimulated the government not only to tax the sector, but also to continuously make investment in agricultural research, extension services and rural infrastructure.

Adding to the positive impacts of investments in rural infrastructure such as roads, railways, and electrification, the Meiji government enforced a number of important reforms. Some were general as, for example, the educational reforms making school attendance compulsory for rural and urban children alike. Others were sector specific. The 1873 land tax reform transformed the feudal sharecrop tax to a fixed-rate cash tax. This meant that the more land a household owned, the more produce it had to market in order to pay taxes in cash. Until the turn of the century, the land tax remained the largest source of state revenue (Nafziger, 1995). It is interesting to note that the tax encouraged or forced the marketing of agricultural products.

In their effort to further promote the diffusion of the *Meiji Noho* production techniques, the government passed the Arable Land Replotment Law in 1899. The new law, hailed by Hayami (1997) as an institutional innovation as ground-breaking as the English Enclosure acts, together with its 1905 and 1909 amendments made participation in land improvement schemes compulsory. In 1923, the government introduced the Rules of Subsidization of Irrigation and Drainage Projects. The previously referred to investments in educational and training capacities were paralleled by those in agricultural research. Starting in the mid-1880s, state run experiment farms, subsequently named experiment stations, were being established across the nation. The gradually evolving scientific capacity became increasingly important in the 1910s when the exploitation and subsequent exhaustion of the potential of the *rono* varieties became evident. Despite the growth in fertilizer use, the rate of increase in rice yield started to decelerate. New varieties were required.

Rice riots and national food self-sufficiency

Around the turn of the century, the domestic supply of rice failed to keep pace with growing domestic demand. The rise in the relative price of rice began to speed up and Japan became a rice importer. Prior to World War I, however, rice imports had been small: approximately 5% of domestic production. Thereafter an increasing share of the urban demand for rice was met by imports. The industrial boom around the time of World War I caused a sharp increase in urban demand and in combination with the increasing centralization of rice trade it led to speculation. Adding to the situation was the long-term deceleration of yield growth reducing the growth in marketable surplus production. The situation exploded in 1918, first in the Toyama prefecture and then riots swept over all the major cities in Japan.

The Rice Riots marked a new era. The inter-war years became a transition period during which Japanese agriculture was unable to meet the demands of an increasingly industrial society. Japan moved from exploitation to subsidization of the agricultural sector. In 1921, the state started to buy and sell rice in the market and gradually, state intervention in the rice market grew as a means to support the small-scale farm households. Japan's first set of agricultural adjustment problems had started to form.

Post-war recovery and growth

While the inter-war years represented a period of stagnation in agriculture, the period of recovery from the devastation of World War II marks the start of a second phase of rapid growth. By 1950, the output of major crops had returned to pre-war levels. The occupation government had given priority to agriculture and with the renewed provision of fertilizer and labour, the sector played a central part in the recovery. Contributing to the fast recovery was the radical land reform implemented 1946–1950. Inspired by Jeffersonian ideals, the occupation authorities designed a reform that would create a class of small-scale rural landowners. The reform transferred all farmland owned by absentee landlords, as well the landholdings of resident landlords exceeding 1 ha, to its cultivators. The subsequent Agricultural Land Law of 1952 placed a ceiling of 3 ha on the scale of landholdings and also further strengthened the rights of tenants (Hayami and Ruttan, 1985).

By turning tenants into owners the US-led land reform had a positive impact on income and asset distribution and thereby contributed critically to the social stability of the rural sector. In spite of its importance in this respect and the stimulating impact it may have had on the recovery, the land reform did not imply a new course for agriculture. The reform strengthened rather than changed the hallmark of Japanese agriculture, the small-scale, family-based, farm household.

Furthermore, we can conclude that the well-trodden technology path also remained intact. As fertilizer use expanded, the rapid adoption of new varieties developed during the inter-war period gave a boost to production. Japan reached the level of scientific maturity in experiment station research and the systems were able to release new improved varieties on a continual basis. Also significant was the spread of small tractors of less than 10 horsepower. This 'minitractorization' was the solution to the relative rise in farm wage resulting from labour migration (Hayami and Ruttan, 1985). By the late 1970s, nearly all farm operations were mechanized.

The war and post-war recovery period also saw the strengthening of cooperation between farmers' organizations and the state. Under the 1947 Agricultural Cooperative Law, village associations were organized into prefectural and national federations. The extension system was modelled after the US system and the agricultural departments of the prefectural governments coordinated extension, experiment stations and agricultural colleges.

In sum, the mentioned post-war developments contributed to the rapid growth of agricultural output. As the post-war recovery took place in Japan, its former colonies Taiwan and Korea also went through years of dramatic change.

The Japanese Pattern Repeated: Taiwan and Korea

As Japan's food problem became apparent during the 1918 Rice Riots, the government turned to its overseas territories for rice imports. This represented a total shift in policy. Previously, in both colonies, rice exports to Japan had been discouraged due to the fear of competition with the domestic rice sector (Hayami and Ruttan, 1985).

The long-term policy, however, shifted to promotion of rice production in the two colonies. In 1920, the government launched a programme titled *Sanmai Soshoku Keikaku* (Rice Production Development Programme). Interestingly this resembled an early version of the modern Green Revolution programmes launched in the late 1960s and 1970s. Under the programme, the Japanese government invested heavily in irrigation and water control and in research and extension. High-yielding Japanese varieties adapted to the ecological conditions of Taiwan and Korea were developed and diffused and chemical fertilizer industries were founded in both countries. Furthermore, investments in transport infrastructure as well as in education were substantial. The resulting rapid agricultural growth in the inter-war period raised the two colonies' share of Japanese rice imports from approximately half before the start of the programme to more than 95% in the 1930s (Ka, 1995). Taiwan and Korea now supplied approximately a fourth of the rice consumed in Japan.

After World War II and the liberation from Japanese rule, the two ex-colonies, both under *de facto* US occupation, experienced dramatic political and social changes. Chiang Kai-Shek's defeated army and the Kuomintang party fled the Chinese mainland and took control of Taiwan. The Korean peninsula was ravaged by war until 1952, when the truce between the United Nations and North Korea led to its division along the 38th latitude. The geopolitical conditions are of course crucial in both cases. Under the threat of an invasion from the North and from China respectively, the South Korean and Taiwanese political elites gained a wide autonomy permitting them to implement reforms that under other conditions might have been successfully resisted by vested interests.

As in the case of post-war Japan, land reform became an important starting point for both countries in the strengthening of an agricultural production system dominated by small-scale owner-cultivators. The land reforms, often referred to as the internationally most successful ones, kept the peasantry politically quiescent. The regimes that were established in Taiwan and South Korea in the

1940s and 1950s were highly statist. The dirigist approach towards development applied by the new regimes did, however, not represent something fundamentally new. Farmers in Taiwan and South Korea had experienced the hard Japanese colonial state, which maintained public order with penetrating administrative systems backed up by a strong police force. As will be shown, in both countries, although with a time lag in the case of South Korea, the regimes set up a mechanism for state intervention in agricultural production that in many respects resembled the one in Japan.

Taiwan

In 1895 Taiwan was occupied by Japan. The new colonial government soon developed local sources of revenues that covered the cost of colonial administration and investment. In order to persuade Japanese capitalists to invest in Taiwan, the colonial state invested massively in infrastructure. And, it focused heavily on the development of agriculture.

Like Japan, Taiwan has experienced two distinct periods of rapid production and productivity growth in agriculture. The first rapid expansion phase, which has been referred to as a Green Revolution by many commentators, started in the early 1920s and lasted until the second half of the 1930s. The second phase is the post-war success story of Taiwan's agricultural development between 1946 and 1970.

Between 1923 and 1937, average rice yields rose by about 50% and the area harvested expanded as a result of both area expansion and a rising multiple cropping index. With rapid growth also in the yield of sugarcane, total agricultural production increased at an annual rate above 4% (Mao and Schive, 1995). This represented a substantial increase compared to the previous (1913–1921) growth rate of 2% annually, which had been achieved mainly through area expansion. Two main factors seem to explain the timing as well as the magnitude of the growth that took place in the 1920s. On the demand side, Japan's sudden need to import large quantities of rice created an opportunity for subsistence farmers to produce beyond domestic demand. By 1937, approximately 50% of the rice and almost all sugar produced in Taiwan were exported to Japan. On the supply side, a real breakthrough occurred around 1925 in the efforts to adapt Japanese varieties to the sub-tropical conditions in Taiwan. In 1926, the *japonica* varieties were officially designated as *ponlai* (heavenly rice).

The new strains were, however, only high-yielding under well-controlled water conditions, requiring long-term investments in irrigation infrastructure. In the case of Taiwan, we have to look back into the early colonial period to see how this development came about.

The 1905 land tax reform, irrigation development and science-based agriculture

Before World War I, the Japanese promoted sugar over rice production. This was partly because sugar imports constituted a drain on Japan's foreign exchange reserves and partly because of the successes in domestic rice production leaving the nation more or less self-sufficient at the time. However, unlike in other sugar-producing colonies, the Japanese in Taiwan refrained from establishing a plantation-type agriculture based on hired labour. Instead the indigenous mode of production, especially family farming, persisted and the Japanese actually strengthened this pattern. In the absence of Japanese private capital, mobilizing local resources to promote agricultural commodity production was the only realistic alternative.

Taiwanese smallholders entered into a system of contract farming under which they delivered sugarcane to Japanese sugar corporations. Although exploitative as such, the system nevertheless contributed to the persistence of a family farm structure that in the mid-1920s could draw benefits from the mentioned rice boom. In 1905, a land tax reform was implemented that annulled the land rights of 40,000 top-level landlords, compensating them with government bonds (Tomich *et al.*, 1995). Tenants became the legal owners and assumed responsibility for land tax payments. The overall aim of the reform, which was preceded by a thorough land survey, was to increase tax revenues. Nevertheless, the fixed land tax reform contributed to the establishment of a unimodal agrarian structure. Although ownership was still skewed and did not become more equitable until the post-war reforms, an agrarian structure of small operational units emerged, more or less from the start of the colonial rule. Contributing to this development was the transfer of Japanese agricultural technologies based on land-saving and labour-using principles.

Of fundamental importance for the intensification of agriculture was the colonial government's heavy investment in irrigation infrastructure. Irrigated area as a share of total arable area roughly doubled between 1903 and 1920 (Barker *et al.*, 1985:98; Ka, 1995:61). The 1920s and 1930s saw continued rapid irrigation development with a growth in irrigated area by some 72% (Barker *et al.*, 1985). The colonial government also invested in agricultural research at an early stage. The Japanese long-term investments in building a research capacity paid off. The development of the *ponlai* rice varieties represented a spectacular success.

Heavenly rice

The *ponlai* series of varieties were developed by cross-breeding of *japonica* varieties or between *japonica* and traditional Taiwanese varieties (*chailai*) belonging to the *indica* subspecies. Early efforts to transfer Japanese varieties to Taiwan had been largely unsuccessful. But the *ponlai* varieties, which suited the Japanese consumers' taste, spread rapidly. Apart from being fertilizer-responsive and high-yielding, the *ponlai* varieties were early maturing, thus permitting double-cropping. By 1935, almost one-half of the paddy field area planted was grown with *ponlai* rice. As in Japan, the existence of well functioning Farmers' Associations was a precondition for the successful development, spread and use of the new technologies, including the growing application of chemical fertilizer after the introduction of the *ponlai* varieties.

As in the 1950s and 1960s, the growth of the agricultural workforce was high in the 1930s (Tomich *et al.*, 1995). During all the three decades, it was the small-scale farms which, due to their successful intensification of land and adoption of land-saving technologies, managed to absorb and keep labour productive, and thereby reduce unemployment. The full potential of this production system was realized with the post-war developments among which the land reforms played the most central role.

Post-war recovery, national food security and the rule of the Nationalists

Taiwan's transformation from a less developed agrarian economy to an industrial high-income economy represents the most cited success story of development in modern times. The links between agricultural production and farmers on the one hand and the growth of rural-based, small-scale industry on the other hand has stimulated many to present Taiwan as a show-case of agriculture-led industrialization. With a closely linked perspective, others place more emphasis on Taiwan's success in promoting equitable growth through its broad-based unimodal strategy of agricultural development.

In the immediate post-war years the country experienced a serious food shortage, but recovery was fast. Access to fertilizers, more labour input, expanded crop area and multiple cropping index and rehabilitation of irrigation facilities contributed to the rapid recovery. On the demand side, farm prices rose quickly as the result of relatively faster increasing food prices during the runaway inflation in the immediate post-war years. In 1951, agricultural output surpassed the pre-war level (Mao and Schive, 1995). The fast recovery of the sector had not been completed, however, by the time that new challenges presented themselves.

In 1949, General Chiang Kai-shek and the Nationalist army retreated from the mainland to Taiwan. An influx of nearly 2 million soldiers and civilians to an island of 6 million people contributed to the acute food scarcity of the period, but also in the medium term the food situation posed a challenge (Moore,

1985). Three factors made it necessary to achieve a continued growth of rice production during the 1950s. The first was the high population growth rate, above 3% per year until 1964. The second was the increasing per capita rice consumption due to rising income levels. Thirdly, Taiwan was heavily dependent on rice (and sugar) exports in a situation of acute scarcity of foreign exchange (Moore, 1985). Consequently, a number of reasons forced the Kuomintang (KMT) leadership to engage in agriculture from the beginning of their rule. Their track record in that field from the mainland was not impressive.

Referred to as a 'catastrophic learning experience' (Brewster, 1967, in Tomich *et al.*, 1995), according to the KMT administration's own self-examination, the expulsion from the mainland of KMT had been due partly to its failure of reducing urban–rural inequality and curbing exploitative relations between landlords and tenants. Having nowhere else to escape to in the case of repeated failure, the reformed KMT party (1950–1952) allowed more room for principled leadership and less room for lobbyists and vested interests. It was determined to obtain the support of the Taiwanese farm population. Thus, as a result of the chances of history, Taiwanese smallholders were suddenly facing a new regime which was eager to get agriculture going. Its first step was land reform.

Redistributive land reform and US support

On the very year of their arrival, the KMT government started to implement a land reform programme. The Land Reform was carried out in three stages between 1949 and 1953. The first step was the Farm Rent Reduction Programme in 1949. The second one was the sale of public land in 1951. The final and perhaps most well-known stage was effected in 1953. This was the so-called Land-to-the-Tiller Act, applying a 3-ha ceiling on rented paddy land. Larger holdings were purchased by the government and paid for with government bonds and shares of government-owned industries.

The reform was a success. Farmers benefited greatly under the programme and the income distribution resulting from the reform brought social justice to the rural areas. The increase in incentives made farmers work longer and harder, adopt new technologies, upgrade water control facilities and participate in community activities. This contributed to the impressive advances in yields after 1952. Between 1952 and 1966 output per unit of cultivated land grew by 4.1% annually.

Like in Japan the land reform was supported by the USA. Strengthened by the success of the Japanese land reforms and challenged by the victory of Mao's agrarian revolution, the US post-war liberal approach to land reform lasted long enough for a US commitment in Taiwan and Korea in the late 1940s and early 1950s. The resistance by US conservatives, who represented the traditional bias of the United States against redistributive reform, gained in strength during the early 1950s and, as a result, the liberal reformists in the US administration were never able to drive through reforms in the Philippines (Putzel, 1992).

The US assistance in the land reform process in Taiwan was administered through the influential Sino-American Joint Commission on Rural Reconstruction (JCRR). Not formally a part of the public service and staffed by both American and Chinese personnel, the organization came to operate as the agricultural arm of the US Mission to Taiwan. Over the 1951–1965 period, its investments in agriculture represented just below 60% of net domestic capital formation in agriculture (Francks *et al.*, 1999). The JCRR achieved a worldwide reputation for its efficiency in channelling the extensive American aid into 'integrated rural development' projects applying a 'Farmers First' approach long before these concepts were discussed in the literature and within the walls of international development agencies.

Functioning as the *de facto* Ministry of Agriculture, and being well financed, the JCRR came to play a central role in the post-war agricultural success. The organization invested heavily in irrigation and other areas of rural infrastructure. It supported and coordinated nationwide research projects, education and extension activities, but perhaps most importantly it supported and strengthened rural institutions and organizations. In this way, it used the previously established Farmers' Associations as the main providers of agricultural services in the villages. With its local level, bottom-up perspective, the JCRR, through the continuous interaction between technical staff and end users, was able to significantly strengthen the capacity of local agencies. Ultimate political control over the JCRR and over key macro-economic policy decisions affecting agriculture remained with the government, but the organization retained a high degree of independent influence over the planning and implementation of agricultural and rural development programmes. Among the different agencies concerned with agriculture, JCRR became the farmers' favourite (Moore, 1985).

Sow and reap: the continued Green Revolution and the intersectoral transfers of resources

In 1956 Taiwanese breeders achieved a breakthrough when the world's first fertilizer-responsive semi-dwarf *indica* rice variety (Taichung Native No. 1) was developed through cross-breeding. It outyielded the best local *indica* varieties and within 2 years of its release it had been adopted by 90% of farmers in a district near the Taichung Agricultural Experiment Station. New varieties, increasing fertilizer use, improved water control and extensive training and extension activities were all factors contributing to productivity growth. Average paddy yields climbed from 2.4 tons/ha in the 1946–1955 period to 4.1 tons/ha in the 1966–1975 period (Barker *et al.*, 1985).

The rapid productivity growth of rice and other crops together with a gradual diversification into high-value crops made agricultural incomes grow steadily. Farm households also increasingly supplemented their farm incomes through non-agricultural employment in the rural areas. Without leaving their farms, household members increased their incomes through full- or part-time employment in the growing rural industry sector and in this way contributed to the supply of labour for industrialization (Andersson, 2003). Rising farm and non-farm incomes seem to account in part for Taiwanese farmers' acceptance of the state's tough direct and indirect forms of taxation.

The government had the right to make compulsory purchases of rice at official prices. Between 1952 and 1968 the official price averaged 70% of the market price (Moore, 1985:146). However, until the early 1970s, the bulk of collected rice was obtained through the rice-fertilizer-barter system, which had begun already in 1948. The government monopoly on fertilizer supply made it possible to require farmers to pay for it in rice.

State intervention was massive. This is seen in, for example, the state's procurement of some 60% of all rice marketed in the 1950s; the state's monopoly of rice imports and exports; its close monitoring of the 'private' rice trade through licensing; making tactic releases of rice on the market as well as restrictive regulations on the movement of rice between 'food zones'. Serving as an important example of a mechanism of surplus transfer out of agriculture, the direct and indirect taxes were hardly unique in comparison with the imaginative methods used by governments in other countries. Rather, it was the magnitude of the transfer that stands out. More surprising than the heavy state intervention *per se* was the way it was administered.

Farmers' Associations

The Provincial Food Bureau (PFB) was the institution responsible for the state's control over the production, collection and marketing of grain. The PFB operated to a large extent through the Farmers' Associations (FAs). Farmers had to hand their procured rice over to local FAs, to which virtually all farmers belonged. The FA collected, stored and milled paddy and distributed much of the rice ration in return for a handling fee from the PFB. The fee seems never to have covered costs (Moore, 1985). The local FAs also stored and allocated fertilizer, almost always in return for rice. Highly regulated by the PFB, which dictated the price and margins used, the FAs provided a cheap and effective way for the state (PFB) to intervene in production and marketing. Through the

FAs, farmers, in a sense, financed an administering system that exploited them while, at the same time, the Nationalist state succeeded in maintaining the appearance of a 'lean state' in personnel terms (Moore, 1985:146).

Farmers' acceptance of the role of the FAs in the surplus transfer system seems to have several explanations. At the broader level, such acceptance must be seen in relation to the generally rising standard of living resulting from increasing both volume and value of agricultural production as well as increasing non-farm incomes. Furthermore, one could speculate that the loyalty of Taiwan's farmers rested also on their liberation from landlordism by the land reforms. The high-level but 'cold' conflict with the People's Republic also may have contributed. At a more specific level, the FAs seem to have remained popular among farmers due to their successful provision of important services, especially the very effective extension service. On the whole, the success behind the FAs, as well as the other significant local-level farmers' organization, namely the Irrigation Associations, seems to be related to the strong local participation. The FAs were organized in every village and elected a chairman among themselves as their operative link to the township Association. However, despite their voluntaristic and participatory forms, neither the FAs nor the Irrigation Associations used their position to act as a pressure group on behalf of farmers. Such efforts would not have been tolerated by the regime.

After the agricultural boom period 1946–1967, growth slowed down. Coinciding with the overall structural transformation of the economy, the agricultural sector gradually began to experience many adjustment problems that other industrializing countries had experienced before them. As the agricultural sector's comparative advantages declined, farmers responded by diversifying their production. To cut labour costs, small-scale mechanization, particularly the use of the so-called 'pedestrian tractors', increased. As concluded by Francks *et al.* (1999:195), from the late 1960s onwards, 'the agricultural authorities had little choice but to abandon the policies through which agriculture had supported the industrial sector and move towards a situation in which the reverse would occur.'

Concluding the Taiwanese section, it can be argued that while the overall importance of the state in Taiwan's miracle development is still being debated, the central role of the state in the agricultural sector during the post-war period of very rapid agricultural growth seems to constitute a less controversial issue that still needs to be underlined.

Korea

While the agricultural sectors of Japan and Taiwan have been broadly described as important contributors to overall economic development, the role of Korean agriculture has been interpreted in less positive terms. According to Mellor (1995) it was the rapidly expanding non-agricultural sector which pulled the lagging agricultural sector into faster growth. For our purposes, however, the case of Korea offers yet another interesting example of how distinct periods of rapid agricultural productivity growth were driven by government interventions and carried through by small-scale family farms applying science-based high-yielding technologies. Like Taiwan, Korea/South Korea has experienced two such growth periods. We start with the one driven by the colonial state during the inter-war period.

Unlike in Taiwan and Japan, the pattern of rainfall distribution through the year in Korea is such that it is possible to grow rice without irrigation. As a consequence, irrigation did not develop to the same extent as it did in Japan prior to the Meiji period. Also, in comparison with Taiwan, where the climate permits double cropping and irrigation, Korea started its 19th century agricultural development with a relatively low share of irrigated area. After the Russo–Japanese war of 1904–1905, Japan made Korea into a Japanese protectorate and in 1910 Korea became a formal colony. Korean rice (*japonica*, round grain) was considered as directly competing with Japanese production and was initially suppressed (Hayami and Ruttan, 1985). On the whole, the Japanese paid less attention to agricultural development in Korea, seeing a much lower agricultural potential there than in the sub-tropical island of Taiwan. Instead, industrial development became more important and the Japanese zaibatsu companies invested in both light and heavy industries, especially in the north where hydroelectric power supplies were available (Francks *et al.*, 1999).

As a consequence of the Rice Riots in 1918, however, the Japanese promoted rice production also in Korea. In spite of the country's climatic similarity to Japan allowing a direct transfer of Japanese high-yielding varieties, yield growth was insignificant. Adding to the puzzle was the Korean farmers' relatively better access to fertilizer, first through their advantageous nearness to Manchuria, from where large quantities of soybean cake could be imported, and later through the large-scale modern nitrogen plants established by Japanese industrialists in North Korea in the 1930s. The explanation is found in the previously mentioned generally low level of irrigation in Korea. In the absence of good water control, the presence of fertilizer-responsive varieties failed to result in higher yields.

The Japanese reaction was to invest heavily in irrigation development. Under the Rice Production Development Programme irrigated paddy area expanded by 52% between 1925 and 1935 (Hayami and Ruttan, 1985:286–290). The time needed for such an expansion and the farmers' subsequent, gradual adoption of seed-fertilizer technology implied a time lag of approximately 10 years compared with Taiwan. Starting in 1935, Korea experienced a 5-year period of remarkable yield growth, catching up with Taiwan's yield levels (single crop) by the end of the decade (Barker *et al.*, 1985; Hayami and Ruttan, 1985). By 1935, rice exports from Korea met 12.9% of Japan's demand for rice, approximately the same share as that provided by the Taiwanese export of *ponlai* rice to Japan.

Thus, a similarity in terms of colonial heritage between Korea and Taiwan is the adoption in both countries of the Japanese-style 'fertilizer consuming rice culture'. As in Taiwan, production was based on small-scale family farm households. A difference in this respect was that ownership was even more skewed in Korea than in Taiwan. While the Japanese authorities tried to control or curtail the traditional power of the Taiwanese landlord class (Ka, 1995:175), the situation in Korea was different. Landlordism spread rapidly under the Japanese rule and the proportion of pure tenant households increased from 38% to 55% between 1918 and 1937. Together with Japanese corporations, the Japanese settlers controlled 15% of all cultivated land in the south including the best agricultural land (Putzel, 1992:78; Francks *et al.*, 1999:107).

Another difference was the relatively harsher administrative practices used by the Japanese in Korea. In post-war independent South Korea the entire agricultural extension system set up under colonial rule was abolished. This was a reaction against the authoritarian agricultural extension practices that had been used and, as a consequence, post-war Korea was left without an effective agency for the promotion of agricultural technology (Moore, 1985:171). In contrast, Taiwanese farmers had accepted Japanese acculturation policies relatively willingly, partly due, it seems, to the intermediary role that farmers' local organizations played in the contact with the colonial authorities.

The post-World War II development – decades of food grain deficits

Almost immediately after the arrival of US forces in September 1945, the US military government began the process of land reform. Land reforms in North Korea and political unrest in the South were forces driving the reform programme in spite of opposition from both the interim Syngam Rhee administration and the US State Department, which initially considered land reforms 'a long-range problem which the Koreans will have to work out for themselves' (Putzel, 1992:79).

In January 1951, still in the middle of the Korean War, Rhee's government, with the full backing of the USA, began land redistribution. By the end of 1952, the reform programme had been more or less completed. No more than 7% of farm families were tenants and the landlord class as such had been wiped out. A 3-ha ceiling was applied and compensation was paid to Korean, but not to Japanese, landlords. The compensation was made in bonds which in practice turned out to be virtually worthless. Like the post-war land reforms of Japan and Taiwan, the Korean reform turned a unimodal agrarian structure of small-scale, family-based 'operational' units' into a unimodal structure also in terms of ownership.

In spite of the redistribution of land ownership and the stronger incentive structure associated with such change, there were no dramatic changes in farm output. Yields were already very high and further productivity growth would not be possible without land improvements and new technologies. Fertilizer use was high as imported fertilizer was being subsidized by the USA. Farms remained extremely small so resources for private investments in land improvements, for example, irrigation and drainage, were limited. On the demand side government policies did not help. The Rhee government, like many other governments following the import-substitution industrial strategy, applied an agricultural squeeze policy but in contrast to the Taiwanese leadership, the Korean government did not reinvest much resources into the sector. Moore (1985:171) points at a number of factors explaining why Korean governments largely ignored agriculture in the 1950s and 1960s.

From a technology point of view, the entire above-mentioned extension system was abolished under Rhee's government; also, no new agricultural technologies were being generated due to the government's decision not to strengthen the agricultural research system. The potential of the transferred and adapted Japanese varieties was already exploited and the new varieties spreading into other parts of Asia were not suitable for the Korean climate. Thus, there were no technologies available that would allow a rapid development.

From a political point of view, the land reform may have created sufficient goodwill in the rural areas for the Korean governments to rely on the loyalty of farmers without making investments in the sector. Finally, and perhaps most importantly, the geopolitical situation contributed to the United States' decision to provide food grains under the PL 480 system.[4] While staving off famine, the large-scale import of American grain between 1955 and 1969 depressed producer prices and negatively affected farmers' terms of trade.

The relative neglect of agriculture did not, however, imply non-exploitation. Contributing to downward pressure on prices caused by the PL 480 imports was the government procurement policies. In the period of 1948–1960, farmers, who were forced by law to sell rice to the government, never received payment covering the cost of production and in 6 of these 13 years the government purchase price was under 50% of the market price (Moore, 1985:172).

Although a double-edged sword, extensive food aid, including the PL 480, was necessary to stave off starvation in the late 1950s. In the aftermath of the Korean War the population was swelled by the influx of refugees from the North and, due to factors discussed, domestic production could not keep up with demand. PL 480 grain imports alone were equivalent to about 9% of domestic production in the late 1950s (Francks *et al.*, 1999:117). Grain imports were necessary and at the beginning of the 1970s the country had to import almost 25% of its grain requirements.

The developmental state and the switch from exploitation to support

With the military coup of 1961, which brought General Park Chung Hee to power, South Korea embarked on its famous export-orientated industrialization strategy. The shift in power also marked the beginning of a more active statism in relation to agriculture. Up until the second half of the 1960s, the shift in the attitude towards agriculture was gradual. President Park himself and others of the new military leadership came from rural backgrounds and although this may have contributed to the enhanced interest for agriculture, the shift can also be seen as an attempt to legitimize the new regime (Moore, 1985:174). The widening gap between rural and urban incomes had become a problem. Shanty towns around Seoul, and, perhaps even more politically dangerous, an erosion of President Park's support in the rural areas, were among the factors that contributed to the need for a shift in policy.

In the early 1960s, a number of changes affecting agriculture were initiated. The Ministry of Agriculture was reorganized and an agricultural research and extension system was re-established in 1962 with US assistance. In 1961, the Law for the Maintenance of Farm Product Prices was introduced and thereafter the government's buying price, although below the market prices throughout the 1960s, always covered production costs (Francks *et al.*, 1999:122). Government procurement increased as the policy focus shifted from providing rations to controlling and stabilizing prices. In 1962, the Fertilizer Control Law placed fertilizer procurement and marketing in the control of the state and the construction of large-scale fertilizer plants was commenced. Two years later, in 1964, a programme to consolidate farmers' fields was launched.

The first fundamental change affecting the economy of farm households did not come until 1968, however, when the government substantially increased the official purchase prices for rice and barley. Subsequent annual increases meant that by the mid-1970s South Korea was applying remunerative prices to farmers well above both domestic and international market prices. This reversal of policy, which would not have passed the test of today's policy pundits in IMF and elsewhere, should be seen in the light of a number of events and processes taking place in the period of 1967–1972.

The South Korean government grew increasingly uncomfortable with the widening food gap, not least in the light of the worsening security situation on the Korean peninsula during the 1960s. As it gradually became clear for the Park administration that the USA was not going to supply food grains on concessionary terms forever and that Washington planned for the PL 480 to be repayable in hard currency, more attention was directed at the growing food deficit. PL 480 terminated in 1970. In this respect, the South Korean case shares several features with the Indian and the South-east Asian cases to be discussed in Chapter 4 (Djurfeldt and Jirström, this volume).

It was at this juncture that the Green Revolution started in South Korea. Korean breeders had difficulties in overcoming the virus and fungus problems that reduced the productivity of the Korean *japonica* rice varieties. As the high-yielding varieties developed by IRRI began to be released, Korean breeders showed an interest in the possibility of crossing the IRRI and South Korean types. After joint efforts, a successful variety named *Tongil* ('Unification') was released on a large scale in 1972. By 1978 over 76% of the rice area was planted with the highly fertilizer-responsive *Tongil*-type varieties. Between 1968, when the policy of remunerative prices was initiated, and 1978 total rice production increased by 67%, equivalent to an annual rate of growth of more than 5% (Moore, 1985:175).

As in the pre-Green Revolutions in Japan and Taiwan, as well as in the case of colonial Korea's own pre-war Green Revolution in 1935–1940, it was not solely the availability of the new high-yielding seeds which produced the rapid growth. As in the contemporary Green Revolution in tropical Asia, the state used a host of instruments to promote the spread of the new technology. One precondition was the existence of well developed irrigation and drainage facilities. Investments in irrigation had been stepped up in the latter half of the 1960s, but even during the period of neglect of agriculture discussed previously, foreign aid was channelled towards irrigation development. As a result, by 1975 only 7% of the paddy acreage remained dependent on rainfall alone (Francks *et al.*, 1999:111).

In contrast to the Taiwanese case, but as we shall see, in consonance with the development in most of the countries of South-east and South Asia, the Korean 'belated' Green Revolution went hand-in-hand with the government's introduction of more favourable price ratios. The Korean government was also increasingly subsidizing the cost of fertilizer, resulting in a 25% increase in fertilizer use between 1968 and 1974 (Moore, 1985:175).

What made the Korean case stand out in terms of a state-promoted diffusion of the new technology was the coercive mechanisms used by the government. In practice, farmers only received the high government buying price for *Tongil* while traditional varieties continued to be handled by the free market. As consumers preferred the taste and cooking qualities of the traditional rice, the government had to rely on what sometimes is referred to as 'considerable encouragement' in order to persuade farmers to accept the new varieties. This included massive propaganda campaigns, privileged credit allocations to adopters, the physical destruction by public officers of non-*Tongil* seed-beds, adoption quotas to be met by public officials in the areas they served, subsidized farm machinery and highly publicized competitions (Moore, 1985).

The government's interventions in the supply and demand side of agricultural production, through the rice production programme and through its willingness to buy grain from farmers at support prices, eventually gave results. Domestic farm production increased while the gap between urban and rural incomes was narrowed. The policies resemble the Common Agricultural Policy in the EU as well as national policies pursued elsewhere in Europe and the USA which all built on price support for essential farm products.

In 1978–1979 South Korea was self-sufficient in rice, its main staple crop. The 'golden age' of the mid-1970s ended, however, rather quickly. The government had to cover the expensive policy of subsidizing both producers and consumers by deficit financing, ultimately, of course, fuelling inflation. Subsidies had to be cut back and for another 5 years, until 1984, Korea had to resort to rice imports. By the mid-1980s South Korea had again become self-sufficient in rice. By then, the transformation into an industrial economy had run a long course and the agricultural sector entered a stage of structural adjustment.

Summary of East Asian Experiences

Before turning to the South-east and South Asian cases, let us preliminarily extract the characteristics of the East Asian model. Tentatively, then, they are:

- The East Asian cases are largely pre-Green Revolution, as conventionally understood. The role of the state in driving the application of science to agriculture is obvious in all three cases. Fertilizer industry was expanded with state investment. Similarly, state intervention was strategic for the expansion and improvement of irrigation schemes and rural infrastructure. Research and extension systems were established by the state and received much attention. Thus the state-drivenness of agricultural development in East Asia is beyond doubt.
- These cases share a political goal, namely self-sufficiency in food grains, which became important *inter alia* due to political factors stemming from the rivalry between states in the international system of states.
- In the Japanese and South Korean cases nationalism had an obvious role in motivating and legitimating agricultural development policies, while in the case of Taiwan, the Cold War and anti-communism played a similar role as an ideological driving force.
- In all three countries, a unimodal agrarian structure based on small-scale, family farms as operational units was in place already before World War II. In all three countries, this unimodal structure was further strengthened as post-war reforms added land ownership to the agrarian structure.
- Albeit driven by authoritarian regimes, the East Asian experiences involve participatory strategies in agricultural development. Farmers' organizations, not least irrigation-based organizations, played an important role in this respect.
- The role of aid was substantial and made a positive difference both in Taiwan and South Korea (See Stein, 1995:33; Bräutigam, 1998:149–152).

One should be wary of generalizing too much from the East Asian model. One question, which we will return to, is to what extent the form and scope of state-drivenness can be explained by specific historical factors, which cannot be generalized to, for example, the contemporary situation in Asia and Africa.

While the productivity gains in staple food production in East Asia were dramatic enough to justify comparisons with the Green Revolution of the 1960s, the development in South-east and South Asia starting in that decade was in at least three ways more revolutionary. First, the magnitude and urgency of the food-population crisis was greater and the production success of the Green Revolution represented a dramatic improvement in food security for millions. Second, breakthroughs in the efforts to breed high-yielding seed varieties of rice and wheat for countries in the tropics represented scientifically much greater achievements. Third, the Green Revolution represented a distinct shift in agricultural policies in the countries involved. In the following chapter we will focus on two South-east Asian cases, the Philippines and Indonesia, and on two South Asian cases – India and Bangladesh.

Notes

[1] As noted in Chapter 1 (Djurfeldt, this volume), China developed and released its own modern varieties ahead of the rest of Asia and managed a large-scale dissemination of a new semi-dwarf *indicia* variety in 1964, 2 years prior to IRRI's release of IR-8.

[2] Before then, during the 1970s, state investments supporting agriculture had been significant, especially the rapid development of the domestic fertilizer production capacity. Farmers' experience of and access to high-yielding seed–water–fertilizer technologies provided an important pre-condition for the rapid take-off in production in 1978.

[3] In 1873, the US Commissioner of Agriculture, Horace Capron, headed an advisory group to Japan. He reported that 'the Japanese farmers have brought the dwarfing to perfection' and that the Japanese claimed that they had shortened the wheat straw so 'that no matter how much manure is used it will not grow longer, but rather the length of the wheat head is increased' (Dalrymple, 1978:11).

[4] Public Law 480 of the USA, the Agricultural Trade Development and Assistance Act of 1954, permitted the export of grain as aid. The Korean government's import of cheap PL 480 food grains, paid for in local currency, is just one example of how American crop surpluses were consciously used in strategic foreign policy making from the 1950s until and throughout the 1970s.

References

Andersson, M. (2003) *Bending the Kuznets Curve – Wrenching and Levelling Forces during Agricultural Transformation in Taiwan and Chile. Lund Studies in Economic History 25.* Almqvist and Wiksell International, Stockholm.

Asian Development Bank (1977) *Rural Asia: Challenge and Opportunity.* Praeger Publishers, New York.

Barker, R., Herdt, R.W. and Rose, B. (1985) *The Rice Economy of Asia.* Resources for the Future, Washington, DC.

Bräutigam, D. (1998) *Chinese Aid and African Development: Exporting the Green Revolution.* Macmillan Press and St Martin's Press, London and New York.

Dalrymple, D.G. (1978) Development and spread of high-yielding varieties of wheat and rice in the less developed nations. Foreign Agricultural Economic Report No. 95. US Dept of Agriculture Office of International Cooperation and Development in Cooperation with US Agency for International Development, Washington, DC.

Francks, P. (1984) *Technology and Agricultural Development in Pre-war Japan.* Yale University Press, New Haven, Connecticut.

Francks, P., Boestel, J. and Choo, H. Kim (1999) *Agriculture and Economic Development in East Asia: From Growth to Protectionism in Japan, Korea and Taiwan.* Routledge, London.

Hayami, Y. (1997) *Development Economics: From the Poverty to the Wealth of Nations.* Clarendon Press, Oxford, UK.

Hayami, Y. and Ruttan, V.W. (1985) *Agricultural Development – an International Perspective.* The Johns Hopkins University Press, Baltimore, Maryland.

Ka, C.-M. (1995) *Japanese Colonialism in Taiwan, Land Tenure, Development, and Dependency, 1895–1945.* Westview Press, Boulder, Colorado.

Lin, J. Yifu (1998) Agricultural development and reform in China. In: Eicher, C.K. and Staatz, J.M. (eds) *International Agricultural Development.* The Johns Hopkins University Press, Baltimore, Maryland.

Mao, Y.-K. and Schive, C. (1995) Agricultural and industrial development in Taiwan. In: Mellor,

J.W. (ed.) *Agriculture on the Road to Industrialization*. The Johns Hopkins University Press, Baltimore, Maryland.

Mellor, J.W. (ed.) (1995) *Agriculture on the Road to Industrialization*. The Johns Hopkins University Press, Baltimore, Maryland.

Moore, M. (1985) Economic growth and rise of civil society: agriculture in Taiwan and South Korea. In: White, G. and Wade, R. (eds) *Developmental States in East Asia, a Research Report to the Gatsby Charitable Foundation*. IDS Research Reports Rr 16, IDS, Sussex, UK.

Morris-Suzuki, T. (1998) Japanese nationalism from Meiji to 1937. In: Mackerras, C. (ed.) *Eastern Asia*. Longman, South Melbourne.

Nafziger, E.W. (1995) Japan's industrial development 1868–1939: lessons for sub-Saharan Africa. In: Stein, H. (ed.) *Asian Industrialization and Africa*. Macmillan, London, pp. 53–86.

Pingali, P., Hossain, M. and Gerpacia, R.V. (1997) *Asian Rice Bowls: the Returning Crisis?* CAB International, Wallingford, UK, in association with IRRI.

Putzel, J. (1992) *A Captive Land – the Politics of Agrarian Reform in the Philippines*. Catholic Institute for International Relations and Monthly Review Press, New York and London.

Stein, H. (1995) The World Bank, neo-classical economics and the application of Asian industrial policy to Africa. In: Stein, H. (ed.) *Asian Industrialization and Africa*. Macmillan, London, pp. 31–52.

Timmer, C.P. (1992) Agricultural diversifcation in Asia: lessons from the 1980s and issues for the 1990s. In: Barghouti, S., Garbux, L. and Umali, D.L. (eds) *Trends in Agricultural Diversification: Regional Perspectives*. World Bank Technical Paper No. 180. World Bank, Washington, DC.

Tomich, T.P., Kilby, P. and Johnston, B.F. (1995) *Transforming Agrarian Economies: Opportunities Seized, Opportunities Missed*. Cornell University Press, Ithaca, New York.

4 The Puzzle of the Policy Shift – the Early Green Revolution in India, Indonesia and the Philippines

Göran Djurfeldt[1] and Magnus Jirström[2]
[1]Department of Sociology, Lund University, Lund, Sweden; [2]Department of Social and Economic Geography, Lund University, Lund, Sweden

Almost simultaneously, but apparently without any connection between them, India, Indonesia and the Philippines made a U-turn in agricultural policies with the introduction of Green Revolution technologies in the mid-1960s. The shift involved the famous package of technology containing new seeds, fertilizer and pesticides as well as other ingredients like credit, improved extension and training, increased investments in infrastructure, irrigation and, what will primarily concern us here, a new price policy. This policy introduced the idea of remunerative prices and built on the presumption that farmers can be stimulated to produce more if they get fair and reasonably stable prices. The Green Revolution, however, was a question not only of remunerative prices but also of increased margins brought about by the increases in productivity made possible by the new technology. That is why we have to define the Green Revolution as a package of technology, inputs *and* policies.

The simultaneous but uncoordinated shifts in policies call for an explanation. How come that, in the mid-1960s, Suharto in Indonesia, Marcos in the Philippines and Shastri in India all abandoned the low-price policies for food grains and instead introduced agricultural policies with a focus on small farmers who were offered a combination of subsidized agricultural inputs and credit on the one hand and price policies set so as to assure the profitability of production on the other? From existing studies it is easy to piece together a credible account for the policy shift taking one case at a time. The question is, however, can the three separate accounts also explain why the shift came simultaneously and, as it seems, against established wisdom among professional economists?

This chapter explores the questions raised above. Apart from the three country cases mentioned and a brief discussion about the global dimensions of the Green Revolution, we call attention to a fourth case, the belated Green Revolution in Bangladesh. The country has experienced rapid growth in agriculture during the past two decades. This means that, a 'Green Revolution like' development has taken place in Bangladesh during a period of structural adjustments and international demands for less regulated grain markets. These preconditions are thus similar to the conditions currently facing African countries trying to get agriculture moving. Presenting the Bangladesh case towards the end of the chapter, we conclude by summing up the common features of all the country cases from this and the previous chapter into an Asian model of agricultural development.

The Philippines

The first new rice variety, IR-8, still unsurpassed in terms of yields, was officially released by the International Rice Research Institute, IRRI, in December 1966, but already in July that year the Philippine government had obtained seeds for testing and multiplication (Mangahas, 1970). In 1967, it was propagated in the country and diffusion was very rapid. In the following year (crop year 1968/69) more than 35% of total rice production stemmed from high-yielding varieties (HYVs) and the new technology had made an impact on national production. The new varieties outyielded the old by over 75% and during 3 consecutive years, 1968, 1969 and 1970, imports ceased and rice was even exported.

The seeds by themselves do not explain the early success. The Rice and Corn Production Council, RCPCC, set up by the Marcos administration, successfully coordinated different government bodies and private actors in road and irrigation construction, extension services, credit facilities and fertilizer supplies (Salas, 1985). Heading the programme was Rafael Salas, Minister of the Cabinet, who surrounded himself with a group of highly educated experts, the 'Salas Boys'. The programme was concentrated in the provinces that ranked highest in terms of past productivity. Within each of these provinces, high-potential villages were designated as programme villages and within these certain farmers – 'cooperators' – were selected for intensive assistance.

The Philippine case must be understood in a political context. With few exceptions, imports of rice as a percentage of total consumption had been modest during the 1950s – on average less than 3% (Bouis, 1982). Increases in area and yield did not keep pace with population growth in the early 1960s and, as prices started to rise in both real and nominal terms, the administration, presumably in response to consumer pressure, began to increase imports. From 1963 to 1967 food grain imports were huge, peaking at 18% of total consumption in 1965. In spite of high imports, Marcos' predecessor, President Macapagal, 1962–1965, was unable to prevent food scarcity. For many Filipinos, the food lines during Macapagal's regime were traumatic experiences which seemed to give the food line a deep symbolical significance in Philippine politics. As one respondent said: 'It was like returning to days of Japanese occupation!' As suggested by another respondent, the food scarcity problem may have been exaggerated for political reasons, but it is generally agreed that the food lines contributed to Macapagal's failure to be re-elected in the 1965 elections. During the campaign, Marcos vowed to ban massive rice imports to encourage national production.

Price policy

Like the governments of several other Asian countries trying to intervene in their rice economies, the post-war governments of the Philippines had to strike a balance between consumer and producer interests. Rice was both an important component of consumer expenditure and a source of income for the large share of the population who were rice farmers. Central to the policies of monitoring the price level through imports were the fundamental ideas of the strategy of industrialization through import substitution (ISI). Low food prices for urban workers formed part of this strategy.

For the Philippine rice farmers, the slow growth in yield throughout the 1950s implied that the policy of low consumer prices was not compensated for by productivity growth on the farms. In the face of stagnant yields, farmers set their hopes on lowering production costs through land reforms. After independence, the idea of raising farmers' income through land reform legislation had been on the agenda but little was accomplished because of the opposition from the politically influential big landowners. The USA had, after half a century of 'colonial rule' over the Philippines, strong political and economic links to the landed elite, something which made the redistributive land reform option politically controversial (Putzel, 1992). Unlike in East Asia, the USA therefore did not support land reform in the Philippines. According to Putzel (1992), US policy clearly promoted agricultural development based on productivity growth as the solution to the food problem and low farmer incomes.

President Macapagal seems to have shared Marcos' striving for self-sufficiency (Bouis, 1982).[1] The long-run profitability of rural investments was recognized before the Marcos presidency but political circumstance did not permit their implementation. In his efforts to make tax-financed productivity-enhancing investments in infrastructure, Macapagal was hampered by the opposition-controlled Congress. Thus, slow implementation of the 1963 land reform bill as well as limited investments in productivity-enhancing infrastructure made increasing imports necessary in order to keep consumer prices low.

In contrast to his predecessor, Marcos received the necessary support from Congress for massive investments in public works expenditure, especially for road building, irrigation and construction of schools. At the same time Marcos and the Congress permitted prices to increase to encourage more production. The government support price for rice was increased by legislation by 33% in 1966. In 1970, a pan-territorial support price was introduced. Marcos combined a sense for the symbolical significance of rice with a strategy combining investments in small-farmer production of rice, in farm-to-market roads and rural schools. The election slogan in the 1969 election – 'rice, road and schools' – drawing on the achievements made during his first term in office, is generally thought to have won Marcos the election. Although the election campaign's promises to put a stop to imports were not kept – rice imports were still necessary during 1966 and 1967 – the bumper crops thereafter put a temporary end to imports.

Self-sufficiency lost and regained

During the 3 years of national self-sufficiency in rice, the Marcos administration made claims of long-run self-sufficiency, although this, according to Mears (1974), represented an unwarranted confidence. The success of the RCPCC programme contributed, in 1970, to a shift in government attention to other crops, principally sugar and coconuts. Simultaneously fertilizer prices jumped by over 50%. Production increases in 1970 were limited and with major pest infestation and extreme weather conditions in 1971 and 1972, the Philippines experienced poor harvests making major rice imports necessary. Agricultural policy was once again focused on achieving food self-sufficiency.

Implementation of a subsidized credit-fertilizer-extension programme, Masagana 99,[2] began in 1973/74 and lasted for 15 years in at least 14 phases with refinements made with each phase. Again, it was areas with better than average production potential that were selected for programme coverage (Mangahas, 1975). During the initial years, the policy relied on an already existing set-up of government and private rural banks and credit cooperatives. A large number of the latter went bankrupt as a consequence of being forced to lend to default-prone farmers. In this respect, then, there was little market mediation, but rather an attempt to socialize rural credit. The financial burden of the distribution of fertilizer, at discounted prices, was reduced by reallocations. The sugar plantation sector, dominated by big estates and the famous sugar barons, whose influence in Philippine policies Marcos confronted, received less priority as subsidies were re-routed to the small-scale farm sector.

During the early years of implementation of the Masagana programme, fertilizer subsidies were high and amounted to 40% of the commercial price in 1975 (Bouis, 1982). After 1976 they were reduced as the world market price of rice came down. The programme covered 40% of the Philippines rice area in 1974 to 1975 – a period during which the world market price of rice peaked. The Philippines reached self-sufficiency in rice a few years after launching the strategy. Throughout the 1970s and early 1980s, national reserves of rice were always sufficient to meet any shortfall in domestic production. Rice scarcity did not occur again until in the 1990s. It has been claimed (Tolentino, 2002) that Marcos and his Secretary of Agriculture – Tanco – through the

Masagana 99 programme saved the country from mass famine.

The Marcos regime continued a long-established policy of state procurement of rice to be used among other things for feeding the army. The role of the food agency (under different names), was mainly the distribution of imports in urban areas. Domestic procurement was limited in the 1960s – less than 2% – but increased significantly under the National Grain Authority (NGA) in the 1970s (Lantican and Unnevehr, 1985). In 1972, the construction of more than 500 buying stations started as the agency was equipped with more funds. Government procurement did not eliminate or replace private trading in rice, but was only an instrument to regulate the rice market. In 1982, 90% of wholesale was handled by private traders (numbering 22,000) and 99% of retail was controlled by 60,000 retailers (Lantican and Unnevehr, 1985).

The launching of the new agricultural strategy was combined with efforts aimed at supporting a remunerative floor price for producers and a ceiling price assuring reasonable prices for consumers. The elements of the price policy were not new, rather it was the mentioned higher level of the floor support price and the resolve to maintain it through investments in storing capacities and transport facilities.

There seems to be solid evidence for Marcos' stronger interest for and commitment to the small scale farm sector. It is more difficult to establish the antecedents and roots of these ideas, which seem to have implied a break with previous policies, dominated by the landowning elite and catering primarily to their interests. Democracy in the form of direct elections of the President is one important factor, because it implies a necessity to respond to the demands and needs of the electorate. Many of our respondents cite the fact that Marcos was the first President to get re-elected as proof of the success of his 'Rice, road and schools' plank.

The oil crisis starting in 1974 paradoxically had a positive effect on Philippine agricultural development. The international banking system was flooded with petrodollars, and credit was easily available for large-scale investments like those made in irrigation (for example, the massive Pantabangan Dam) and in the construction of feeder roads in the rural areas. US aid was almost freely available: you could get support for any well-designed project. This again must have been conditioned by the Cold War context, the ongoing war in Vietnam and the ever-present threat of rural unrest. Our respondents differ somewhat in their interpretation of the significance of agricultural policy as a counter-insurgency tactic and some definitely disputed that it was high on Marcos' agenda. On the other hand it was definitely high on the US agenda, and played a role in channelling aid resources to the country.

US educational support to its allies in the form of scholarships for higher studies in the USA was important. It included economists trained at Cornell in agricultural economics, at Harvard in business administration and at Stanford. US support also included guest teachers and professors at the University of the Philippines. Influential scholars like Vernon Ruttan, Leon Mears, Peter Timmer and J.W. Mellor were actively participating in studies of agriculture and agricultural policies in the Philippines.

Indonesia

There are many similarities between post-independence food policy in Indonesia and the Philippines. In both countries agriculture had developed according to the traditional vent for surplus pattern characteristic of much of South-east Asia (Hayami, 2000). The area planted to food staples increased in response to the expansion of export crop production. Growing world demand for export crops such as sugar and copra stimulated growth in income and population and as long as large unused land areas were available, it was possible to increase staple food production through area expansion. To keep wages low in the labour intensive plantation sector, the colonial governments pursued a food policy emphasizing a low price of rice, the most important consumption item.

While colonial government intervention had been limited and mainly had relied on temporary changes of import and export regulations, the depression of the 1930s changed the picture. A long period of declining prices started due to the Asian overproduction of rice and the world monetary crisis. In this situation, the Dutch in Indonesia decided to intervene actively. Ending a long period of free imports, the colonial government was forced to set up a system guaranteeing a steady and regular supply of rice in all parts of the vast archipelago. The prime notion held by the Dutch, a notion shared by succeeding Indonesian governments, was that rice was too important to be left outside government control.

In Indonesia, as in the Philippines, the low wage policy remained intact as agricultural exports continued to be the most important source of foreign exchange revenues. The policy tuned in well with the import industrialization strategies adopted in the 1950s. Inherited from the pre-independence era were, however, not only the ideas. Also several of the colonial policy instruments were relied upon, including trade barriers, floor and ceiling prices as well as the institutional apparatus in the form of food agencies and the physical apparatus in the form of rice mills, transportation and communication networks. The food policy, however, had to take another factor into account. In order to reduce the impact of high inflation on the real incomes of civil servants and the military, rice rations had been introduced. Consequently, avoiding too high prices for consumers or too low for producers was not enough. Making sure that these politically strategic groups received a fixed part of their monthly salaries in the form of rice implied a commitment that had to be honoured (Timmer, 1981). Thus, rice as a commodity was further politicized.

The food policies became ever more expensive as imports increased. Rice imports to Indonesia tripled in the second half of the 1950s, but in spite of the growing imports, prices doubled in 1957–1958 (Timmer, 1981). President Sukarno then turned to the farmers for help and in 1959 launched a 3-year self-sufficiency campaign. In spite of its innovative introduction of 'village padi centres' providing seeds, fertilizer, training and credit, the programme failed apparently due to the lack of incentives for the farmers. Still at that time the idea of remunerative prices for the farmers seemed distant. Imports in 1962 exceeded 1 million tons and in 1963 Sukarno launched a personal campaign urging the population to substitute maize for rice. The outcry was great and the plans to make maize a part of the rations to the military and civil servants soon were abandoned.

The first rice crisis

As the economic and political crisis of Indonesia was building up during the first half of the 1960s, the rice economy crumbled. Rice production dropped by 14% on Java between 1960–1964 (Timmer, 1981). In 1965, production was only 2% higher than in 1954, the year marking the recovery to the pre-World War II level of production (Mears and Moeljono, 1981). With a rapidly increasing population this implied that the availability of rice per capita was very low, only 92 kg in 1965 compared with the 120 kg recommended on nutritional grounds.[3] In 1963/64 parts of Java experienced a serious drought and Reuters reported in February 1964 that a million people were starving in Central Java (Bresnan, 1993). Imports peaked at 1.7 million tons in 1964, but this did not stop the runaway inflation and rice prices jumped from 200 rupiahs/kg to 1800 rupiahs/kg during 1965 and continued to rise in 1966 (Barker *et al.*, 1985). The mounting food crisis formed part of the overall economic and political crisis.

The fall of Sukarno in late 1965 and early 1966 and a transfer of leadership marked the beginning of a new era – the Suharto era. During General Suharto's and the new leadership's first 2 years in power official attention was redirected to the agricultural sector. Having cut the ties with China and the Soviet Union after banning the Communist party, Suharto had no choice but to turn to the West. Measures supported by the West to curb inflation made a steady impact but the instability in rice price and production continued. In 1967, a below average dry season rice crop again caused severe food shortage and prices doubled during 1967 and had redoubled by early 1968. Coinciding with the domestic difficulties was a problematic world rice market situation. China, Japan and the Philippines bought large quantities, which implied that there was not enough rice on the market.

It was at this point that the government decided to pay farmers an incentive price for their surplus rice. In its early form the strategy was based on the Rumus Tani (farmer's formula), which meant that the prices of milled rice and urea fertilizer should be about equal for the farmer. This was a first attempt at setting an incentive price for rice and it was to be followed by others. The price incentive was combined with a number of other measures aimed at accelerating production by promoting the adoption of HYVs and fertilizer technology. The BIMAS (Bimbingan Massal or Mass Guidance) programme[4], which had been in operation since 1965, was supplemented by the establishment of village-level branch banks of BRI (Bank Rakyat Indonesia) serving BIMAS participators.

In 1968, shortages of fertilizer supplies and domestic credit prompted the government to start the BIMAS Gotong Royong (BGR or 'mutual self-help programme') in which foreign manufacturers of fertilizers and pesticides were invited to participate directly in supplying credit and distributing inputs and management advice to rice farmers and extension staff in certain locations. The programme contributed to the spread of knowledge of the fertilizer technology, but had serious drawbacks. Widespread defaults in the credit programme and capacity problems of individual companies in supplying chemicals for specific problems were some of the major problems. The use of heavy-handed methods such as aerial spraying of large areas without farmers' consent also contributed to the failure of the BGR programme, which was discontinued in 1970 after only four seasons (Barker *et al.*, 1985).

The most important and perhaps well-known institutional change resulting from the new agricultural policy under Suharto was the establishment in 1967 of BULOG (Badan Urusan Logistic), the new food logistic agency directly responsible to the President. Over the years BULOG developed into one of Asia's most powerful food agencies. In spite of early criticism and reports of financial scandals in the organization, Suharto decided to broaden BULOG's functions and strengthen its organization. To a large extent it was BULOG that implemented the new rice price policy.

In 1969 a first attempt was presented to develop a comprehensive operational rice price policy, which, in turn, formed a central part of the First New Order Five-Year Development Plan (Repelita I). According to the policy, BULOG's task was to: (i) support a floor price high enough to stimulate production; (ii) protect a ceiling price assuring consumers a reasonable price; (iii) make sure that the range between the two prices was large enough to allow traders and millers a reasonable profit; and (iv) keep appropriate price relationships both within Indonesia and between national and international rice markets (Mears and Moeljono, 1981).

In 1970 and 1971, BULOG was successful in implementing the new policy. By mid-1972 the new rice programme with its strengthened BIMAS component and remunerative prices looked like major success stories. The new policies were indeed needed. In spite of the rapid spread of the new rice technology and impressive growth rates in production, imports continued to be high. Still, in the early 1970s, serious deficiencies remained. The average supply of calories was only 80% of the requirements and Indonesia continued to rank among the poorest countries of Asia.

The 1972–1973 rice crisis

The South-east Asian drought in 1972 (the El Niño effect) also hit Indonesia and as production suffered the acute rice crisis returned. BULOG was unable to prevent steep increases in retail prices and in some parts of Indonesia rice prices doubled. Efforts to increase imports failed as adequate supplies simply could not be found at any price. In 1973, the continued lack of rice on the world market made imports of more than a million tonnes expensive. The world market price jumped from US$125/tonne in 1971 to US$630 in 1973 before the crisis ended. A political dimension was added to the crisis, when Jakarta's students hit the streets in mass demonstrations against inflation and gradually launched a more general criticism of foreign capital and the direction of economic policy.

The 1972–1973 rice crisis 'galvanized the Indonesian government to a full-scale commitment to rice self-sufficiency' (Bresnan, 1993:118). Adding to the resolve was the clear message from the 1974 UN sponsored World Food Conference in Bandung: there would be no international grain reserve administered by an international agency. The importing countries would have to be responsible for their own food security.

One effect of the rice crisis was Suharto's decision to increase the authority of civilians in the Cabinet. Already, from the start in 1966, Suharto had strengthened the planning capacity at the government level by establishing a long-term cooperation with economic experts including academics – especially the so called Berkeley Mafia, a number of young Indonesian economists trained at Berkeley and other (mainly) US universities. The influence of experts formed part of what Bresnan (1993) describes as the rise of the technocrats – a phenomenon that we will return to as it represented more than an isolated Indonesian development.

Suharto's personal commitment to food self-sufficiency was strong. The one achievement Suharto himself regards as his greatest was Indonesia's self-sufficiency in rice in the mid-1980s.[5] His background as a 'country boy', growing up far from the urban elites, may have been one factor that brought him closer to the everyday realities of the Java peasantry. In any case, the Suharto regime clearly realized the necessity of economic and social rehabilitation of rural Java in order to remove the main cause of the earlier growth of communist influence. The rural vision of the regime resulting in two decades of sustained rural bias included not only the protection and support of agriculture but also substantial spending programmes to increase the provision of physical infrastructure and social services in rural areas.

The strong pro-small-scale agriculture policy manifested itself also in the government's organization and management. Widjojo Nitisastro, one of the 'Berkeley mafia', was head of the planning agency. In weekly meetings, his famous 'rice team' consisting of top civil servants were under considerable pressure to answer questions about the implementation of rice policies (Bresnan, 1993:121). There are apparent similarities with the role played by Marcos' Minister of the Cabinet, Rafael Salas, the 'rice Tsar' of the Philippines in the late 1960s.

Although the visions and plans pre-dated the oil crisis set off by the 1973 Middle East war, the sudden oil price hike gave the oil-rich country economic licence to deal with the rice problem across a broad front. The creation and expansion of a national fertilizer industry formed part of the strategy to become self-sufficient in rice. Support to farmers during the 1970s and early 1980s was mainly indirect through subsidized inputs – fertilizers (30–40%) and credit (25%). In retrospect, the efficiency of the subsidies has been questioned. The fact remains that Indonesia, by the first half of the 1980s, had reached food self-sufficiency and thereby had – given the magnitude of food crises of the late 1960s and early 1970s – achieved a development goal of historic proportions.

India

While the Green Revolution and the food policies surrounding it in the Philippines and Indonesia largely revolved around rice agriculture, the Indian case was different. In comparison, Indian agriculture is very diverse and wheat and other cereals play a significant role making it necessary to consider food grains as a group when trying to understand the country's agricultural policy. Consequently, the early Green Revolution in India was a question of both wheat and rice production, and it was the early successes in the wheat sector that explained most of the growth, especially in the early phase from the late 1960s.

The first national-scale programmes to increase food production started after World

War II as a direct consequence of the 1942–1944 Bengal famine during the last few years of British rule. The colonial state had all too late abandoned its complete *laissez-faire* food policy and at least 1.5 million and possibly up to 3 million or more people perished in Bengal due to the mismanagement of food supplies (Sen, 1981).

After independence in 1947 India was unable to feed its population without importing wheat from Pakistan – a part of India before partition. Western Punjab had for long been a net-producer of wheat for the rest of India and, after the outbreak of hostilities in 1947 over the control of Kashmir, the Indian food security situation worsened further.

Europe, including the former colonial power Great Britain, was not in a position to export cereals to India as the continent itself faced serious food security problems after the war. Apart from imports from Canada and Australia, the USA was the main supplier of grain to India. In 1949, the USA was reluctant to supply the large quantities of grain needed, arguing that this would facilitate a food rationing system, which in turn would create a permanent need for mandatory procurements (Perkins, 1997). Such procurements, it was feared, would work as a disincentive for producers and reduce the chances of India becoming self-sufficient in food grains. However, the rapid escalation of the Cold War – accelerated by the USSR's first nuclear bomb test and Mao's victory in China in 1949 – transformed US policy. Grain supplies for India at reduced prices were suddenly no longer too much of a problem.

The Nehru era

During the first half of the 1950s, food grain production increased at a satisfactory speed (2.5%/year) outgrowing population growth (approximately 2%), mainly due to area expansion and some investments in irrigation (Barker *et al.*, 1985). However, by the mid-1950s it became clear that India was going to follow a strategy of de-emphasizing agriculture in favour of industrial development. As for agriculture, Nehru and others on the Planning Commission were inspired by China's agriculture in which substantial yield increases were claimed to have been achieved through altering the socio-economic structures of agriculture. The dedication to social rather than technical reform had already been emphasized through the launching of the Community Development Programme in the early 1950s.

During the latter part of the decade production stagnated and large imports were again needed. India combined commercial imports with PL 480 grains from the USA. Consequently prices remained low for the domestic producers, leaving no incentives for productivity increases. As food shortages grew and prices threatened to increase, Nehru decided to substantially increase the domestic grain procurement at low prices. Agricultural producers realized that further downward pressure on prices was to be expected and unrest was growing.

At this point, Nehru agreed to let a team of American experts organized by the Ford Foundation prepare a study on the problems of Indian agriculture.[6] Their 1959 report, *India's Food Crisis and Steps to Meet It*, called for a new approach to agricultural development. The central outcome of the recommendations of the report was the setting up of the Intensive Agricultural District Programme (IADP) in 1961. The IADP was based on a 'package' approach to increase India's agricultural yields. It consisted of a combination of institutional, economic and technical innovations to be implemented at the district, block, village, farm and field level. On the basis of one pilot district in each of seven states 'India would attempt to marshal all of the inputs, to be made available to capable farmers, needed for intensive high-yielding practices' (Perkins, 1997:182). Apart from the technical components – improved seeds, fertilizer, irrigation, and pesticides – the package approach also stressed the importance of adequate credit facilities, technical advice and a guaranteed price providing an incentive to accept the risk of trying a new technology. However, according to Barker *et al.* (1985), the attempt to implement special product price supports in project districts was unsuccessful because of the larger market forces at work. Thus, remunerative prices did not really form part of the implementation. In 1963 the programme was expanded to seven new districts and in 1964–1965, a new programme was announced covering 100 districts. The new programme – the Intensive Agricultural Areas Programme (IAA) – had far less financial backing than its predecessors.

In spite of the launching of the intensification programmes, agriculture continued to stagnate. Careful programme evaluations were unable to show faster growth in either the IADP or the IAA districts as compared to non-programme districts. According to an expert committee, the main obstacle to improved performance was the low yield response to fertilizer and other inputs of the then recommended varieties of food grains, which were locally developed and without the dwarfing genes of the high yielding varieties to come (Barker *et al.*, 1985). In 1965 and 1966 severe droughts were to cut production further and India was not well prepared.

The Shastri interregnum and the policy shift

Nehru died in 1964 and was succeeded by Lal Bahadur Shastri. During the Shastri interregnum, before the emergence to power of Indira Gandhi, important changes took place in India's development policies, including those in agriculture. Three authors, Frankel (1978), Varshney (1995) and Perkins (1997), have dealt with the shift in policies. Frankel and Varshney both build on interviews with a large number of key persons, while Perkins adds important material based on US records.

For Frankel, the Shastri interregnum is important because it led to a quiet dismantling of Nehruvian policies and principles, among them those in the fields of agriculture and planning. Nehruvian price policies were of the standard variety pursued in many countries in the 1950s and early 1960s, stressing import substitution.

Frankel and the other authors stress the important role played by C. Subramaniam, who during his term in office as Minister of Agriculture completely changed the price and procurement policies. In one of his speeches he describes the situation he met with when moving from the Department of Industries to that of Agriculture:

> . . . no industrial unit can progress and succeed unless it is a profitable concern. If it is a losing concern, no industry can prosper. I looked at agriculture from a similar point of view and, after study and analysis, came to the conclusion that Indian agriculture was a losing concern for the farmer. He did not receive a return commensurate with his labour, or with the investment he was prepared to make. This was mainly because of the price policy which had been adopted since independence.[7]

True to this spirit, Subramaniam took the initiative to form the Food Corporation of India, a major player in the rice and wheat markets. From the late 1960s the organization had the mandate to buy at the prices proposed by the Agricultural Prices Commission (APC), also formed by Subramaniam. The Food Corporation procured between 10% and 20% of the marketed production. The APC was instructed to collect data on farm economics and to suggest procurement prices to the Minister. These prices were meant to ensure a fair level of profit and to facilitate the adoption of the new technology.

The introduction of remunerative prices was combined with the emphasis on the agronomic component of the package concept earlier introduced. Narrowing down the objective and de-emphasizing the institutional support in the credit and cooperative fields, the new strategy was embodied in the High Yielding Varieties Programme (HYVP) and involved a concentration of seeds, fertilizer, and extension in areas with high quality irrigation conditions. By betting on the potentially productive areas at the expense of others, the shift in 1966 thus represented a turnaround from the egalitarianism under earlier programmes.

Shipments of semi-dwarf wheat seeds from Mexico and rice seeds from IRRI were rapidly supplied to the promoted areas and in 1967 2 million ha were planted under the HYVs. In 1971, the area under HYVs had increased to 15 million ha and by 1975, 27 million ha were covered with the new varieties. It has been estimated that HYVs supplied

62% of India's total cereal output in 1975 as compared with 6% in 1967 (Barker *et al.*, 1985:245).[8]

With imports falling from 10 million tons in 1966 to about 3 million tons between 1971 and 1973, the more comfortable situation brought about a return to the problems of poverty and equity. However, when import costs jumped during the 1973–1974 world food crisis, the focus again shifted to reach national self-sufficiency in food grains. As a result gains in production continued during the 1970s and by the early 1980s a satisfactory situation with modest imports, or even exports, each year became the norm. Although it is possible to talk of an early success of the Green Revolution also in the case of India, the benefits of new seed technology actually played a more important role after 1975 than they did in the early years (McKinsey and Evenson, 2003). From the late 1970s to 2000, a period during which the Indian population doubled, food production more than doubled, much as a result of the spread of the Green Revolution within the country (see Otsuka and Yamano, Chapter 13, this volume).

Returning to the shift in policies in the mid-1960s, Frankel (1978) sees them as a consequence of developments inside the ruling Congress Party. For the mobilization of voters, the party was dependent on the rural elite. The interests of this very elite, however, were hardly considered at central level under Nehru's rule. He had the charisma to impose his moderate socialism onto a party, whose rank and file were conservative rather than socialist in their inclinations. The rural elite was more influential at state level, which explains a great deal of the dilemma in Nehruvian policies, namely that the plans remained paper products; when it came to implementation at state level the plans and policies of the Centre were actively or passively sabotaged.

After Nehru's death, Shastri was the choice of the party bosses, a group of leaders with their main backing in their respective states and among the rural elites. Frankel interprets the change in agricultural policies as the result of this shift of power. Higher prices for agricultural produce were in the economic interests of the rural elite. Varshney (1995) makes the important point that remunerative prices are beneficial to all agricultural producers who sell at least part of their crops, i.e. to the vast majority of Indian farmers. His interpretation is slightly different, *viz.* that Shastri and Subramaniam signified a first step in the democratization of the Indian polity, in which the majority of the population, i.e. farmers, started to make their voice heard, not only at state level, but also at the national level. Frankel and Varshney share what may be termed a nationalistic interpretation of the new agricultural strategy. Perkins' focus is slightly different: he sees the Green Revolution as a global process and he regards the role of the USA as fundamental.

To conclude, compared with the cases previously discussed, the Indian Green Revolution has two specific features, which must be mentioned. The first one concerns the slower growth rates, at least initially, and the second one the limited impact on poverty.

The Indian Green Revolution is symbolized by the early breakthroughs in Punjab and Haryana. Here we had the heartland of the Indian and Asian wheat revolution with varieties directly imported from Borlaug's breeding programme in Mexico. But the IRRI varieties of rice also proved suitable to growing conditions in the two states, which traditionally were not so big in rice. These two states for a long time stood for a disproportionate share of the food grains – i.e. of both wheat and rice – procured and used in the public distribution system. The slower impact in the rest of the country was due, among other reasons, to lesser suitability of the IRRI varieties. Not until the national breeders had developed cross-breeds of domestic high-yielders and the IRRI varieties could advances be made in the traditional rice growing areas in East and South India. This accounts for the much higher rates of growth in the late 1970s and in the 1980s (Bhalla, 1997, 2001).

The persistent poverty in India, finally, would seem to contradict our argument and although the distribution effects are not central in this analysis, they deserve a comment. It was not until the late 1970s that the Indian Green Revolution had any impact on the poverty rates; and much of the progress on this front can be attributed to overall development in agriculture (Datt, 1998, 1999) (see Otsuka and Yamano, Chapter 13, this volume).

There are obviously many reasons why poverty continues to be widespread in India, despite agricultural growth. A prominent reason can be found in the deficiencies in the public distribution system for food grains. Very briefly, this system is far from effective in reaching the target group of the poorest and most vulnerable sections of the population. So, for example, the system works much better in relatively better off states than in, for instance, states like Bihar, Orissa, Madhya Pradesh and Rajasthan. Moreover, the system is not immune to discrimination against minorities, like scheduled castes and tribes. Partly for this reason, India is in the paradoxical situation of combining a food surplus and large grain reserves with rampant poverty.[9]

The Global Dimension and the Role of the USA

Ecologically the Green Revolution was a global process; it institutionalized a process of global diffusion of genemass that is as old as agriculture itself. The Mexican dwarf wheat, for example, which was used for the Indian wheat revolution in the Punjab, was the result of a crossing of Mexican varieties with Japanese plant material, brought to the USA after World War II and made available to the breeders employed in Rockefeller's Mexican Agricultural Program.

Similarly, the new rice varieties were crossings of plant material from different lines of *Oryza sativa* that had developed separately over the centuries, i.e. the *japonica* and *javanica* families. The breeding programme was carried out at the International Rice Research Institute, IRRI, in the Philippines. IRRI was and is part of the Consultative Group on International Agricultural Research, CGIAR, a donor-funded international research organization, which in the 1960s was funded largely by Rockefeller and Ford Foundations.

Geopolitically there are also important global dimensions to the Green Revolution. As was earlier the case in Japan, Taiwan and South Korea, the US input to the modernization of agriculture in South-east and South Asia was very much shaped by the Cold War and the anti-communist agenda. To these concerns, Perkins (1997) shows that, from the 1960s and onwards, a new element was added, namely the concern with the growing population. Among American academics, neo-Malthusianism gained increasing influence, having started in the late 1940s but becoming politically influential only during the Kennedy administration.

One reason why the apocalyptic visions of the propagation of the teeming millions in Asia gained political influence was the apprehension that overpopulation and ensuing food scarcities would fuel the communist movement. The urgency of these concerns increased with the war in Vietnam and led, Perkins argues, to a change in American policies. From an earlier stress on food exports, nicely tailored to the domestic concerns with overproduction of wheat, US policy moved to stress export of technology, rather than export of surplus grain. This again led to the two big Foundations investing considerable resources in developing new technologies for rice farming (Perkins, 1997).

Concluding Remarks on the Policy Shift

The problem with the nationalistic interpretations of the change in policies taking place roughly at the same time in the three countries under discussion is, of course, that they would have difficulty in explaining the simultaneous developments. For this reason, and disregarding the nationalistic sentiments that may be hurt, it is difficult to avoid the conclusion that the Green Revolution was a global process, to a large extent engineered and steered by the USA.

What is interesting, especially compared with the current situation, is that, driven by concerns with the communist threat and fed by visions of a neo-Malthusian apocalypse, US policies stressed the export of technology and the need to make countries technologically capable of attaining self-sufficiency in food grains (Perkins, 1997). This is much in contrast to the current situation where the CGIAR system is starved of resources and where crucial technologies in plant breeding are controlled by private companies rather than by institutions in the public domain. Today export of technology is increasingly based on commercial principles, while US and also European policies have reverted to the dumping of surplus grain on the world market. This again is creating difficulties for late-comers to the Green Revolution while, fortunately, countries like India have gained the necessary competence largely to pursue their own development of agricultural technologies.

The global and geopolitical dimensions of the Green Revolution do not invalidate the nationalistic explanations expounded by authors like Frankel and Varshney in the case of India. On the contrary, the two types of explanations complement each other. There is much to Varshney's argument that the New Agricultural Policy of C. Subramaniam was a step in the democratization of agricultural policies, and that it reflected the interests of the agrarian and rural population in Indian politics. Something similar can be said about both Indonesia and the Philippines.

When Marcos was elected it was on a plank confronting the sugar barons and other landlord interests, while flirting with the ordinary rice farming electorate with the slogan 'Rice, roads and schools'. Similarly, although Suharto did not come to power via elections, his overriding concern in establishing the legitimacy of the military regime was to win the Javanese population consisting of a majority of small rice farmers. This again had a background in the crushing of the PKI, the Communist Party of Indonesia and the pogroms against its supporters in 1965. The new agricultural strategy was part of the effort to undermine agrarian radicalism and to build up rural political support (Rock, 2002).[10]

One of the most persistent myths about the Green Revolution was that it mainly benefited large farmers and that it contributed to a concentration of landownership, massive proletarianization and pauperization. As is increasingly realized, this is far from what happened (see for example Lipton and Longhurst, 1989; Hazell and Ramasamy, 1991; David and Otsuka, 1994; Pingali *et al.*, 1997).

Since the distributional effects of the new technology were not as foreseen by contemporary leftist and radical critics, but more like an all-win game, the strategies formulated in the late 1960s bore fruit. With large sections of the agrarian and rural population gaining from the new policies, the Suharto regime won widespread legitimacy. Something similar can be said about the Philippines. Although our interviewees on the whole doubted that anti-insurgency was part of Marcos' strategy, it remains a fact that the militancy of the Huq subsided, not to re-emerge on a large scale until the late 1980s.

The puzzle of the role of the economists and the academics remains. As stated in the introduction, the policies adopted by the three governments in the late 1960s were clearly against contemporary economic orthodoxy. In the case of India, this is especially clear. Planning was at the core of the Nehruvian strategies; the Planning Commission was not only at the high command of the Indian development strategy, it was also the vehicle for left-leaning economists and statisticians to put their stamp on the policies pursued. A major adversary of the New Agricultural Strategy was the Planning Commission and the left within the Congress Party. They favoured a low-price policy for food grains, and they were opposed to the new technology because it required large investments in fertilizer and agricultural and rural infrastructure which would be a drain on the resources committed to industrialization.

Shastri and Subramaniam did not belong to the Congress left; they were not economists and they did not try to give a theoretical underpinning for their priorities. To implement their policies they had to move against both the Congress left and the supremacy of the Planning Commission in development policy.

Frankel interprets the Shastri interregnum as a first step towards the liberalization of the Indian economy, while Varshney, as already mentioned, sees it as a result of the increasing political influence of the farming population. We need not resolve this difference in interpretation, but instead stress that the academic economists seem to have had more influence under the old order than under the new.

When it comes to the influence of intellectuals and of the economists, in Indonesia, the role of the Berkeley mafia is well-known (Bresnan, 1993:83ff.). They made up what in modern parlance would be a think-tank with regular meetings with Suharto and which undoubtedly had great influence on the formulation of the new policies, including the price policies. What is puzzling is that the advocated policies were not in line with contemporary mainstream economics. Something similar can be said about the Philippines where the 'Salas Boys' had a reputation and standing similar to the Berkeley mafia's.

There are two possible explanations for the fact that the academic think-tanks gave advice to their respective governments which was not in line with mainstream economics. One reason may be that they were influenced by new debates in the 1960s. *Transforming Traditional Agriculture* by T.W. Schultz was first published in 1964, but gained widespread influence only later (Schultz, 1964). Similarly, J.W. Mellor published an influential book in the mid-1960s (Mellor, 1966). As already mentioned, Mellor visited the Philippines, as did V. Ruttan, whose important work co-authored with Hayami appeared in 1971 (Hayami and Ruttan, 1971). Others can be added to the list. These personalities may have had an influence on the Berkeley mafia and Salas Boys, because their views of agricultural development were more in tune with the new strategies pursued than with established orthodoxy.

This first explanation, stressing the influence of American academics, would add to the global dimensions of the Green Revolution which, as we have seen, are indisputable. An alternative explanation would tone down the American intellectual influence and concentrate instead on the intellectual abilities of economists and other social scientists associated with the early stages of the Green Revolution. It would stress their ability to independently analyse the agrarian problems of their own countries and, confronting both mainstream and Marxist economics, stress that a resolution of the food crises in their respective countries required a different price policy than had hitherto been dominant.

Which of the explanations is most credible? The intellectual histories of the three countries have not yet been written in such detail that a final answer is possible. It would have to await the work of future generations of historians.

Before making an attempt to sum up the two chapters on Asian experiences and to single out the significant common features making it possible to suggest an Asian model of Green Revolution development, we focus briefly on Bangladesh, a latecomer to the group of Asian countries which, supposedly against all odds, seems to have reached national food self-sufficiency.

Bangladesh

Bangladesh does not belong to the group of Asian countries in which the Green Revolution of the 1960s and 1970s represented a shift in agricultural policy and a leap in domestic grain production. Instead, rapid spread of modern varieties took place only during the 1980s and 1990s. The country, which still in the 1980s was referred to as a hopeless case in terms of food security, has moved from a situation of chronic food deficiency to one of relative stability in which production in a normal year suffices to meet the demand of the population. In 1998, the worst floods in living memory hit the country; still, the performance of the crop sector in 1998–1999 was one of the best since independence.

The belated Green Revolution of Bangladesh is of interest for this study precisely because of its relatively late occurrence. Bangladesh's Green Revolution can be claimed to have taken place during a period when calls and demands were made for structural adjustments, liberalization of markets and a de-emphasis of the role of the state in the development process. In other words, preconditions were quite different from those facing the Philippines, Indonesia and India in the late 1960s and 1970s and more similar to those presently faced by many African states.

A market liberal interpretation of Bangladesh as a latecomer in the Green Revolution would emphasize the positive effects of the policy changes in the agricultural input markets in the 1980s. The most important ones are undoubtedly those related to irrigation management. They represent the final steps in the gradual metamorphosis from domination of publicly owned and bureaucratically managed systems to a situation where a substantial share of facilities are privately owned and managed (Zohir *et al.*, 2002). By removing all restrictions on importation, standardization and placement of tubewells and pumps, a virtual 'tubewell revolution' took place. The increase in the number of shallow tubewells fielded was spectacular – between 1987 and 1996 their number grew from 183,000 to 624,000 (Zohir *et al.*, 2002). Destandardization and reduced import duties made less expensive Korean, Chinese and Indian engines available to farmers and the increased competition caused a general fall in the prices of tubewells.

The improved availability of cheaper irrigation equipment resulted in a rapid expansion of groundwater irrigation. Still in 1980, less than 13% of cultivated land was irrigated. The share had risen to 30% in 1990 and to more than 50% by the turn of the century (Zohir *et al.*, 2002). Improved control of water facilitated the adoption of modern varieties of rice along with an increasing use of fertilizer. While the proportion of cultivated land covered by modern varieties increased from approximately 20% in 1980 to 65% in 2000, the use of NPK fertilizer rose from 30 kg/ha to 99 kg/ha during the same period (Zohir *et al.*, 2002).

If a more intensive application of Green Revolution technology explains the rapid production achievements made during the past two decades, the questions that follow are why this development came later than in most other Asian countries,[11] and moreover, if it was delayed due to bad policies.

The non-take-off period

It was not a lack of access to the new technology that blocked the early development. High yielding varieties of rice, IR-8, IR-5 and IR-20, were introduced in East Pakistan (Bangladesh in 1971) at about the same time as in most of rice growing tropical Asia.[12] Their impact was, however, very limited and at independence in 1971, still less than 5% of the total rice area was planted with high-yielding rice varieties (Hossain, 1989).

There are several intertwined explanations of why the Green Revolution never took off in the initial stages. During the second half of the 1960s, the food situation was still not considered alarming by the central government. Since the partition from India in 1947, East Pakistan had become dependent on West Pakistan for food imports. During the 1960s grain imports from West Pakistan and foreign countries increased and in 1969–1970 reached the level of 1.3 million tons (Faaland and Parkinson, 1976). However, food grain imports from West Pakistan formed part and parcel of the overall strategy of development of Pakistan. Faaland and Parkinson (1976) describe the two-pronged strategy according to which, on the one hand, large-scale irrigation and drainage systems and even major river flow control schemes were to improve the conditions for agriculture; and on the other hand, large and growing quantities of food grain imports were to be financed 'through industrialization, import-saving and export-earning developments, and by foreign assistance' (Faaland and Parkinson, 1976:129). It is well known that not much of these grand plans was realized during the 1950s and 1960s.

Although imports were necessary and had been so for long, a rather rapid growth in rice production during the 1960s reduced the overall pressure. Rice production grew annually by nearly 3%, mainly as a result of more intensive land use. The cropping intensity rose from 130% to 148% during the decade (Hossain, 1989). By the end of the 1950s the land frontier had closed and as population growth was picking up during the 1960s, the pressure on the land increased and farmers explored ways of planting two crops per year.

Gradually, however, the options to increase production through intensification narrowed. Yields per crop increased only marginally and as concern for the future food situation grew, modest measures were taken by the East Pakistan government to accelerate rice production through the introduction of the new Green Revolution technologies.[13] The central government of Pakistan was fully engaged in the spread of Green Revolution wheat and rice technologies in West Pakistan. As for its eastern wing – East Pakistan – the central government considered the lack of water control and the small size of farm holdings, as compared with the large-scale production systems in West Pakistan, as a serious constraint and consequently, according to the views held by some of our interviewees, never invested much in the development of a Green Revolution in East Pakistan.

By independence in 1971, the Green Revolution had not had any significant impact on production. The situation deteriorated during the first half of the 1970s. An opinion forwarded by several of the interviewees was that several factors combined to make this a period of stagnation rather than one of Green Revolution development. The war of liberation with Pakistan and the resettlement of ten million refugees who had fled to India during the war represent one factor. The very serious droughts and floods from 1972 to 1974 and successive crop failures made the situation disastrous, with thousands of people starving to death in late 1974 and early 1975. On the top of this, there was the coup of 1975 in which the Prime Minister Sheik Mujibur Rahman was killed. It was only in 1976 that production recovered to the pre-independence level.

In spite of all these disasters, the Green Revolution technologies spread, although slowly, during the decade with almost a fifth of the cultivated area under modern varieties in 1980. In 1979, the Medium Term Food Production Programme (MTFPP) was launched. Although self-sufficiency for long had been an elusive goal, the MTFPP has been claimed to be the first national programme set up to reach the goal (interview). Still in the late 1970s, however, the major constraint to a more rapid Green Revolution was the flooding of land during the rainy season and the lack of irrigation facilities during the dry season. Investments in irrigation development had continued along the large-scale strategy and only gradually did small-scale groundwater technologies grow in importance. The semi-governmental organization responsible for the procurement and distribution of modern irrigation equipment was the Bangladesh Agricultural Development Corporation, BADC. Set up as a parastatal in 1963 (East Pakistan Agricultural Development Corporation) in line with World Bank recommendations, BADC had sole control over the procurement and distribution of not only irrigation equipment but also fertilizer, improved varieties and other types of agricultural machinery.

In the late 1970s, private sector participation in the input markets began as a result of policy changes and BADC, which still exists in spite of pressure from international donors to the contrary, gradually lost its complete control over these markets. Representing a step-by-step development during the 1980s with a strong push in 1988 when standardization requirements were removed, the liberalization of the agricultural equipment markets resulted in a substantial acceleration of the area under irrigation in Bangladesh, especially during the 1990s. As a direct effect, a more intensive application of Green Revolution technologies took place.

Late or delayed?

Could the last two decades' achievements have been reached 10–20 years earlier? Departing from our proposed three-actor model – state–market–farmers – it can be argued that relative to the role of the state, the private market forces were given little emphasis in the young country's first decade of development. As for the role of the small-scale farm sector, it can be argued that the Bangladeshi farmers have, in spite of difficulties, proved to be as eager to participate in and drive agricultural development as farmers elsewhere. Institutional changes, including the growth of, for example, cooperatives and the NGO-sector, have been important. Farmers' willingness to adopt new technologies has been great, not least when taking into account that it has taken place during periods of great political and economic uncertainty.

So, the question remains of whether a more market liberal approach including deregulation of markets could have been successful at a much earlier stage. According to most of the persons interviewed, the answer is clearly negative. As pointed out by one respondent, in the early stages of development 'there were hardly any markets at all – output or input.' Markets had to be developed, farmers had to learn to apply new technologies – irrigation, fertilizer and seed technologies – and rural infrastructure, such as roads and electrification, was very weak. Still in the late 1980s, only 30% of farmers had a marketable surplus and the rather slow growth in effective demand for food created a rather drawn-out development of the markets for surplus producers. Although rarely remunerative, the government's control of cereal prices through procurements and regulation not only benefited net consumers, including a substantial share of farm-households, but also reduced price fluctuations, an important factor for the small-scale grain producer.

We end this glance at the belated Green Revolution in Bangladesh with a quote addressing the question whether market liberalization could/should have been implemented earlier. Pointing at the positive effects of liberalization and privatization of input markets, Ahmed (2001:48), however, also reminds us of the role of the government:

> While the measurement of the impact of liberalization warrants a distinction between pre- and post-reform periods, the government's role in the pre-liberalization period (creating fertilizer and irrigation markets and nurturing them to maturity) must not be underrated. The efforts of research and extension departments and the construction of rural infrastructure accumulated over time to build a foundation for growth in agricultural productivity. The impact of market liberalization could have been smaller without the cumulative effects of market development in the pre-liberalization period, as is observed in many developing countries, particularly in Africa.

Asian models: conclusions

Let us now try to round up this exercise of comparative analysis. Obviously, it is not entirely misplaced to speak of an Asian model of agricultural development, because some factors recur in all the cases that we have discussed. Forgetting the specificities for a moment, then, the common features are:

- State intervention was strategic for the expansion and improvement of large-scale irrigation schemes and rural infrastructure, for expanding capacity in fertilizer industry, and for the national agricultural research and extension systems, which played a prominent role in the process. This holds for all cases, from the early Japanese development to the Asian latecomer, Bangladesh.
- We use the terminology of *state-drivenness* and *market mediation* to render the fact that a state-driven process of development does not necessarily imply that market mechanisms have no influence. Administratively regulated markets are an outstanding characteristic of the Asian model of agricultural development, but within the framework of these markets, private commercial activities were significant.
- A third characteristic of these state-driven, market-mediated processes of development is that they were small or family farmer based, and that the unimodal character of agrarian structures grew even more pronounced in the process.[14]
- A price policy assuring profitability to smallholder agriculture seems to be a common feature: we have given much emphasis to the U-turn in agricultural price and trade policies which occurred at about the same time in the mid-1960s in India, Indonesia and the Philippines. We have stressed that this was an essential, although often neglected, part of the Green Revolution policy package, and a precondition for the spurts in production. It was also a means of assuring that the new technologies became small farmer-based. Moreover, it seems

sensible to view this as an indication of an *incipient democratization*, mirroring the greater influence of the rice farmers in national politics. Similar, although not so dramatic, developments can be traced in Japan and Taiwan, where economic carrots and sticks were among the instruments used in inducing agricultural development. South Korea finally had a U-turn much like the South-east and South Asia cases and at about the same time.

- These cases share a political goal, namely self-sufficiency in food grains, which became important due to political factors stemming, *inter alia*, from the rivalry between states in the international system of states. Achieving self-sufficiency became important for regime survival, but it was also a goal promoted by the donors, especially by the USA.
- In all the cases, except perhaps Taiwan, nationalism had an obvious role in motivating and legitimizing agricultural development policies. In the case of Taiwan, the Cold War and anti-communism played a similar role as an ideological driving force.
- Foreign aid played an important role in the process, not only motivated by strategic considerations during the Cold War, but informed by the neo-Malthusian and anti-communist agenda (Perkins, 1997), which motivated an export of technology crucial for making the Asian economies independent of food aid and import.
- Finally, industrialization, although not discussed in the two chapters, seems a common factor. Even if agricultural growth did not everywhere lead that in industry, as Mellor maintains it did in Taiwan, there is no case of agricultural growth unaccompanied by industrial growth, with the possible exceptions of the Philippines (Mellor, 1995). It is conceivable that a process of industrialization is a necessary precondition for a dynamic process of agricultural development. Although the role of industry in the development of agriculture is not part of our research questions, it must evidently be kept in mind when discussing agricultural development potentials, in Africa and elsewhere.

The specificities have to do with, among other things, timing:

- In East Asia, Japan and Taiwan are largely pre-Green Revolution cases, while in the rest of Asia the movement towards self-sufficiency in rice starts later. There is a common background to the later start in the rest of Asia having to do not only with the breakthrough in seed technology, but also with the food crisis at the beginning of the 1970s, adding to the shock of the oil crisis. This also explains the divergent timetable in the case of Korea where, as we have seen, the Green Revolution is a drawn-out process, culminating in the late 1960s and early 1970s, largely in tandem with the development in South and South-east Asia.
- There is also a shift in the financial policies at about the same time. In the 1950s and 1960s most countries had followed a 'squeeze agriculture' policy, trying by all means to keep down farm gate and food prices in order to mop up a surplus for industrial growth. After the delinking of the dollar from the gold price in the early 1970s, a fundamental structural change in the Bretton Woods system, many countries turned to deficit financing as a means of driving agricultural development. In a number of countries this meant subsidies for farm inputs, remunerative farm gate prices and subsidized food prices – policies obviously fuelling at least a moderate inflation.
- Equally important, the U-turn of price policies in South and South-east Asia signifies a first partial break with the import substitution industrialization strategies followed since decolonization. Conventional accounts sees export-led industrialization as the break with ISI occurring only in the 1970s (except in Taiwan and South Korea where it came earlier), but here we see that this break is antedated by the revamped agricultural development

strategies. Thus one can say that the shift to an export-led strategy of industrialization was preceded by a shift to import-substitution in food grains.

- In moving from East to South Asia, agricultural development strategies become more top-down and less participatory. Concurrently, the balance between national and global institutions changed, with the Bretton Woods institutions and CGIAR institutes, especially, playing a crucial role both in financing and in supplying specialized inputs to national programmes.
- The specificities of the South Asian cases have to do with the character of their states and with their social structures, but there are agronomic specificities as well. While Green Revolution technologies had an early breakthrough in Indian Punjab, the more general breakthrough in Indian rice production came only when the IRRI varieties had been crossed with national improved varieties in the late 1970s. Similarly, in Bangladesh the breakthrough came even later with improvements in small-scale irrigation technologies which made it possible to expand the double cropped area. Both the Indian and Bangladesh states are far from the efficient machineries that we associate with development in East and parts of South-east Asia. In the Indian case, persistent poverty is largely explained by the impotence of the national executive in battling various sectional interests corrupting the attempt to target the food distribution system to the most needy and discriminated against parts of the population. India and Bangladesh also had a much weaker position of family farming, which seems another important factor to account for their more sluggish performance. The same institutional factors probably account at least partly for the slower and later industrialization in South Asia.

The specificities notwithstanding, the model of the Asian Green Revolution that emerges is one of a market-mediated, small farmer-based, state-driven process. It is conditioned by geopolitical and institutional factors and part of industrialization. However, it has no direct causal links to demographic factors. Finally, technology is not a driving force, but a necessary, although not sufficient, factor.

Returning finally to the causal model outlined in the introduction, we hope that the above analysis substantiates our contention that this model explains the Green Revolutions in Asia. Obviously, a historical analysis based on secondary data, like the one we have made, does not constitute a rigorous test of the model. Such a test could be done only with more solid comparative data.

The reader may recall the alternative hypotheses spelled out in Chapter 1 (Djurfeldt, this volume): (i) the counter-factual hypothesis that a continuation of the low-price policies and ISI strategy would have delivered the same end result of national self-sufficiency in food grains; (ii) the rival explanation that the Green Revolution was and is a capitalist project, not a state-driven one; and (iii) the neo-liberal counter-factual hypothesis that a free-market would have achieved the same results more efficiently. Although we do not claim to have rigorously tested our model, we continue to contend that it accounts for the secondary data we possess far better than any of the rival hypotheses. We leave it to the reader to evaluate this claim.

As we said already in Chapter 1, we regard the question *of transferability of the Green Revolution* as less interesting and fruitful: it tends to lead to an argument of the type 'This feature is common . . .', 'This feature differs between the cases . . .', 'So, on the one hand . . .' and 'On the other hand . . .'. Few controversies are resolved by this type of analysis.

We want to use our causal and explanatory model differently. In the following chapters the overarching question is: can the causal model developed on the basis of the Asian cases *explain* what *is* happening and what *is not* happening in sub-Saharan Africa? Can these explanations in turn be used for formulating strategies for agricultural development in Africa?

Notes

[1] In his memoirs, Macapagal wrote: 'It is imperative to undertake measures to attain increased rice production and productivity and to seek self-sufficiency. It is equally imperative that in the process, the largest number of people must not be allowed to be without rice – even if the road to self-sufficiency thereby becomes longer and even if recourse is made to importation.' (Macapagal, 1968 quoted in Bouis, 1982).

[2] Sometimes translated as 'productive 99', the programme name signifies the nominal target of 100 cavans (5 tons) per ha (Barker *et al.*, 1985).

[3] The 1968 Workshop on Food in Jakarta (Mears and Moeljono, 1981) recommended the quantity 120 kg per capita as an average figure for the entire Indonesian consumption. Considering that rice was not the main staple in all parts of the country, while being the dominant one in, for example, Java, the recommended average did not reflect regional nutritional needs.

[4] BIMAS was developed from the DEMAS (Mass Demonstration) programme, a student extension service started in 1963 by the Faculty of Agriculture at Bogor as a response to the national food situation. The students first tried to gain the confidence of farmers and then attempted to teach them to use selected seeds and fertilizers. Simultaneously they made sure that input supplies to the targeted number of farmers were timely (Mears and Moeljono, 1981; Barker *et al.*, 1985).

[5] Tomich *et al.* (1995) quote a review article of Suharto's autobiography in the *Far Eastern Economic Review*, 19 January 1989, p. 17.

[6] The Ford Foundation, on Nehru's invitation, had been active in India since the early 1950s supporting the work of the community development programme and Ford Foundation representatives had a good working relationship with Nehru and other members of the Planning Commission and the ministries (Perkins, 1997).

[7] *A New Strategy in Agriculture: a Collection of Speeches* by C. Subramaniam. Indian Council of Agricultural Research, New Delhi, 1972, p. 4, quoted by Varshney (1995:54).

[8] Barker *et al.* (1985) seem to draw on official statistics which may contain some exaggeration as to the use of modern varieties in the period 1967–1975. Cf. Bhalla and Singh, 1997.

[9] The central reference on this point is Tyagi (1990), complemented with a whole string of articles in the leading Indian journal *Economic and Political Weekly*.

[10] Rock has a related analysis of the Malaysian agricultural policies where the aim to win the Malay rice farmers for the regime was an important part of the strategy.

[11] Here we have to keep in mind that also in India, as previously mentioned, it is during the last two decades that the spread and impact of Green Revolution technologies have been the largest. The difference in comparison with Bangladesh is of course that Green Revolution technologies also had a major impact on India's food security in the early phase.

[12] In 1965, a set of IRRI varieties and lines was imported to East Pakistan (Bangladesh) to support an accelerated rice production programme sponsored by the Ford Foundation. The famous IR-8 together with IR-5 and in 1970, IR-20 – the latter representing the first modern variety suitable for the summer-fall crop (*Aman*) – spread in locations with superior water control (Faaland and Parkinson, 1976; Dalrymple, 1986).

[13] An important institution in the early efforts was the Comilla Academy under the leadership of Aktor Hamid Kahn. The experience of integrated rural development that had developed in Comilla constituted an important building block for the future launching of technology packages demanding local level organization and cooperation structures.

[14] Longitudinal data on rice farm size and farm distribution supporting this claim is provided through the World Rice Statistics compiled by IRRI (IRRI, 2004).

References

Ahmed, R. (2001) *Retrospects and Prospects of the Rice Economy of Bangladesh*. The University Press, Dhaka.

Barker, R., Herdt, R.W. and Rose, B. (1985) *The Rice Economy of Asia*. Resources for the Future, Washington, DC.

Bhalla, G.S. and Singh, G. (1997) Recent development in Indian agriculture: a state level analysis. *Economic and Political Weekly* (Review of Agriculture, March 29), A1–A14.

Bhalla, G.S. and Singh, G. (2001) *Indian Agriculture: Four Decades of Development*. Sage, New Delhi.

Bouis, H. (1982) *Rice Policy in the Philippines*. Food Research Institute. Stanford University, Stanford, California.

Bresnan, J. (1993) *Managing Indonesia: the Modern Political Economy*. Columbia University Press, New York.

Dalrymple, D.G. (1986) *Development and Spread of High-yielding Rice Varieties in Developing*

Countries. Bureau for Science and Technology, Agency for International Development, Washington, DC.

Datt, G. (1998) *Poverty in India and Indian States: an Update*. International Food Policy Research Institute, Washington DC.

Datt, G. (1999) Has poverty declined since economic reforms? Statistical data analysis. *Economic and Political Weekly* (December 11–17).

David, C. and Otsuka, K. (1994) *Modern Rice Technology and Income Distribution in Asia*. Lynne Reinner, Boulder, Colorado.

Faaland, J. and Parkinson, J.R. (1976) *Bangladesh: the Test Case of Development*. University Press, Dhaka.

Frankel, F.R. (1978) *India's Political Economy, 1947–1977: The Gradual Revolution*. Princeton University Press, Princeton, New Jersey.

Hayami, Y. (2000) An ecological and historical perspective on agricultural development in Southeast Asia. Policy research working paper; 2296. World Bank, Washington, DC.

Hayami, Y. and Ruttan, V.W. (1971) *Agricultural Development: an International Perspective*. The Johns Hopkins University Press, Baltimore, Maryland.

Hazell, P.B. and Ramasamy, C. (1991) *The Green Revolution Reconsidered: the Impact of High-yielding Varieties in South India*. The Johns Hopkins University Press, Baltimore, Maryland.

Hossain, M. (1989) *Green Revolution in Bangladesh: Impacts on Growth and Distribution of Income*. University Press, Dhaka.

IRRI (2004) World Rice Statistics, Area planted (or harvested) to modern varieties, selected Asian countries 1965–1999. Table 34, http://www.irri.org/science/ricestat/index.asp

Lantican, F. and Unnevehr, L. (1985) Rice price and marketing policy. Paper presented at the PIDS-CPDS-PCARRD-IRRI-ADC Workshop on Agricultural Policy, College, Laguna, The Philippines, 3–4 May.

Lipton, M. and Longhurst, R. (1989) *New Seeds and Poor People*. The Johns Hopkins University Press, Baltimore, Maryland.

Mangahas, M. (1970) An economic analysis of the diffusion of new rice varieties in central Luzon. PhD thesis. The University of Chicago, Chicago, Illinois.

Mangahas, M. (1975) *The Political Economy of Rice in the New Society, 14:3,1975*. Food Research Institute, Stanford University, Stanford, California.

McKinsey, J.W. and Evenson, R.E. (2003) Crop genetic improvement impacts and Indian agriculture. In: Evenson, R. and Gollin, D. (eds) *Crop Variety Improvement and Its Effect on Productivity – the Impact of International Agricultural Research*. CAB International, Wallingford, UK.

Mears, L.A. (1974) *Rice Economy of the Philippines*. University of the Philippine Press, Quezon City.

Mears, L.A. and Moeljono, S. (1981) Food policy. In: Booth, A. and McCawley, P. (eds) *The Indonesian Economy During the Soeharto Era*. Oxford University Press, Kuala Lumpur.

Mellor, J.W. (1966) *The Economics of Agricultural Development*. Cornell University Press, Ithaca, New York.

Mellor, J.W. (ed.) (1995) *Agriculture on the Road to Industrialization*. The John Hopkins University Press, Baltimore, Maryland.

Perkins, J.H. (1997) *Geopolitics and the Green Revolution: Wheat, Genes and the Cold War*. Oxford University Press, New York.

Pingali, P.L., Hossain, M. and Gerpacio, R.V. (1997) *Asian Rice Bowls: the Returning Crisis?* CAB International, Wallingford, UK.

Putzel, J. (1992) *A Captive Land – the Politics of Agrarian Reform in the Philippines*. Catholic Institute for International Relations and Monthly Review Press, London and New York.

Rock, M.T. (2002) Exploring the impact of selective interventions in agriculture on the growth of manufactures in Indonesia, Malaysia, and Thailand. *Journal of International Development* 14, 485–510.

Salas, R.M. (1985) *More than the Grains: Participatory Management in the Philippine Rice Sufficiency Program, 1967–1969*. Simul Press, Tokyo.

Schultz, T.W. (1964) *Transforming Traditional Agriculture*. Repr., University of Chicago Press, Chicago, Illinois.

Sen, A. (1981) *Poverty and Famines: an Essay on Entitlement and Deprivation*. Clarendon Press, Oxford, UK.

Timmer, P.C. (1981) The formation of Indonesian rice policy: a historical perspective. In: Hansen, G.E. (ed.) *Agricultural and Rural Development in Indonesia*. Westview Press, Boulder, Colorado.

Tolentino, V.B.J. (2002) Governance constraints to sustainable rice productivity in the Philippines. In: Sombilla, M., Hossain, M. and Hardy, B. (eds) *Developments in the Asian Rice Economy*. IRRI, Los Banos, Philippines.

Tomich, T.P., Kilby, P. and Johnston, B. (1995) *Transforming Agrarian Economies. Opportunities Seized, Opportunities Missed*. Cornell University Press, Ithaca, New York.

Tyagi, D.S. (1990) *Managing India's Food Economy: Problems and Alternatives*. Sage, New Delhi.

Varshney, A. (1995) *Democracy, Development, and the Countryside: Urban–rural Struggles in*

India. Cambridge University Press, Cambridge, UK.
Zohir, S., Shahabuddin, Q. and Hossain, M. (2002) Determinants of rice supply and demand in Bangladesh: recent trends and projections. In: Sombilla, M., Hossain, M. And Hardy, B. (eds) *Developments in the Asian Rice Economy*. IRRI, Los Banos, Philippines.

5 Spurts in Production – Africa's Limping Green Revolution

Hans Holmén
Department of Geography, Linköping University, Linköping, Sweden

At the turn of the millennium, which incidentally also happens to be the early phase of Africa's post-SAP period, reports abound about deepening poverty, enhanced food deficiencies, and flight from the countryside. Many express a familiar and accentuated concern about the subcontinent's ability to feed itself. Some expect the recently implemented SAP reforms to be a long-overdue cure. Others assume they will just enhance the misery. Is it a question of 'short-term pains for long-term gains' or rather a renewed process of underdevelopment? To present an unquestionably 'true' picture of sub-Sahara's food and livelihood situation on a few pages is, of course, an impossible task. The following section therefore merely attempts to summarize a few aspects of the process, which we believe will shed light on vital aspects of the intensification problematics.

This chapter has several objectives. It introduces the reader to the present food situation in sub-Saharan Africa. It gives examples of historical agricultural intensification experiences and of some attempted sub-Saharan Green Revolutions. This is followed by an overview of food production development in eight countries based mainly on *Afrint* country studies and official FAO data. The purpose of this exercise is to determine if (and when) food crop intensification has taken place in Africa south of the Sahara or if, rather, processes of extensification and/or involution have been more prominent. This analysis further gives special emphasis to development before, during and after the implementation of Structural Adjustment Policies (pre-SAP, SAP, post-SAP).

The Present Food and Livelihood Situation in Sub-Saharan Africa

Natural resource base

Africa, accounting for about one-fifth of the earth's land surface and some 12% of world population, is a vast continent which is often said to be under-populated. However, its frequently leached and depleted soils, and often adverse climatic conditions pose comparatively difficult constraints for agriculture. Moreover, the size of sub-Saharan Africa, in combination with its low population densities, renders modernization problematic because infrastructure investments (and utilization) tend to become costlier and slower to realize than in, for example, Asia's more densely populated major Green Revolution regions.

Population

Sub-Saharan Africa's population has quadrupled since 1950. It reached some 650 million

(eds G. Djurfeldt, H. Holmén, M. Jirström and R. Larsson)

in the year 2000 and is projected to more than double to some 1.5 billion in 2050 (medium variant) (UNPP, 2003). In aggregate, in 2000 some 410 million people, or almost two-thirds of the population, were classified as rural (FAOSTAT data, 2004). Naturally, this figure varies regionally. In the eight countries included in this study, the share of rural in total population ranges from around 60% (Ghana, Kenya, Nigeria, Zambia) to 85% (Malawi, Uganda, Ethiopia), with Tanzania falling neatly between at nearly 70% (FAOSTAT data, 2004).

Population densities

Populations in the countries under study have tripled or quadrupled in the last 40 years. During the same period, expansion of arable land has been of a much lesser magnitude and is actually stagnating or has even declined in recent years, indicating that land frontiers may have been reached (FAOSTAT data, 2004). The result is a rapidly accentuated population pressure and declining ratios of arable land to agricultural population. In all countries studied man/land ratios (agricultural land/rural population) have been reduced by half during the last 4 decades (FAOSTAT data, 2004).[1] Farm sizes are generally shrinking and 'land distribution in [the] small-farm sectors appears to be becoming . . . comparable to those of many Asian countries at the time of their green revolutions' (Jayne *et al.*, 2003:253).

Poverty

Poverty is widespread. Whereas in the year 2000 the world GDP/capita (PPP) was US$7446, in sub-Saharan Africa it was only US$1690. In 1999, almost half (47%) of the population in sub-Saharan Africa (more than a quarter of the world total) eked out a living on less than US$1/day (UNDP, 2002). A better assessment may be obtained by looking at the share of people falling below the respective country's national poverty line. Of those countries for which data are available, on average 48% of the population in Africa south of the Sahara in years 1987–2000 had an income under the national poverty lines. Differences are great, though, ranging from 31% in Ghana to 68% in Zambia (UNDP, 2002).

Food supply

Although in global terms the world food situation has improved, in sub-Saharan Africa it has worsened. In the mid-1960s, 57% of the world population were estimated to live in countries with an average daily food consumption under 2200 kcal. In the late 1990s, only 10% of a much larger world population lived in countries with food consumption under 2200 kcal/person/day. Sub-Saharan Africa, excluding Nigeria, stands out as the only region that failed to make any progress in raising average per capita food consumption between the 1960s and today (FAO/Earthscan, 2003). In 1999 there were 30 countries in the world with average food consumption below 2200 kcal/person per day. Of these, 22 were in Africa south of the Sahara (FAO/Earthscan, 2003). The prevalence of poverty and under-nourishment is being aggravated by HIV/AIDS as adult labour is affected and the young and the elderly will have to assume the role of providers.

Food imports and food aid

Whereas population in Africa south of the Sahara has quadrupled since 1950, cereal imports have increased 10-fold between 1961 and 2001 (FAOSTAT data, 2004). For seven of the eight countries included in this study,[2] food imports accounted for, on average, 15% of the value of merchandize exports in the year 2000 (Hunger, 2003) – a high figure for agriculturally based economies. With declining per capita food production and little with which to pay for imports, food aid to sub-Saharan Africa has been a persistent phenomenon during the last decades. For example, seven of the eight countries included in the study[3] have received food aid (cereals) in *all* years since 1970 (FAOSTAT data, 2004). In 2003, 25 countries in sub-Saharan Africa faced food emergencies due to drought, floods and refugee problems (FAO/GIEWS, 2003). Six of these countries (Ethiopia, Kenya, Malawi, Tanzania, Uganda, Zambia) are included in this study.

Extensive agriculture

The above résumé gives at hand that intensification of (food) agriculture 'ought' to take place in contemporary sub-Saharan Africa. Most commentators, however, seem convinced that it does not. Apparently, extensive agriculture dominates, as it seems always to have done. And, apart from some sporadic and short-lived outbursts, the Green Revolution is said 'never to have happened' there. Instead, it is commonly argued that yield growth has been of minor importance and that growth of agricultural production has been almost entirely based on extending the area under cultivation (Kydd *et al.*, 2002; Evenson and Gollin, 2003). At the same time sub-Saharan Africa is deemed to have a 'vast agricultural potential' (Eicher, 2001:3), which, apparently, is not being made use of.

It can be argued that, in such a situation, intensification of (food) agriculture 'ought' to take place. When there is an 'objective' need for it, lack of intensification leads to catastrophe. Several commentators have, for example, tried to explain the Rwanda genocide in 1994 in purely Malthusian terms of population growth, accelerating environmental degradation and acute competition over land (see e.g. Andre and Platteau, 1998; Ohlsson, 1999; Gasana, 2002). We are extremely puzzled by this. Is it true that intensification is absent or insignificant under these circumstances? And, if it is, how can this be explained – and changed?

Potential for Agricultural Intensification in Sub-Saharan Africa

A possible, or at least partial, explanation could be the low population densities. Following the logic of Boserup (1965) and many others, intensification is not likely to occur until possibilities to expand extensive farming are exhausted. Compared with other regions in the world, sub-Saharan Africa has sometimes been deemed under-populated (Amin, 1972). Historically, the main limiting factor in sub-Saharan African agriculture, apart from environmental constraints, has been labour, while land until recently has been much less of a problem. In large parts of Africa, it is only now that the land frontier has been reached – or is about to be reached. The pressure to change established ways of production (and accompanying social institutions, etc.) has been low compared with more densely settled regions such as those where the Green Revolution first took off in Asia.

Intensive farming, however, is not a new phenomenon in sub-Saharan Africa. Nor is the agricultural history of the subcontinent without successes (Haggblade, Chapter 8, this volume). There are a number of well-documented historic examples of long-lasting successful agricultural intensification, i.e. 'islands' of agricultural intensification (see e.g. Widgren and Sutton, 1999; Börjesson, 2000), as well as archaeological evidence of now 'defunct' large-scale intensive farming systems (Soper, 1999; Sutton, 1999). We do not know exactly why these 'islands' of intensive agriculture disappeared. However, even if population pressure, perhaps, was not the only factor behind intensification in these areas, Netting (1993) has shown how, for example, the Kofyar people in Nigeria abandoned their intensive farming systems in favour of more extensive but less labour demanding practices when more land became available to them.

Today, increased pressure on land due to rapid population growth over the past 30–40 years, and in some areas since the early 20th century, has produced an uneven intensification process in sub-Saharan Africa (Turner *et al.*, 1993). In many places, however, the gains of population growth in terms of increasing aggregate output have been offset by negative effects from continuous cropping in the form of land degradation and fertility loss, causing declining per capita output and impoverishment of the rural population (Lele and Stone, 1989; Turner *et al.*, 1993; Scoones and Toulmin, 1999). However, where farmers have had better access than elsewhere to (urban) markets and where opportunities of income diversification have been present, population growth has more often been found compatible with sustained agricultural growth and rising per capita incomes without negative environmental side effects (see e.g. Turner *et al.*, 1993; Tiffen *et al.*, 1994; Larsson, 2001).

It should be noted, though, that also where high population densities and markets are present, the direction and pace of intensification are not self-evident (Berry, 1993). As clearly spelled out by the authors of success stories, such as that of Machakos in Kenya, the crucial importance of market access also depends on a range of other factors (Tiffen *et al.*, 1994; Scoones and Toulmin, 1999). Markets do not operate in a vacuum, but are, above all, dependent on government policies as regards, for instance, institutions, infrastructure and production incentives.

Hence, the role of the state in providing favourable conditions for market-orientated production is obvious. The question of agricultural intensification in Africa is made urgent through unprecedented rates of population growth. Its solution, however, is intimately linked to policy factors. The inadequacy of spontaneous forces (population and markets) to solve present problems of sustainability and low productivity underscores what Lele and Stone (1989) deem a need for 'policy-led agricultural intensification', i.e. a Green Revolution.

Agricultural intensification and the Green Revolution, thus, are not the same thing. Perhaps too simplistic, we can define intensification as a process whereas the Green Revolution is a project. Intensification entails agricultural productivity enhancement, which can take place in various ways and for different reasons. It can be a slow process or happen more abruptly depending on the circumstances. The Green Revolution is a particular form of intensification. It is not, generally, a spontaneous process. Rather, it is purposely initiated, involves a high degree of support and takes place over a comparatively limited period of time.

Early Experiences of the Green Revolution in Sub-Saharan Africa

The problem with the African food crisis is neither technology (e.g. wrong crops) nor nature (e.g. Africa's limited irrigation potential). Nor is it that African governments have been reluctant to 'interfere' in the agricultural sector. On the contrary, there has been no scarcity of attempts at state-led intensification. However, during the last decades, experiences of attempted Green Revolutions in sub-Saharan Africa have been episodic events, resembling short-lived 'spurts of production' rather than lasting productivity improvements. The question, therefore, rather than 'Why have Green Revolutions been absent in Africa?' should be 'Why have they not been sustained?' A few short examples are provided below.

Zimbabwe

Zimbabwe (Rhodesia) is (besides Kenya) one of the two centres in sub-Saharan Africa from where 'indigenous' maize-based Green Revolution technologies have spread over the subcontinent. Two distinct stages have been identified. The foundation of Zimbabwe's first maize-based Green Revolution (1960–1980) was laid early in the 20th century (Eicher and Kupfuma, 1997). In response to demands from the white commercial farmers and their organizations, the government established research stations early in the 1900s. Research on hybrid maize was initiated in 1932 and in 1949 the first HYV (SR1) was released. An improved high yielding, long duration hybrid (SR52) was successfully released in 1960. When tobacco exports declined due to sanctions against the illegal Rhodesia regime, attention was shifted to the need to develop a short-duration maize hybrid that could replace tobacco in low-rainfall areas. Such varieties (R200, R201, R215) were successfully released in the early 1970s (Eicher and Kupfuma, 1997). The success at this stage was the result of, on the one hand, strong farmer organizations able to articulate their needs and, on the other hand, a government that, in response to these demands, invested in relevant research. A further stimulus was provided by the isolation of Ian Smith's regime and, hence, the need to attain national self-sufficiency in production of basic food crops.

The second Zimbabwean Green Revolution (1980–1986) had already begun in the 1970s when smallholders began adopting the above-mentioned short-duration HYVs. The new government reorientated basic agricultural institutions (credit, research, extension) to serve the majority of farmers – the black smallholders – who rapidly adopted hybrid maize and fertilizer. In production terms, the second Green Revolution was overwhelmingly successful. However, by the mid-1980s, transport and storage capacities were exhausted and grain silos were overflowing (Eicher and Kupfuma, 1997). While grain prices fell and produce could not be disposed of, the government's budget was severely strained. Hence, the smallholder support programme was quickly scaled down and in 1991 major economic reforms were launched in order to reduce subsidies for maize and fertilizer (Eicher and Kupfuma, 1997).

Kenya

Like in former Rhodesia, white commercial farmers had already initiated maize inbreeding and hybridization programmes in the 1930s (Harrison, 1970). In the early 1950s attention was redirected to breeding for drought tolerance, pest resistance and early maturity (Harrison, 1970). The first synthetic (Kitale2) was released in 1961 and in 1964 the first hybrids (K611, K621, K631) were launched.

As in Rhodesia/Zimbabwe, the Kenyan Green Revolution in maize evolved in stages. During the first phase, 1964–1974, large-scale farmers in the high-potential areas rapidly adopted the new hybrids together with inorganic fertilizer. A second phase of maize technology diffusion, 1975–1984, was characterized by a surge in smallholders' adoption of improved seeds, especially in the high-potential areas where their adoption rates eventually equalled those of large-scale farmers. During the third phase, 1985–1991, smallholders in low-potential areas adopted improved seed, but fertilizer use remained low (Hassan and Karanja, 1997). Contrary to the case in Zimbabwe, the spread of the technology to large numbers of smallholders was not facilitated by extensive and costly subsidies. Instead, quantitative restrictions on fertilizer imports and an import licensing system were introduced in the 1970s (Lele *et al.*, 1989).

Since the 1980s the Kenyan Green Revolution in maize has slowed down if it has not been halted. Initially, donors began to finance fertilizer imports, albeit on an *ad hoc* basis (Lele *et al.*, 1989). However, the reform of fertilizer trade tended to benefit large-scale farmers and cash-crop producing smallholders while leaving small-scale food crop growers without access to inorganic fertilizer (Lele *et al.*, 1989). Also, austerity measures introduced as part of Structural Adjustment Programmes have virtually dried up public credit facilities for small-scale peasant farmers and, hence, their ability to afford improved seeds and inorganic fertilizer (Oluoch-Kosura and Karugia, Chapter 10, this volume). Likewise, 'the scientific and institutional co-operation that created the maize success story of the 1960s and 1970s collapsed in the 1980s due to weakened public financial support for research with a subsequent general decline in overall maize production' (Hassan and Karanja, 1997:90).

Nigeria

Nigeria is a complicated case. On the one hand it has been the home of Africa's most ambitious attempt to implement a Green Revolution (Akande, Chapter 9, this volume). On the other hand, Nigerian agriculture has reportedly been neglected, especially after the oil boom in the 1970s (Nzimiro, 1985). One indication would be that after the modernization programmes of the 1970s, 'traditional' smallholders, using simple technologies and the bush-fallow system of cultivation, account for around two-thirds of the country's total agricultural production and some 90% of food consumed (van Buren, 2001). For many years and under different governments the approach has been top-down with little influence for smallholders (Nzimiro, 1985; Bienen, 1987). However, 'several recent studies . . . suggest that at least some areas of the country have experienced dynamic agricultural change over the last two decades' (Goldman and Smith, 1995: 250; see also Turner *et al.*, 1993; Smith *et al.*, 1997).

Maize became an important crop following the introduction of an improved open-pollinated variety of maize (TZB) in the mid-1970s (Goldman and Smith, 1995). High-yielding maize was the focus of projects in the mid-1970s and was adopted by smallholders alongside their traditional cereal crops (Lawrence, 1988). Hence, even if the initial adoption rate for maize was slow, by the late 1980s the new high-yielding varieties had become a major food and income crop (Goldman and Smith, 1995). In the mid-1990s, maize was widely grown in the Northern Guinea Savannah, where it accounted for 30% to 40% of area under agricultural production (Smith *et al.*, 1997). By the year 2000, maize accounted for 25% of Nigeria's total cereal harvest (van Buren, 2001).

Also, a breakthrough in cassava breeding was made in 1977 when the high-yielding, mosaic-resistant TSM-varieties were released from the International Institute of Tropical Agriculture (IITA). This made possible the transformation of cassava, Africa's second most important food crop, from a low-yielding, famine-reserve crop into a high-yielding cash crop for both rural and urban consumers (Nweke *et al.*, 2002).

Whatever else can be said about the Nigerian Green Revolution (see especially Akande, Chapter 9, this volume), these two experiences contradict the claims that agriculture in this country has been neglected. Rather, it appears that it is the smallholders who have been neglected. The Nigerian Green Revolution appears to have been rather exclusive and directed primarily at politically well-connected large farmers (Akande and Kormawa, 2003). Besides this, the problem in Nigeria seems not to be in the first place to increase production, but rather to handle and distribute what is already being produced. Van Buren (2001:757) reports that inadequate transport and storage facilities cause massive postharvest losses.

Comment

A number of tentative conclusions can be drawn from these examples. One is that the African experiences of Green Revolutions are quite varied. Both successes and failures can be variously explained. Partly, the delay and limited successes of Green Revolutions in sub-Saharan Africa are due to the fact that few suitable varieties were available until the 1980s, a situation that has since been remedied (Evenson and Gollin, 2003). Also, production potentials may not be as limited as is often suggested, which is evidenced by problems with 'over-production', e.g. in Zimbabwe and Nigeria. Despite this, it may of course be claimed: 'attempts to introduce the green revolution *on a large scale* in Africa [have] failed' (emphasis added, Asiema, 1994:17). But then it should also be noted 'green revolution crops nevertheless gained importance. Although maize is not indigenous to Africa, it is the single most planted cereal on this continent' (Asiema, 1994: 17). More could, however, be done both to increase yields and to handle them postharvest.

The above few examples show that the problem is not primarily one of how to raise food production and productivity. Sub-Saharan Africa has seen a number of 'spurts in production'. Rather, the problem is how to turn these spurts into sustainable levels of production. Another conclusion is that an African Green Revolution apparently should give more attention to extra-technological supportive measures, to market development and postharvest handling of agricultural produce than to boost production *per se*. This highlights our earlier definition of Green Revolutions as being much more than technology. A third conclusion is that African governments, at least in the above cases, have been neither as disinterested nor as inactive as is often claimed. The problem rather has to do with how – and why – agricultural policies have been implemented (the way they have). Before penetrating such issues (Chapter 6), we need to take a closer look at the issue of intensification, regardless of whether it has been associated with Green Revolution attempts or not.

Post-independence Food Crop Performance in Sub-Saharan Africa

Country selection procedure

The purpose of this study is *not* to give a representative picture of the overall agricultural and food situation in Africa south of the Sahara. Instead, it is *a study of intensification* and of the conditions under which it takes place – or not. We selected eight countries within what can be called the 'sub-Saharan maize and cassava belt' as these crops are of major importance in the region. Maize and cassava (and sorghum and rice) are also 'typical' crops around which an African Green Revolution most likely will have to be built. The countries chosen for study – Ethiopia, Ghana, Kenya, Malawi, Nigeria, Tanzania, Uganda and Zambia – have different (current and historical) profiles in terms of population pressure, land availability, market orientation and histories of state involvement in agriculture and food production. As mentioned, the analysis is further divided into pre-SAP, SAP and post-SAP periods. It is our belief that this spatial as well as temporal spread of the analysis will allow us to better assess conditions under which food crop intensification does and does not take place in contemporary Africa south of the Sahara. Thereby it may enable us to identify critical preconditions as well as real constraints for the successful launching of the 'delayed' sub-Saharan Green Revolution.

A word of caution

The analysis is based on statistics concerning population, land use, production (tonnes) and productivity (yields/ha) of major food crops, and adoption rates for various inputs (seed, fertilizer). Statistics are obtained from the FAO and, through the *Afrint* studies, from official statistical bureaus and/or ministries in the countries concerned. It is well known that such statistics in sub-Saharan Africa often are of mixed quality. The costs of tight monitoring are often prohibitive, if not insurmountable, especially now that governments' resources are shrinking. Agriculture in sub-Saharan Africa is not a stable undertaking and the acreage actually utilized for food production varies from season to season, as does crop composition. Weather conditions are highly variable (e.g. early rains, late rains, small rains, big rains or no rains at all) and influence both area planted and area harvested. Estimations of yield and crop acreage are further complicated by varying degrees of subsistence production, by the prevalence of shifting cultivation in certain areas, and by the shifting contents and frequencies of intercropping.

Not only must figures be dealt with cautiously, the number of unexplained underlying factors remains 'uncomfortably large' and 'growth literature has had difficulty in coming to grips with the particular character of this continent' (Deininger and Okidi, 2003:481). Even if there are differences sometimes between e.g. FAO statistics and country data, these tend to concern details and/or specific years but they generally point to the same trend. Nevertheless, much research on African development (or its absence) rests on a fragile base. At the aggregate level available data are often scarce, inconsistent and of disputable reliability. Hence, 'any objective assessment [of production or production potentials] is far from easy to make' (Scoones and Toulmin, 1999; see also Lele and Stone, 1989; Lipton and Longhurst, 1989; Alexandratos, 1995).

Periodization of SAP

As a matter of convenience, the period during which Structural Adjustment Policies were implemented in sub-Saharan Africa is usually defined as *c.* 1985 to 1995. Friis-Hansen (2000:9) finds that 'by the mid-1990s, large parts of the agricultural adjustment policy agenda had been implemented in most Sub-Saharan African (SSA) countries' (with some delay for francophone Africa). As can be seen from Fig. 5.1, there are some deviances from this supposed generality in the countries included in this study and for some of them it is questionable whether we can presently talk of a post-SAP period.

In Nigeria an effort to introduce economic stabilization policies was made in 1982 but failed and was replaced by a full-fledged SAP in 1986 (Akande and Kormawa, 2003). Similarly, in Uganda an earlier introduced SAP in 1981 collapsed in 1984 due to 'contradictions within SAP, deep-seated problems in the economy . . . and civil war' (Bazaara and Muhereza, 2003). It took until a new government was established in 1986 for Uganda's SAP to become manifest. Ethiopia had a late start and did not adopt these policy measures until the early 1990s, following a change of the political regime; and the process can hardly be said to have finished. Zambia's 'stop-and-go' implementation has lasted beyond the year 2000. Also, in Malawi, SAP was partly interrupted in the late 1980s and it is being debated whether SAP has been abolished or given a new start since the mid-1990s (Harrigan, 2003). Not only have SAPs been implemented with highly varying degrees of enthusiasm (Oluoch-Kosura, 2003) and mixed results (Msambichaka *et al.*, 1995; Hunger, 2003), severe set-backs – variously interpreted either as a 'failure of adjustment' or as a 'failure to adjust' – have led to disruptions, 'time-out' and even reversals to more active state involvement in some countries.

Indicators of Intensification

In contrast with widespread views of stagnant food production in sub-Saharan Africa, a look at FAO's index of food production reveals a slow but steady increase in aggregate food production between 1961 and 1985 and a faster growth from 1985 to 2002 (FAOSTAT data, 2004). This seems to confirm Sender's argument that most 'pessimistic' assessments of African agriculture are not supported by the data, e.g. from the FAO, on which they are supposedly based (Omamo, 2003). With few exceptions, the *Afrint* country studies,[4] however, present a less optimistic picture of recent trends, indicating that the FAO might paint too rosy a picture. It is, however, not only trends in aggregate food production that are of interest but,

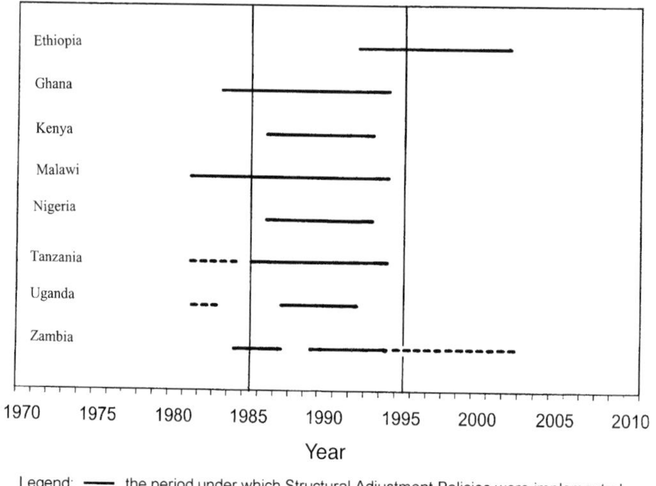

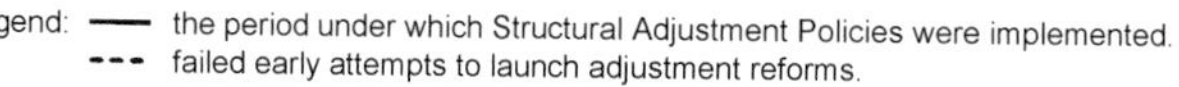

Fig. 5.1. Periodization of SAP in eight African countries. (Sources: *Afrint* macro papers; Harrigan, 2003.)

even more so, differences in yields and preconditions for production and productivity improvements among different categories of peasant farmers. These issues will be looked into more closely based on the commissioned country studies (see Holmén, Chapter 6, this volume).

Also, an interesting observation from the aggregate FAO production statistics is that although growth rates appear to have increased for most countries in recent years, present patterns are unstable and display greater year-to-year fluctuations than did those observed for earlier periods. This observation is not only confirmed but also strengthened when looking at trends in maize production in the individual countries under study (Fig. 5.2).

This indicates that general improvement in aggregate food production post-SAP appears to be accompanied by a greater insecurity in national food self-sufficiency. Unstable, and possibly worsening, weather conditions may play a role here but 'the major culprit may be policy related, particularly market reforms' (Oluoch-Kosura, 2003:10; see also Saasa, 2003). Since, in these cases, there is only a weak correspondence of drought years and low maize harvests, it seems as if both peasants' vulnerability and national food insecurity have increased during SAP and post-SAP (see also Oluoch-Kosura and Karugia, Chapter 10, this volume).[5]

Other studies reveal that from 1971 to 1996/97, aggregate production increased at more than 2.0%/year for all the major food crops in Africa (Mareida *et al.*, 2000). Hence, the problem is not so much a lack of production increases but rather that population growth has outstripped food production. Higher production levels have various causes, which differ among crops, countries and regions within countries. For SSA as a whole, 'almost three quarters of the increment in rice production has resulted from an increased area under rice cultivation . . . For maize . . . yields increased at a rate of 1.0%/year, contributing about one-third towards production increments' (Mareida *et al.*, 2000:534f). As for roots and tubers, the authors find an annual growth rate of 2.7%, half of which was due to yield improvements (Mareida *et al.*, 2000:534f).

Maize production

Maize is the prime staple crop and deserves some special attention. It has been a favoured crop by governments trying to increase food production in all the countries studied. It dominates cereal production in Ghana, Kenya, Malawi, Tanzania, Uganda and Zambia. Only in Ethiopia and Nigeria is the maize dependency less prominent (Figs 5.3 and 5.4). The countries investigated show a general upward trend in maize production. Although trends have been fluctuating and uneven, present levels of production are on average 400% to 500% higher today than they were 40 years ago in Ghana, Nigeria, Tanzania and Uganda. Total annual maize production in Kenya and Malawi is about

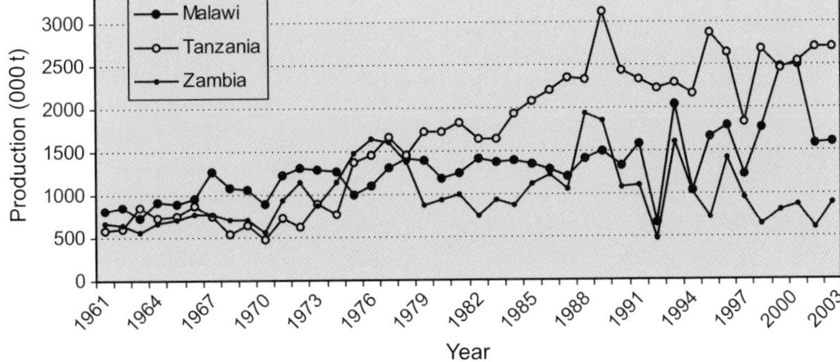

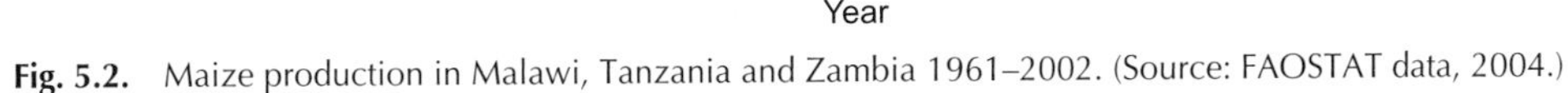
Fig. 5.2. Maize production in Malawi, Tanzania and Zambia 1961–2002. (Source: FAOSTAT data, 2004.)

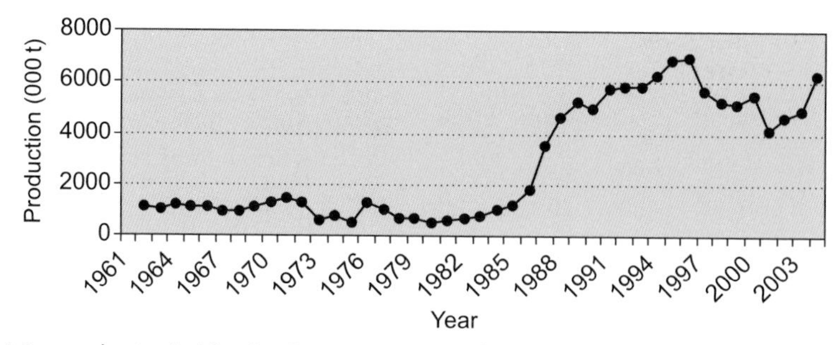

Fig. 5.3. Maize production in Nigeria. (Source: FAOSTAT data, 2004.)

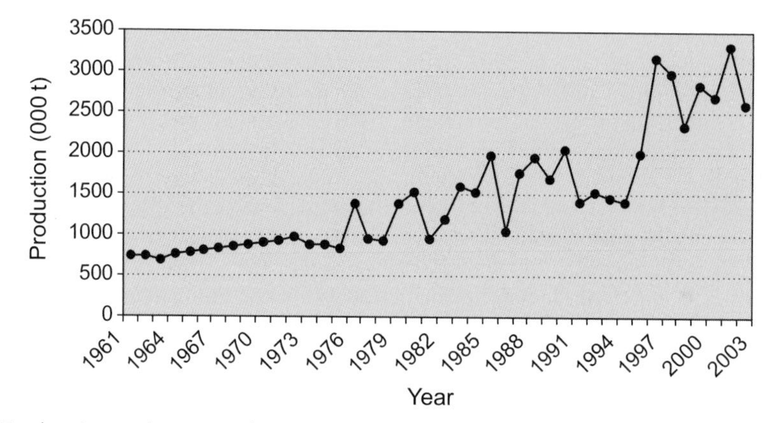

Fig. 5.4. Total maize production in Ethiopia. (Source: FAOSTAT data, 2004.)

double that of the early 1960s. The real laggard, according to FAO statistics, is Zambia, where present levels of production are only some 12% over those of the mid-1960s (FAOSTAT data, 2004).

These improvements in maize production have occurred at different points in time in different countries and at least three broad patterns of growth can be identified. Maize production in Ghana, Nigeria and Uganda show stagnation pre-SAP and continuous growth through SAP. This development was sustained through post-SAP in Ghana and Uganda whereas Nigeria suffered from a decline in maize production in the post-SAP period. So did Zambia, where there was general stagnation before and during SAP, interrupted by a brief jump when SAP was abandoned 1987–1988 and then decline as SAP was given a second chance. In contrast, maize production in Kenya and Malawi shows a slow and steady upward trend throughout the whole period since independence. Tanzania's growth in maize production began in the mid-1970s but appears to have stagnated during SAP and post-SAP. A further observation is that the range of annual fluctuations in total maize production has started to oscillate wildly since the introduction of SAP in a number of countries (Kenya, Malawi, Tanzania and Zambia).

Yields

Maize yields have improved throughout the period under study but yield levels remain low – on average 1.5 t/ha in the early 2000s (Table 5.1). Within-country variations are great, however, ranging from 0.8 t/ha to 2.7 t/ha between regions in Zambia, and from 0.25 t/ha to 2.2 t/ha in Ethiopia in recent years (FAO/WFP, 2003:2004). Of greater interest here are the differences in when improvements occurred (Table 5.1). For Tanzania and Malawi, there appears to be no big influence of SAP on yields. There is a

Table 5.1. Maize yields (t/ha) 1965–2001 (3-year average). (Source: FAOSTAT data, 2004.)

	1965	Pre-SAP*	Post-SAP**
Ethiopia	1	1.5 (1991)	1.6
Ghana	1	0.8	1.4
Kenya	1.1	1.6	1.6
Malawi	1	1.2	1.3
Nigeria	0.8	1.1	1.1
Tanzania	0.8	1.3	1.6
Uganda	1	1.2	1.8
Zambia	0.9	1.8	1.5

*Average yield in the 3 years immediately preceding SAP.
**2000–2002.

slow and steady increase since the 1960s. In Ghana and Uganda, to the contrary, yields have made rather big improvements during SAP and post-SAP. As for the remaining countries (Ethiopia, Kenya, Nigeria and Zambia), they all experienced yield improvements pre-SAP but these trends were broken during SAP. In Ethiopia, Kenya and Nigeria yields stagnated whereas Zambia has faced declining yields post-SAP.

Maize area harvested

Whereas yield improvements have been small and fairly similar since the 1960s – a general increase from 0.95 to 1.5 t/ha (Table 5.1) – trends in maize area are more diverse (Fig. 5.5). According to FAO figures, half the countries studied (Ghana, Kenya, Malawi, Tanzania) have seen a slow but steady growth in maize area throughout the period. In Ethiopia, Kenya and Tanzania a levelling off occurred during SAP and post-SAP. Uganda had a very unstable pattern until the mid-1980s when PRM's[6] political take-over coincided with the introduction of SAP, after which maize area more than doubled in 15 years. In Zambia, area expansion came to a halt in the mid-1970s, when it dropped about 50% and has thence remained at the same low level. The opposite trend is found in Nigeria where maize area declined slowly between 1961 and 1982 and then expanded dramatically (about 450%) through SAP and then dropped somewhat post-SAP (Table 5.2).

A different way to measure the importance of yield versus acreage behind growth in maize production is to compare growth rates between these two factors during different periods, i.e. on the one hand the whole period since the early 1960s (Table 5.3) and, on the other hand, from the introduction of SAP until the present (Table 5.4).

Over the whole period under study – from independence to the early third millennium – both area expansion and yield improvement have evolved at roughly similar rates in Kenya and Tanzania (Table 5.3). However, in Tanzania the relative importance of yield improvement was somewhat higher than that for area expansion whereas the opposite was true for Kenya. In Zambia, maize area decreased by almost a third whereas yields rose by three-quarters. In Ghana, Nigeria and Uganda, area expansion has been

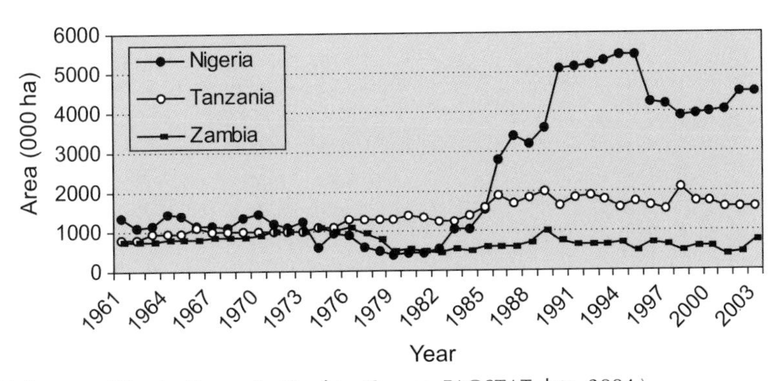

Fig. 5.5. Maize area: Nigeria, Tanzania, Zambia. (Source: FAOSTAT data, 2004.)

Table 5.2. Increase in maize area (%) pre- and post-SAP (000 ha). (Source: FAOSTAT data, 2004.)

	Pre-SAP to end-SAP*	End-SAP to post-SAP**
Ethiopia***		100 (1961–2002)
Ghana	41	100
Kenya	14	15
Malawi	24	7
Nigeria	438	–20
Tanzania	27	–8
Uganda	67	23
Zambia	28	–17

*1961–1963 to average last 3 years of SAP.
**Average last 3 years of SAP to 3 year average, 2000–2002.
***In Ethiopia, where SAP was introduced late, it is still too early to talk about post-SAP.

Table 5.3. Growth in maize area and yield 1961–63 to 2000–2002 (3-year average). (Source: FAOSTAT data, 2004.)

	Growth (%)	
Country	Area	Yield
Ethiopia*	158	74
Ghana	247	53
Kenya	51	33
Malawi	82	31
Nigeria	245	20
Tanzania	73	84
Uganda	280	54
Zambia	–30	77

*1963–1965.

Table 5.4. Growth in maize area and yield since SAP (3-year average). (Source: FAOSTAT data, 2004.)

	Growth (%)	
Country	Area	Yield
Ethiopia*	47	17
Ghana	105	80
Kenya	33	–10
Malawi	34	11
Nigeria	132	–9
Tanzania	16	21
Uganda	111	58
Zambia	7	18

*1993–1995.

relatively more (i.e. 5 to 12 times) important for production increases. The same holds for Ethiopia and Malawi, but at lower (i.e. about double) levels.

Also in the period since the introduction of SAP area expansion has been more, sometimes much more, important than yield improvements as a factor behind increases in maize production in six out of eight countries investigated. Broadly speaking, SAP did not significantly alter the relative importance of area versus yield growth seen in the pre-SAP period, although differences have been reduced in most cases. In Tanzania, the trend was sustained and there was no big difference between the two variables even though yield improvements continued to be slightly more important than area expansion. In the case of Zambia, maize area remained rather stable whereas yields continued to improve. More remarkable is that average yields fell in Kenya and Nigeria post-SAP and that despite (because of) this, area expansion continued to be great (more than 100%) in Nigeria post-SAP.

The *Afrint* papers confirm this picture and stress that in all countries (except Zambia), area expansion is the main source of growth in gross maize production. In some cases, post-SAP area expansion is seen as a response to market liberalization (e.g. Seini and Nyanteng, 2003). Others point out that area expansion proves that there are now (finally) available seed varieties adapted to the savannahs (Fakorede *et al.*, 2003). Generally, however, this is interpreted as a sign of declining use of improved technologies after SAP was introduced (Akande and Kormawa, 2003; Bazaara and Muhereza, 2003; Mulat and Teketel, 2003; Oluoch-Kosura, 2003; Saasa, 2003; Seini and Nyanteng, 2003). This, to some extent, contradicts the findings of both Evenson and Gollin (2003) and Maredia *et al.* (2000). The former found that the contribution of modern varieties to yield growth has been higher since 1980 than before, but add that even then their contribution was low (Evenson and Gollin, 2003:6). In contrast, Maredia *et al.* (2000:539) state that the 'overall rate of adoption of improved maize varieties in the early 1990s was estimated to be about 37% of the total maize area in Africa, which is quite comparable with an estimated 42% in Asia and 41% in Latin America'. They also claim that in some countries (e.g. Kenya and Zambia) adoption rates were as high as 70–80% (Maredia *et al.* 2000:539).

None or limited intensification of maize farming is revealed everywhere. Yields remain low and have declined in some cases. If average maize yields are low (e.g. in Ghana 47% below potential), this hides seasonal variations, which are sometimes great (Mulat and Teketel, 2003; Seini and Nyanteng, 2003) as well as regional yield differences (Isinika *et al.*, 2003; Saasa, 2003). Large tracts still attain low yields whereas well-endowed areas may have double that amount. Hence, many smallholders only reach maize yields of about or below 1 t/ha whereas in 'isolated cases' large, commercial farmers may reach maize yields of 5 t in Ghana (Seini and Nyanteng, 2003) or even 9 to 10 t in Zambia (Saasa, 2003).

Rice

Among the *Afrint* countries, rice is grown in Ghana, Kenya, Malawi, Nigeria, Tanzania and Uganda. In most cases it is a minor crop representing only a few per cent of maize production, the only exceptions being Ghana and Nigeria. Production is increasing in all countries studied but yields are generally low (about 1.5 t/ha). Ghana and Kenya do better at 2.2 and 3.5 t/ha, respectively. Yield trends are increasing only in Ghana whereas in the other countries they are either stagnant or declining (Fig. 5.6; FAOSTAT data, 2004). In Nigeria and Tanzania particularly, yield decreases are a post-SAP phenomenon (Akande and Kormawa, 2003; Isinika *et al.*, 2003; FAOSTAT data, 2004).

In Kenya, Nigeria and Tanzania rice production has grown since the 1960s but at a more stable pace before SAP. Ghana and Uganda have seen rapid growth in rice production since SAP and gross production has increased some 500%. Spectacular growth, albeit on a lesser scale, is also the case in Malawi post-SAP but here rice production collapsed just before SAP and production is now back on the level obtained in the late 1970s (FAOSTAT data, 2004). In Ghana, growth in rice production is due to both area expansion and yield improvement but more so to the former (Seini and Nyanteng, 2003) whereas growth in Malawi and Uganda is directly proportional to area expansion (FAOSTAT data, 2004).

Sorghum

Like rice, sorghum is a minor crop in most countries investigated, the production of which amounts to only a few per cent of maize output. However, it is important in Ethiopia where it accounts for 20% of cropped acreage (Mulat and Teketel, 2003), and it is a major crop in Ghana and Nigeria where production equals 30% and 140% of maize production, respectively (Fig. 5.7; FAOSTAT data, 2004). Output trends differ in different countries with increases occurring in Ghana, Nigeria and Tanzania, and stagnation in Ethiopia, Kenya, Uganda and Zambia (Mulat and

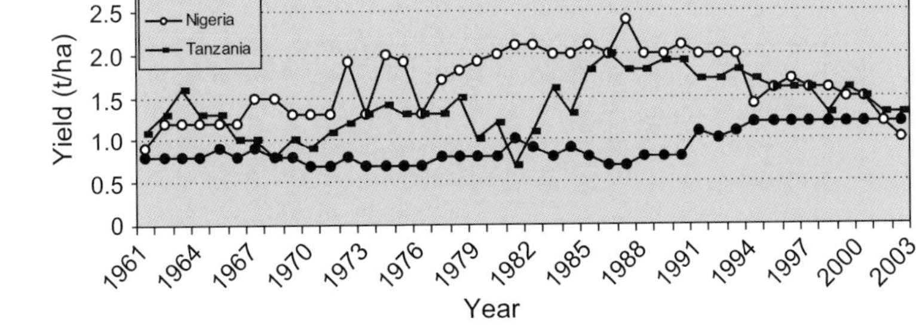

Fig. 5.6. Rice yields: Ghana, Nigeria, Tanzania. (Source: FAOSTAT data, 2004.)

Teketel, 2003; FAOSTAT data, 2004). It has been stated that, in sub-Saharan Africa, 'unlike other cereal crops, most of the adoption in farmers' fields of improved sorghum . . . varieties has occurred in recent years' (Mareida *et al.*, 2000:540). This, however, is not reflected in either the FAO or *Afrint* reports. Yields are generally low and stagnating around 1 t/ha, except in 'isolated cases' where they may reach 2 t/ha (Seini and Nyanteng, 2003). Changes in output are therefore a direct reflection of changes in area harvested. Ghana has seen a slow and steady increase since the 1960s. In Nigeria and Uganda, area harvested decreased steadily until SAP, after which it just as steadily increased and is now more or less equal to what it was in the 1960s (FAOSTAT data, 2004). In Zambia, due to the recent frequency of severe droughts, substituting maize, sorghum is facing renewed interest among smallholders (Saasa, 2003).

Cassava

Cassava has for long been regarded as a famine reserve crop but in recent years it has become more of a market product. Its importance, if measured in area harvested compared with that of maize, varies greatly among countries. In Kenya cassava acreage is equal to only 5% of the maize acreage and in Malawi it equals 11%. At the other end of the spectrum are Uganda (61%), Nigeria (76%) and Ghana (93%) (FAOSTAT data, 2004) but in these cases maize is not as dominant as in, e.g. Malawi or Zambia. Output has grown steadily in all countries studied but growth patterns are quite diverse.[7] In Kenya and Tanzania, a slow growth until the 1970s has become more fluctuating since the 1980s and production is levelling off in Tanzania and decreasing in Kenya since the introduction of SAP (FAOSTAT data, 2004). Only in Zambia has cassava output followed an even

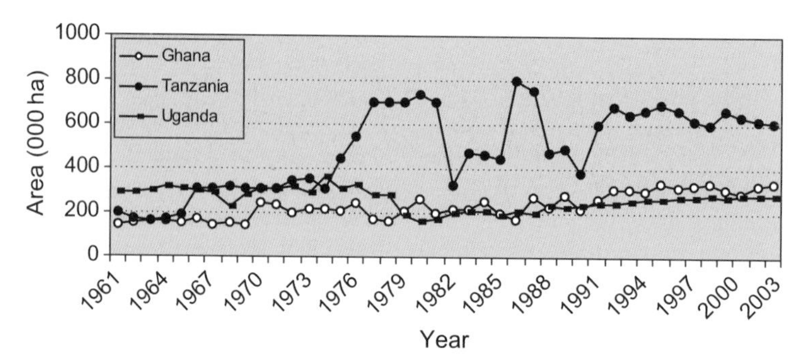

Fig. 5.7. Sorghum: area harvested: Ghana, Tanzania, Uganda. (Source: FAOSTAT data, 2004.)

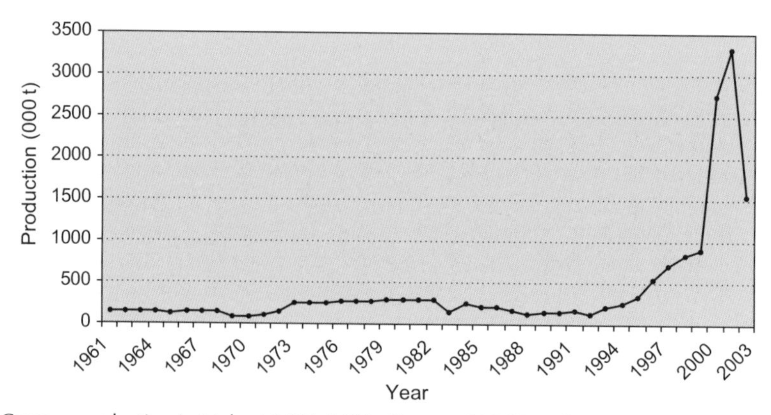

Fig. 5.8. Cassava production in Malawi 1961–2002. (Source: FAOSTAT data, 2004.)

growth path, albeit at a very low level. In Ghana, Nigeria and Uganda, growth in cassava production has coincided with SAP and spectacular growth rates have occurred post-SAP in Ghana, Malawi and Uganda with production increases of some 300% to 500% taking place in only a few years' time (FAOSTAT data, 2004).

In spite of the alleged cassava revolution in sub-Saharan Africa (Nweke *et al.*, 2002), yields have doubled post-SAP in Malawi and Uganda but remained rather stable in the other countries. In all other countries, increased output is a result both of increasing hectarages and of improved yields, but much more of the former (Akande and Kormawa, 2003; Seini and Nyanteng, 2003; FAOSTAT data, 2004). The span in average yield levels realized in different countries is great, ranging from 6–7 t/ha in Zambia and Kenya to 16 t/ha in Malawi in recent years (Figs 5.9 and 5.10; FAOSTAT data, 2004). Averages may be misleading, however. Saasa (2003) reports yields of only 2 t/ha in Zambia whereas Seini and Nyanteng (2003) mention yields as big as 28 t/ha in isolated cases in Ghana.

Summary of food crop performance 1961–2002

Throughout the period under study, food production has increased in all countries under study, albeit at a slow pace. This has mainly occurred through area expansion even if some productivity improvements have also emerged. The importance of area expansion has been accentuated since the introduction of SAP and yields remain low, much below their potential. Hence, it is yet too early to talk of a sub-Saharan green revolution. Yields of maize and sorghum have improved slightly in Ethiopia PDR, but in a

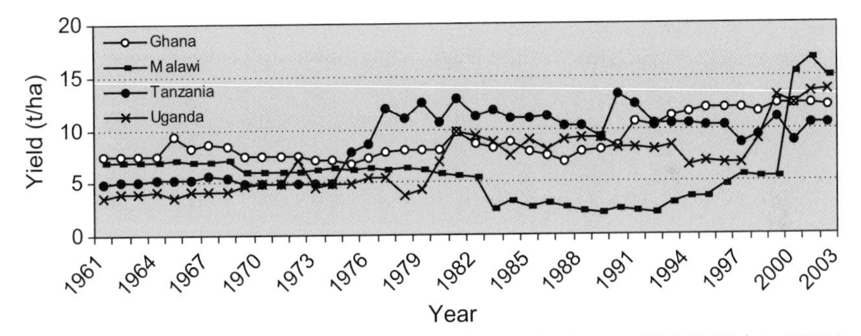

Fig. 5.9. Cassava yields in Ghana, Malawi, Tanzania and Uganda. (Source: FAOSTAT data, 2004.)

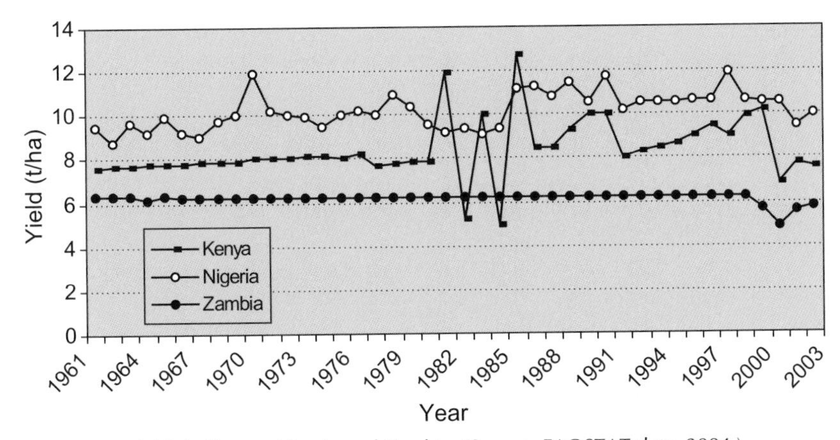

Fig. 5.10. Cassava yields in Kenya, Nigeria and Zambia. (Source: FAOSTAT data, 2004.)

very unstable manner. With the new government in 1992 came stagnation rather than increasing yields. With the imposition of SAP, general food-crop productivity improvements (cassava, maize, rice, sorghum) occurred only in Ghana, but post-SAP stagnation then set in for cereals. Other countries have not fared so well through recent reforms. In terms of productivity, SAP appears to have made no difference in Kenya. In the other countries, trends are inconsistent with some temporary yield improvements immediately after SAP (Ghana, Uganda) or stagnating and even declining productivity for most crops (maize in Zambia, rice in Nigeria and Tanzania). In terms of yield improvements, Tanzania actually performed best before SAP. The country experienced yield-growth for cereals before SAP but has since faced stagnating and declining yields for all crops. So far SAP cannot be said to have correlated with increased food productivity generally, the exception being cassava, the productivity of which has increased, sometimes dramatically, in recent years (post-SAP).

Apart from the recently enhanced importance of cassava in several countries, and the replacement of millet by rice as the third largest food-crop in Ghana (Seini and Nyanteng, 2003), from the information available there appears to have been very little structural change and no major changes in cropping patterns seem to have occurred in the smallholder sector. However, in recent years, large, commercial farmers in Zambia have abandoned maize in favour of export crops whereas, among the smallholders, drought tolerant crops are gaining renewed interest, albeit on a small scale (Saasa, 2003). Generally, maize remains the overwhelmingly dominant crop in most countries concerned (Isinika *et al.*, 2003; Milner *et al.*, 2003; Mulat and Teketel, 2003; Oluoch-Kosura, 2003; Saasa, 2003) and, if anything, the importance of maize (area harvested) appears to be accentuated (FAOSTAT data, 2004). The use of modern technologies and improved seeds appears to have either stagnated or declined in recent years and in several countries agriculture is pushed into the margins (Engel, 2001; Bazaara and Muhereza, 2003; Carlsson, 2003; Isinika *et al.*, 2003; Milner *et al.*, 2003; Mulat and Teketel, 2003; Oluoch-Kosura, 2003), resulting in enhanced fluctuations in yields and increased vulnerability for the smallholders.

Intensification?

Akande and Kormawa (2003:3) state: 'in most African countries . . . land scarcity, in a way that severely constrains agricultural production, is an isolated case rather than a widespread occurrence. Consequently, . . . intensification, in most cases, will be a choice rather than a [necessity]'. In this sense, extensification should not automatically be seen as a 'failure'. So far, intensification of food crop agriculture in sub-Saharan Africa has been limited both in time and in space and extensification has been a more 'practical' option in large parts of the subcontinent. Nevertheless, 'high level' intensification (e.g. HYVs, mixed farming, fertilizer, specialization on high-value crops) is practised primarily by a small group of large farmers. To a lesser extent, it does take place also among smallholders in heavily populated areas and/or near market centres and communication networks (see Larsson, Chapter 7, this volume). In other areas, a 'lower level' of intensification takes place as 'the shifting cultivation and bush fallow system is fading out because of population pressure' (Akande and Kormawa, 2003:6; see also Saasa, 2003), and when 'land pressures are forcing smallholder farmers to practise continuous cropping, often in cereal monoculture' (Milner *et al.*, 2003:9). Often such soil-mining practices go hand in hand with encroachment into marginal or unsuitable lands (Milner *et al.*, 2003; Mulat and Teketel, 2003; Oluoch-Kosura, 2003) with, sometimes, serious effects on the quality of land.

The picture that emerges is a rather mixed one. Spurts in production, particularly for maize (and in recent years cassava), have occurred frequently but have generally not been sustained. In most countries studied, yield improvements were better pre-SAP than thereafter. However, trends are inconsistent and point in different directions.

Contrary to widespread beliefs, there was a slow but steady increase in aggregate food production in Africa pre-SAP. However, population growth outstripped production and on a per capita basis the result was disappointing. Moreover, most of the increased production was due to area expansion, namely extensification was more important than intensification pre-SAP. For maize productivity (yields), implementation of SAP had no great impact in Malawi and Tanzania, whereas it appears to have been beneficial in Ghana and Uganda. In Ethiopia, Kenya, Nigeria and Zambia, the positive trend was broken during SAP. Also since the implementation of SAP, maize area expansion was more important than yield improvements. As suggested, this may well be a response to market liberalization. However, a previous trend of (slow) intensification appears to have been interrupted during SAP when inputs became more expensive while produce prices deteriorated, leading to widespread de-adoption of previously purchased inputs (see Holmén, Chapter 6, this volume). As for the other crops investigated (rice, sorghum, cassava) no consistent trends were found among countries, except that cassava yields made impressive growth in recent years in some countries. A first impression is that no or very little intensification has taken place in sub-Saharan Africa, neither before SAP nor since it was introduced.

However, focusing singularly on aggregate level statistics draws attention away from all the interesting differences. Average yield levels were found to be low for all crops studied (maize 1.5 t/ha, rice 1.5 t/ha, sorghum < 1 t/ha, cassava 2–7 t/ha). But for all of them we found reports of yields substantially above average – for maize 10 t/ha, for rice 3.5 t/ha, for sorghum 2 t/ha, and for cassava 16 and even 28 t/ha. Yield levels differed both between countries and within countries. These figures tell us that intensification of food staple production definitely has taken place in Africa south of the Sahara – in some regions and among certain categories of farmers. It may be pedestrian, but intensification is taking place also among staple food producers.

However, yields generally remain low and for many smallholders the shifts, e.g. from maize to cassava, may be indicators of stagnation (involution) just as well as signs of diversification and growth. Smallholders literally farm *small* farms. In Malawi the average size of a household is less than 1 ha and 41% of smallholders farm less than 0.5 ha/household (Milner *et al.*, 2003). This is not sufficient to provide them with food for the whole year. 'Most people in the rural areas run out of food three months before the next harvest' (Government of Malawi, 2002:41). Malawi is one of the poorest countries in the world with a high export dependency. Export crops (tobacco, tea, cotton, sugar) are primarily estate-crops. Smallholders primarily produce food (maize) even if some diversification has taken place in recent years. 85% of the workforce is in agriculture (Milner *et al.*, 2003) and more than 60% of Malawians live below the poverty line (Owusu and Ng'ambi, 2002).

Moreover, with increasing population pressure on land, plots are declining (Owusu and Ng'ambi, 2002). There is a considerable rural-to-urban migration of men looking for jobs, leaving women to do the farming (Milner *et al.*, 2003). When the government stopped subsidizing inputs, there was a 'drop of 43% in the usage of fertilizer [and] seed sales . . . declined by 56%' (Owusu and Ng'ambi, 2002:25). These circumstances have led, on the one hand, to overuse, declining soil fertility and yield-losses, and on the other to the breakdown of traditional social security systems and civil unrest, e.g. land disputes, robberies, theft of livestock and crop produce, and violence against women and alleged witches (Engel, 2001; Government of Malawi, 2002). Small-scale peasants presently have few opportunities to invest in agricultural modernization and those who do are afraid to do so.[8]

The circumstance that almost 1 million smallholders in Malawi engage in micro- or small-scale enterprises is quite possibly a sign of involution rather than development – if everyone is poor, where are the customers? Instead of being signs of positive livelihood diversification and growth, many of these micro-activities tend to represent desperation, distress sales rather than development. The fact that many smallholder households have (partly) replaced maize with cassava can in many cases be explained on similar grounds. Cassava, usually, is not fertilized and therefore is cheaper to produce. Harvesting can be done outside peak seasons, which is particularly important for overburdened female-headed households. But it is also a less nutritious crop and therefore this shift does not necessarily mean improvement.[9] The recent spread and commercialization of high-yielding cassava may, however, alter this picture.

This has happened at the same time as Malawi's estate sector continued to grow during SAP, from averaging some 750,000 ha in 1980–1989 to 1,150,000 ha in 1990–93. Its GDP share increased from 13% in 1973 to 30% in 1994 (Engel, 2001:38). It is to be noted that a large proportion of the estate sector is not cultivated since estate owners have found more lucrative income sources elsewhere. Due to enclosures, smallholders are struggling under an 'artificial Malthusian pressure' and, in the midst of chronic under-nutrition, an astonishing 2.6 million ha of suitable land is not cultivated (Engel, 2001:38). Malawian smallholders, unfortunately, are not alone in facing such situations. The practice of elites' land grabbing is frequently reported. In Uganda, although land-hunger is rampant, it is estimated that, due to enclosure and large-scale absentee landownership, only 27% of the country's arable land is presently cultivated (Bazaara and Muhereza, 2003).

The issue of intensification versus extensification, obviously, is a rather complicated one, and even more so when looking for the driving forces behind such processes. Both population pressure and market forces are at play here – in different combinations in time and space. But, as the above paragraph has shown, also other factors – enabling as well as constraining – are involved. What we can document in contemporary Africa south of the Sahara, therefore, is no clear-cut trends of either intensification *or* extensification. Rather, simultaneous processes of intensification, diversification and development occur parallel to processes of extensification, de-agrarianization and involution. So far, and in sharp contrast with the Asian experience, attempts to launch Green Revolutions and/or to improve the productivity of African food agriculture – especially small-scale agriculture – have been quite exclusive, involving (by design or default) only limited sections of the farming populations. In other words, they have hardly been smallholder-based.

Chapter 6 aims to shed further light on processes involved and, particularly, to assess the potential for a sustainable food crop intensification in sub-Saharan Africa in the early third millennium.

Notes

[1] Ethiopia has had its borders changed and is left out of this calculation.

[2] Zambia is excluded in these statistics due to incomplete data.

[3] Nigeria excluded due to incomplete data.

[4] Akande and Kormawa, 2003; Bazaara and Muhereza, 2003; Isinika *et al.*, 2003; Milner *et al.*, 2003; Mulat and Teketel, 2003; Oluoch-Kosura, 2003; Saasa, 2003; Seini and Nyanteng, 2003.

[5] Time series for maize production in Malawi, Tanzania and Zambia were compared with national precipitation data obtained from http://www.cru.uea.ac.uk/~timm/cty/obs/TYN_CY_1_1.html

[6] Popular Resistance Movement.

[7] There is no information on cassava for Ethiopia in either FAOSTAT or the *Afrint* country study.

[8] Malawi is not alone in experiencing increasing rural unrest and escalating criminality. Leaving collapsing states such as Congo and Somalia aside, even in countries where civil war has ended, notably Uganda, reports abound of theft, banditry and cattle-raiding, and the number of people reporting being affected by such unrest has doubled during the 1990s (Bazaara and Muhereza, 2003; Deiniger and Okidi, 2003).

[9] See also Bazaara and Muhereza (2003) for similar comments on signs of involution in Uganda.

References

Akande, T. and Kormawa, P. (2003) *Afrint* case study of Nigeria: Macro Study Report (final) (mimeo.). International Institute of Tropical Agriculture, Ibadan, Nigeria.

Alexandratos, N. (1995). *World Agriculture: Towards 2010. An FAO Study.* FAO/John Wiley & Sons, Chichester, UK.

Amin, S. (1972) Underpopulated Africa. 1972 (*Maji Maji No. 6*).

Andre, C. and Platteau, J.P. (1998) Land relations under unbearable stress: Rwanda caught in the Malthusian trap. *Journal of Economic Behaviour and Organization* 34, 1–55.

Asiema (1994) Africa's Green Revolution. *Biotechnology and Development Monitor*, 17–18.

Bazaara, N. and Muhereza, F. (2003) *Afrint* Macro Study Report – agricultural intensification and food security issues in Uganda (draft) (mimeo.). Centre for Basic Research, Kampala.

Berry, S. (1993). *No Condition is Permanent. The Social Dynamics of Agrarian Change in Sub-Saharan Africa*. The University of Wisconsin Press, Madison, Wisconsin.

Bienen, H. (1987) Domestic political considerations for food policy. In: Mellor, J.W., Delgado, V. and Blackie, M.J. (eds) *Accelerating Food Production in Sub-Saharan Africa*. The Johns Hopkins University Press, Baltimore, Maryland.

Boserup, E. (1965) *The Conditions of Agricultural Growth*. Aldine, Chicago, Illinois.

Börjesson, H.C. (2000) Geography and history of an intensive farming system in Tanzania. Department of Human Geography, Stockholm University, Stockholm.

Carlsson, E. (2003) *To Have and to Hold. Lund Studies in Economic History 28*. Lund University Press, Lund, Sweden.

Deininger, K. and Okidi, J. (2003) Growth and poverty reduction in Uganda, 1999–2000: Panel Data Evidence. *Development Policy Review* 21, 481–509.

Eicher, C.K. (2001) Africa's unfinished business: building sustainable agricultural research systems. AEC Staff Paper 2001–10. Michigan State University, East Lansing, Michigan.

Eicher, C.K. and Kupfuma, B. (1997) Zimbabwe's emerging maize revolution. In: Beyerlee, D. and Eicher, C.K. (eds) *Africa's Emerging Maize Revolution*. Lynne Rienner Publishers, Boulder, Colorado, pp. 25–43.

Engel, U. (2001) *Prospects of Crisis Prevention and Conflict Management in Mulanje District, Malawi*. GTZ, Eschborn, Germany.

Evenson, R.E. and Gollin, D. (2003) Assessing the impact of the Green Revolution, 1960 to 2000. *Science* 300, 758–762.

Fakorede, M., Badu-Apraku, B., Kamara, A., Menkir, A. and Ajala, S. (2003) *Maize Revolution in West and Central Africa: An Overview*. International Institute of Tropical Agriculture, Ibadan, Nigeria.

FAO/Earthscan (2003) World agriculture: towards 2015/2030. An FAO perspective. http://www.fao.org/docrep/005/y4252e/y4252e00.htm

FAO/GIEWS (2003) Food supply situation and crop prospects in sub-Saharan Africa. FAO Global Information and Early Warning System on food and agriculture (GIEWS). Report No. 1, May 2003. http://www.fao.org/docrep/005/ac977e/ac977e01.htm

FAO/WFP (2003) *Special Report: Crop and Food Supply Assessment Mission to Zambia. 10 June 2003*. FAO/WFP, Rome.

FAO/WFP (2004) *Special Report: Crop and Food Supply Assessment Mission to Ethiopia, 12 January 2004*. FAO/WFP, Rome.

FAOSTAT data (2004) Food and Agriculture Organisation of United Nations, http://www.fao.org/waicent/portal/statistics_en.asp

Friis-Hansen, E. (ed.) (2000) *Agricultural Policy in Africa after Adjustment*. CDR Policy.

Gasana, J. (2002) En ministers lärdomar av folkmordet i Rwanda 1994, http://www:omvarldsbilder/2002(021010.html

Goldman, A. and Smith, J. (1995) Agricultural transformations in India and Northern Nigeria: exploring the nature of Green Revolutions. *World Development* 23, 243–263.

Government of Malawi (2002) *Agricultural Sector Priority Constraints, Policies and Strategies Framework for Malawi*. Ministry of Agriculture and Irrigation. Malawi Sector Investment Programme.

Harrigan, J. (2003) U-turns and full circles: two decades of agricultural reform in Malawi 1981–2000. *World Development* 31, 847–863.

Harrison, M.N. (1970) Maize improvement in East Africa. In: Leakey, C.L.A. (ed.) *Crop Improvement in East Africa*. Commonwealth Agricultural Bureau, Farnham Royal, UK, pp. 21–59.

Hassan, R.M. and Karanja, D.D. (1997) Increasing maize production in Kenya: technology, institutions and policy. In: Byerlee, D. and Eicher, C.K. (eds) *Africa's Emerging Maize Revolution*. Lynne Rienner Publishers, London.

Hunger (2003) *Agriculture in the Global Economy. Hunger 2003. 13th Annual Report on the State of World Hunger*. Bread for the World Institute, Washington, DC.

Isinika, A., Ashimogo, G. and Mlangwa, J. (2003) *Afrint* country report – Africa in transition: Macro Study Tanzania (final). Dept of Agricultural Economics and Agribusiness, Sokoine University of Agriculture, Morogoro, Tanzania.

Jayne, T.S., Yamano, T., Weber, M.T., Tschirley, D., Benfica, R., Chapoto, A. and Zulu, B. (2003) Smallholder income and land distribution in Africa: implications for poverty reduction strategies. *Food Policy*, 253–275.

Kydd, J., Dorward, A., Morrison, J. and Cadish, G. (2002) Agricultural Development and Pro-poor

Economic Growth in Sub-Saharan Africa: Potential and Policy. ADU Working Paper 02/04.

Larsson, R. (2001) *Between Crisis and Opportunity: Livelihoods, Diversification, and Inequality among the Meru of Tanzania*. Department of Sociology. Lund Dissertations in Sociology 41, Lund, Sweden.

Lawrence, P. (1988) The political economy of 'The Green Revolution' in Africa. *Review of African Political Economy* 15, 59–75.

Lele, U. and Stone, S.B. (1989) Population pressure. The environment and agricultural intensification. Variations on the Boserup Hypothesis. MADIA Discussion Paper 4. The World Bank, Washington, DC.

Lele, U., Christiansen, R.E. and Kadiresan, K. (1989) Fertilizer policy in Africa: lessons from development programs and adjustment lending, 1970–87. MADIA Discussion Paper 5. The World Bank, Washington, DC.

Lipton, M. and Longhurst, R. (1989) *New Seeds and Poor People*. Johns Hopkins University Press, Baltimore, Maryland.

Mareida, M.K., Beyerlee, D. and Pee, P. (2000) Impacts of food crop improvement research: evidence from sub-Saharan Africa. *Food Policy*, 531–559.

Milner, J., Tsoka, M. and Kadzandira, J. (2003) *Afrint* country report: Malawi Macro Study (draft) (mimeo.). Centre for Social Research, University of Malawi, Zomba, Malawi.

Msambichaka, L.A., Kilindo, A.A.L. and Mjema, G.D. (1995) *Beyond Structural Adjustment Program in Tanzania: Successes, Failures and New Perspectives*. Economic Research Bureau, University of Dar es Salaam, Dar es Salaam, Tanzania.

Mulat, D. and Teketel, A. (2003) Ethiopian agriculture: macro and micro perspective. *Afrint* Macro Study (final) (mimeo.). Dept of Business Administration, Addis Ababa.

Netting, R.M. (1993) *Smallholders, Householders. Farm Families and the Ecology of Intensive, Sustainable Agriculture*. Stanford University Press, Stanford, California.

Nweke, F.I., Spencer, D.S.C. and Lynam, J.K. (2002) *The Cassava Transformation – Africa's Best-kept Secret*. Michigan State University Press, East Lansing, Michigan.

Nzimiro, I. (1985) *The Green Revolution in Nigeria or the Modernization of Hunger*. Zim-Pan – African Publishers, Oguta.

Ohlsson, L. (1999) *Environment, scarcity and conflict – a study of Malthusian concerns*. Department of Peace and Development Research, Göteborg University, Göteborg, Sweden.

Oluoch-Kosura, W. (2003) *Afrint* country report: Kenya Macro Study. College of Agriculture and Veterinary Sciences, Department of Agricultural Economics, Nairobi.

Omamo, S.W. (2003) Policy research on African agriculture: trends, gaps, and challenges. ISNAR Research Report 21. International Service for National Agricultural Research, The Hague, The Netherlands. http://www.isnar.cgiar.org/publications/catalog/rr.htm

Owusu, K. and Ng'ambi, F. (2002) *Structural Damage: the Causes and Consequences of Malawi's Food Crisis*. World Development Movement, London.

Saasa, O. (2003) *Afrint* country study: Agricultural intensification in Zambia. Macro Study (final) (mimeo.). Institute of Economic and Social Research, University of Zambia, Lusaka.

Scoones, I. and Toulmin, C. (1999) *Policies for Soil Fertility Management in Africa*. IDS/IIED.

Seini, W. and Nyanteng, V. (2003) *Afrint* macro study: Ghana Macro Report (final) (mimeo.). Institute of Statistics, Social & Economic Research, ISSER, Legon, Ghana.

Smith, J., Weber, G., Manyong, M.V. and Fakorede, M.A.B. (1997) Fostering sustainable increases in maize productivity in Nigeria. In: Byerlee, D. and Eicher, C.K. (eds) *Africa's Emerging Maize Revolution*. Lynne Rienner Publishers, Boulder, Colorado, pp. 107–124.

Soper, R. (1999) *African Agricultural Systems and the Terraced Landscape of Nyanga, Zimbabwe*. The World Archeological Congress 4.

Sutton, J. (1999) Engaruka: the success and abandonment of an integrated irrigation system, c. 15th–17th. In: Widgren, M. and Sutton, J.E.G. (eds) *'Islands' of Intensive Agriculture in the East African Rift and Highlands: a 500–year Perspective*. EDSU, Stockholm.

Tiffen, M., Mortimer, M. and Gichuki, F. (1994) *More People, Less Erosion. Environmental Recovery in Kenya*. John Wiley & Sons, New York.

Turner, B.L., II, Hydén, G. and Kates, R. (eds) (1993) *Population Growth and Agricultural Change in Africa*. University of Florida Press, Florida.

UNDP (2002) *Human Development Report*. Oxford University Press, Oxford, UK.

UNPP (2003) *World Population Prospects*. The 2002 Revision Population Database. http://esa.un.org//unpp/p2k0data.asp

van Buren, L. (2001) Nigeria: economy. In: *Europa Yearbook: Africa South of the Sahara*. Europa, London.

Widgren, M. and Sutton, J.E.G. (1999) *'Islands' of Intensive Agriculture in the East African Rift and Highlands: a 500-year Perspective*. African environments past and present, Department of Human Geography, Department of Physical Geography, Stockholm University, Stockholm.

6 The State and Agricultural Intensification in Sub-Saharan Africa

Hans Holmén
Department of Geography, Linköping University, Linköping, Sweden

This chapter analyses agricultural intensification – or its absence – in eight countries in Africa south of the Sahara (Ethiopia, Ghana, Kenya, Malawi, Nigeria, Tanzania, Uganda and Zambia). Special emphasis is directed at the triad state–market–smallholders and their respective roles in staple food intensification. The analysis is based on the findings of *Afrint* macro-level studies.[1] These reports have been supplemented with 'external' statistics, more general literature and relevant research papers. The purpose is to analyse macro-level processes unfolding with or without, thanks to or despite state intervention, and the effects of these processes on food-crop production and productivity.

Whether intensification takes place or not is a consequence of individual peasants' production choices. These circumstances are in turn influenced by a complex interplay of 'external' factors such as population pressure, land scarcity, access to alternative income sources, availability of credit and technology, markets for inputs and produce, and on the priorities (constraints) set by the political apparatus. Presumably, the priorities made by the state and market actors are not primarily determined by peasants' needs and world-views but rather by how state and market actors perceive their own situations and survival options.

This chapter focuses on the behaviour of state and market – and the preconditions for and effects of their behaviour – in relation to agricultural intensification. The analysis is divided into three periods: pre-SAP (from independence to introduction of SAP – Structural Adjustment Programmes), SAP (the relatively short period under which SAP where implemented), and post-SAP (the most recent years following SAP) – if, indeed, these economies can be said to have become sufficiently transformed to allow us to talk of a post-SAP period.

The State and National Food Self-sufficiency Pre-SAP

The *Afrint* country studies reveal that national food self-sufficiency has been a prioritized objective in all countries but one (Uganda), at least from the 1970s. From independence and until the mid-1970s, food generally was not a big problem even though fluctuating harvests caused local and/or temporary difficulties. For example, Kenya enjoyed substantial agricultural growth until 1980 (Oluoch-Kosura, 2003) and Malawi was considered food self-sufficient until 1980 (Milner *et al.*, 2003), as were Nigeria and Zambia until the 1970s (Akande and Kormawa, 2003; Saasa, 2003). Uganda never had a deliberate food policy since national food insecurity has not been perceived as a problem (Bazaara and Muhereza, 2003). Only Ghana embarked upon a 'bold attempt

 The African Food Crisis
(eds G. Djurfeldt, H. Holmén, M. Jirström and R. Larsson)

to transform . . . agriculture in the 1960s' (Seini and Nyanteng, 2003). With agriculture still capable of providing the necessary foreign exchange to pay for imports and virgin land still available in most cases, there seemed to be no great need to pay special attention to the food sector (Seini and Nyanteng, 2003). In the 1960s and 1970s, 'the idea of green revolution . . . was unknown in Africa' (Akande and Kormawa, 2003:21).

In the 1970s, the situation changed dramatically. Rapid population growth and droughts increasingly strained food security at the same time as a major drop in copper price in 1974 adversely hit Zambia and a quadrupling of the oil price in 1973 negatively affected most governments' budgets (and proved to be a mixed blessing for Nigeria). A number of governments embarked upon ambitious programmes to boost food production. In Tanzania, the Ujamaa policy,[2] launched in 1967, was intensified in the mid-1970s. In Ethiopia the new, radical *Derg* government embarked upon a massive agrarian reform programme. Most countries made efforts to support food production through providing credits, subsidies, inputs, extension services, and marketing facilities to smallholders. In Nigeria (backed by increasing oil revenues), this represented the 'golden age of green revolution when all known strategies were employed to change the food production landscape of the country' (Akande and Kormawa, 2003).

No doubt nationalism played a role here. For quite a few governments, it was important to show that the country was indeed independent and capable of satisfying its food needs by indigenous means (Milner *et al.*, 2003). 'Food production, obviously, was the priority objective' (Akande and Kormawa, 2003:23). Whereas in Ghana in the 1960s the need to provide food for a fast growing urban population led to experiments with state farms and mechanization programmes, in the 1970s policies directed at smallholders were introduced (Seini and Nyanteng, 2003). In various ways governments committed themselves to developing food-crop agriculture (Isinika *et al.*, 2003; Oluoch-Kosura, 2003) and, hence, assumed a leading role in agricultural development (Teketel, 1998; Akande and Kormawa, 2003; Mulat and Teketel, 2003; Saasa, 2003). Investments in the agricultural sector were generally high (Isinika *et al.*, 2003; Milner *et al.*, 2003). In Kenya agriculture grew by 5% annually in the 1970s (Oluoch-Kosura, 2003) and in Nigeria, direct government spending on agriculture increased 20-fold between 1970/71 and 1975/76 (Akande and Kormawa, 2003).

Great expectations

A number of strategies were implemented – state-farms in Ghana, Nigeria, Ethiopia and Zambia, collectivization in Nigeria, Ethiopia and Tanzania – in combination with attempts to reach out to smallholders with campaigns, extension and inputs. In some countries (e.g. Ghana, Nigeria, Kenya), ambitious, large-scale irrigation programmes were launched. Generally, the state provided credit and assumed responsibility for supplying inputs and handling produce through state-led peasant associations, cooperatives and marketing boards. Efforts were made to strengthen agricultural research and to expand extension services. Crop research programmes were initiated in a number of countries (Byerlee and Eicher, 1997) and new high-yielding maize varieties were released (Smale, 1995; Hassan and Karanja, 1997). Private traders were constrained or eliminated as part of the effort to ensure that the peasants got fair, stable and competitive prices, and to establish a degree of parity between agricultural and non-agricultural commodity prices. State monopolies in the handling of agricultural input and products became the rule. This enabled governments to regulate prices, to offer minimum price guarantees and pan-territorial pricing, and to provide inputs like seed and fertilizers at subsidized prices to a largely subsistence orientated smallholder peasantry. In many places, the smallholders found that they suddenly had access to external resources as well as 'markets'.

Missed opportunities?

The outcome of these measures was often interpreted as positive and sometimes they were. A number of countries were self-sufficient in food-crop production during this period. Often enough, production did increase, especially for preferential crops like maize (Ethiopia, Kenya, Malawi, Tanzania, Zambia) and rice (Nigeria). Yields, however, made only modest increases. Production improvements were almost invariably based on area expansion and, in the aggregate, there was no or very limited intensification of food-crop agriculture. One reason, obviously, is that as long as unused land is available it is easier to expand spatially than to intensify. But this is far from the whole story. The deterioration in Africa's terms of trade for its export crops after 1975 obviously plays a role (Friis-Hansen, 2000). So do the experiments with state farms and large-scale irrigation projects (usually orientated towards export crops rather than food staples), which proved uneconomic and absorbed a large part of investments (Akande and Kormawa, 2003; Oluoch-Kosura, 2003; Seini and Nyanteng, 2003). Often enough the governments' approaches to smallholders were relatively benign, which is indicated, for example, by the frequently accepted low loan-repayment ratios and debt-cancellations for agricultural debts (Saasa, 2003; Kelly *et al.*, 2003a; Seini and Nyanteng, 2003). In some cases, however, coercive measures were resorted to, for example the forced movements of people into new settlements in Tanzania and Ethiopia. In the latter case, the peasantry literally became 'tenants of the state' (Teketel, 1998).

Extension services have been criticized for less than optimal performance (Isinika *et al.*, 2003; Milner *et al.*, 2003; Saasa, 2003). States and 'developers' often did not have much faith in the smallholder peasants' abilities to enhance productivity or to develop their production for the market (Netting, 1993; Tomich *et al.*, 1995). Smallholders were also often perceived to be tradition-bound and to lack 'achievement orientation' (Mabogunje, 1980). Thus emerged a preference for 'modern' and 'scientific' models of agricultural development. Top-down management practices and negative attitudes towards the peasantry resulted in very simplified messages being transmitted to the smallholders. The ambition to boost wheat production in Nigeria (Andræ and Beckman, 1985; Akande and Kormawa, 2003) and the 'maize-only' programmes in Malawi and Zambia are cases in point (Smale, 1995; Saasa, 2003).

It has often been claimed that these state-controlled modernization programmes exploited the smallholders (e.g. Bates, 1983). This is definitely true in some countries. However, a number of studies have argued that, in general, 'while there was net taxation of the export crop sector in sub-Saharan Africa (SSA), during the 1970s food crop agriculture was actually subsidized rather than taxed' (see Friis-Hansen, 2000:12). This, however, is not to say that subsidization greatly improved the livelihood of smallholders, nor that it was effective from a food-crop productivity point of view. Through the monopolization of 'markets', farm-gate prices were suppressed and yield improvements, generally, were modest. Fixed prices squeezed the margins between costs of production and revenues from sale of produce for both smallholders and traders and, hence, reduced the incentive to produce a marketable surplus. With governments' priorities increasingly emphasizing low (urban) consumer prices rather than improved (rural) producer prices, the result was maintaining the *status quo* rather than agricultural development. Surplus production under these circumstances was not always attractive and where conditions deteriorated too much, smallholders were reported to be withdrawing into subsistence production (Hyden, 1983; Isinika *et al.*, 2003).

Moreover, with parastatals and marketing boards operating at a loss and the costs of subsidies mushrooming, this policy became economically unsustainable. In Nigeria, the cost of subsidies stood at 10% to 33% of the state's annual budget in the 1970s and early 1980s (Akande and Kormawa, 2003). In Zambia in the 1980s, maize subsidies ranged between 21% and 145% of the total budget deficit (Saasa, 2003).

It is commonplace today to 'explain' this lack of development by stating that African political leaders have been (and perhaps still are) crooks and cleptocrats, who do not care about development and whose only ambition is to enrich themselves by appropriating public resources (Bayart *et al.*, 1999; Tangri, 1999). While such malpractices no doubt occur(ed), sometimes, possibly, on a grand scale, this is not a satisfactory (and definitely not a sufficient) explanation. As we shall see, there are more convincing ones.

Poverty, i.e. a general lack of resources, contributes to explaining the dismal record pre-SAP. In contrast to Asia, it could be argued that the task that African governments set for themselves in the 1970s was just too large. For example, at independence Tanzania had only 16 university graduates (Economist, 2004). This, of course, was quite insufficient as a base for any broad development programme, let alone as foundation for the public sector. Lack of resources and insufficient administrative and managerial capacity could explain the top-down approaches and the simplified messages resorted to. It could also, at least in part, explain the frequent policy shifts and administrative reshufflings that took place (Isinika *et al.*, 2003). The circumstance that policies were poorly coordinated (Akande and Kormawa, 2003), that policies often were not implemented (Lawson and Kaluwa, 1996; Oluoch-Kosura, 2003; Seini and Nyanteng, 2003) and that they were often changed before results could be evaluated (Akande and Kormawa, 2003; Isinika *et al.*, 2003) could also be explained on similar grounds. But this is also not a sufficient explanation and it does not clarify why these efforts failed in sub-Saharan Africa while they succeeded in Asia. After all, policy shifts, bureaucratic awkwardness, top-down approaches, coercion and 'one-message-only' policies were not uncommon in Asia either. Additional explanations are needed.

As outlined below, besides lack of resources and opportunities, the desire for control has probably been at least as strong a motivation for policies and approaches resorted to, as has concern about widespread enhancement of agricultural productivity (Scott, 1998).

A double-edged agenda

Political elites are constrained, *inter alia*, by history, social structures and prevailing power relations (Hettne, 1973). Social institutions determine the obligations of the political elite, limit their autonomy and strongly impact upon what can be done and how. Hydén (1983, 2001) points out that post-independence political elites in sub-Saharan Africa enjoyed much less autonomy than did their counterparts elsewhere. Governments did not (and still do not) represent a corporate class whose development purpose they could be instrumental in fulfilling. Instead, with small and undeveloped middle classes and a majority of the population made up of smallholders who depend(ed) little on either state or market for their daily existence, the state became a target in factional struggles rather than a tool to be used for the realization of policy. Rather than being guided by a forward-looking purpose and executing corporate power, African states – permeated by neo-patrimonial rule and the informalization and personalization of power – were typically looted by their servants in order to honour obligations towards sub-national communities such as tribe and kinship. To express it more politely: 'African leaders do not seem to be giving [growth] the attention it deserves. Most seem keener on redistribution.' (Economist, 2004:16). In Asia, states consolidated their strength and expanded their room for manoeuvre by imposing their project and 'disciplining' factional interests. In sub-Saharan Africa, governments, much less successfully, tried to gain strength and legitimacy by allying with them. This also meant that, with the exception of Nyerere's Ujamaa policy in Tanzania and Mengistu's China-inspired agrarian reform in Ethiopia, there were few if any efforts to transform agriculture or prevailing socio-economic systems in sub-Saharan Africa.

The various African programmes for agricultural development released in the 1970s had a double function. Partly they were aimed at development and partly at nation-building, i.e. the consolidation of power. By providing agricultural inputs (and at the same time eliminating alternative suppliers) and by guaranteeing 'fair treatment' in the form of pan-seasonal and pan-territorial pricing for inputs and produce, the state could show its good intentions and, possibly, gain widespread legitimacy. At the same time, the servants of the enlarged state apparatus became a substitute for a social base that the state did not (yet) have and the state tended to turn a blind eye to malpractices and inefficiencies. Parallel to this, in order to reach out and extend the 'controlled' territory, local bosses, clan leaders and village headmen were co-opted into the clientelistic networks of the state. Also in these cases, malpractices, nepotism and diversion of resources from their intended use were often tolerated (e.g. Bates, 1983).

Favouritism took many forms. While credit and inputs were supposedly distributed evenly and fairly among the peasantry, experience tells a different story. Friis-Hansen (1994:13) mentions that, in Tanzania, 'a politically well-connected village could receive more than it demanded [of scarce hybrid maize seed], while other villages received only a fragment of their requirement'. In Malawi, the Banda government maintained the skewed colonial system where smallholders were confined to growing unprofitable maize, mainly for subsistence, while the lucrative tobacco production was a privileged undertaking by the estate sector, made up by loyalists and the regime's cronies (Orr, 2000). During this period, Malawi's estate sector grew significantly, from occupying 2% of total cultivated land in 1970 to 13% in 1981 (Smale, 1995). In Uganda, efforts to improve agriculture benefited influential people and political followers 'who had nothing to do with farming' (Bazaara and Muhereza, 2003:8). In Nigeria, credit and subsidized inputs were appropriated by 'absentee farmers, retired civil servants, and soldiers' (Olayide and Idachaba, 1987).

On many occasions, development aid added to these adversities. In the 1970s, aid was often given as budget support and could often be allocated at the receiving government's own discretion (Wohlgemut, 1994). Although politically functional, from a development point of view this often turned out to be counter-productive. The externally supplied resources came in handy as assets to be used in patronage and co-optation policies. More important, they functioned as an artificial lifeline reducing the pressure on receiving governments to develop the new countries' internal resources. Not uncommonly, funds were channelled away from agriculture through manipulations by officials and the well-connected, while credits, subsidized inputs and extension largely by-passed the smallholders (Pletcher, 2000; Akande and Kormawa, 2003; Milner *et al.*, 2003; Saasa, 2003). This was probably inevitable. With no perceived Malthusian squeeze, no external threats, artificially fixed territories and the African states lacking a corporate purpose and 'hanging in the air with no structural roots in the societies they were supposed to govern' (Hydén, 1983), the then young (urban) governments had to buy their ways into the countryside. And delayed development was the price the benign approach entailed.

Polarization, external shocks and economic collapse

The result was the establishing of a dual structure comprising, on the one hand, a small group of 'modern', often well-connected, and sometimes absent, commercial farmers and estate owners and, on the other hand, a vast majority of low-productivity, semi-subsistence-orientated smallholders growing traditional varieties and using only small amounts of fertilizers and improved seeds. Whereas the Green Revolution technologies as such have been found to be scale-neutral and, in Asia, have benefited smallholders as well as larger farmers (see Chapter 3, 4 and 13, this volume), in sub-Saharan Africa the beneficial effects of the technology have been much more restricted.

In Tanzania, maize yields almost doubled between 1971 and 1987 (Isinika *et al.*, 2003) and in Kenya agriculture grew rapidly until 1980 (Oluoch-Kosura, 2003). Kenya and Tanzania, however, appear to be exceptional cases. In Kenya, this trend was broken, partly due to external shocks such as oil-price increases and collapsing coffee prices on the world market (Oluoch-Kosura, 2003). The Zambian government, which had more or less neglected agricultural development, faced financial difficulties when copper prices plummeted in 1974–1975. This, in combination with a progressive decline in marketed maize, which the state could no longer afford to import, led to the establishment of an extremely costly programme of state-farming in order to guarantee staple food at low cost to the urban population. By 1984, Zambia became the most indebted country in the world, relative to its GDP (Saasa, 2003). Moreover, on the eve of SAP, Zambia had evolved into a '*rentier* economy driven to the brink of implosion' (Pletcher, 2000:129).

It has been suggested that pre- and post-1974/75 would, perhaps, be more appropriate periods (than pre- and post-SAP) if we want to trace differences in development-related behaviour in sub-Saharan Africa. There is no question that the above-mentioned external shocks had strong impact on governments' finances. However, from the variables investigated in this study (production, yields and area harvested of four major food-crops), the events of 1974/75 appear not to have had any major impact on food-crop intensification. In Zambia maize production fell some 30% and in Kenya fluctuations in maize yields were accentuated. In Tanzania, production of maize, sorghum and cassava increased (in the case of cassava due to yield improvements). In Ethiopia, an ongoing upward trend in maize production was maintained but annual production starts to fluctuate strongly, as do both yields and annually cropped areas. In Tanzania, production increases may be related to an already ongoing Ujamaa-programme and the impact of external shocks is disputable. In Ethiopia, the regime of Haili Selassie was toppled and a new government came into power. In the Ethiopian case, this probably has a greater explanatory value. Otherwise, the 1974/75 shocks are not reflected in the data studied – a possible illustration of the great distance between governments and (subsistence orientated) peasants at the time.

Also, Nigeria, with abundant oil-incomes, had neglected agriculture (despite the large sums allocated to the sector) and witnessed agricultural decline during the 1971–1980 period (Akande and Kormawa, 2003). In Ghana, 'the production of staple crops declined in a steady fashion from 1970 to very low levels in 1983 (pre-SAP)' (Seini and Nyanteng, 2003). In Uganda, production of most cereals stagnated pre-SAP and actually declined for maize (FAOSTAT data, 2004). In Ethiopia, where the radical government had eliminated most incentives for peasants to invest and produce beyond subsistence requirements, and therefore had to resort to forced deliveries of grain, the unsustainability of the system was exposed when the country was hit by severe drought and major famine in the mid-1980s (Mulat and Teketel, 2003).

Like Kenya, both Ethiopia and Uganda suffered from falling coffee prices on the world market. In Malawi, 'a series of external shocks at the end of the 1970s, namely a 35% collapse in the terms of trade, drought in 1979/80 and civil war in Mozambique, which disrupted external trade routes' (Harrigan, 2003:849), forced the government to turn to the IMF and the World Bank for financial assistance. In countries such as Ethiopia, Kenya and Malawi, the land-frontier had been reached and extensification of agriculture no longer seemed a viable option. Hence, for various reasons, external as well as internal, governments in sub-Saharan Africa had to implement Structural Adjustment Policies in order to access continued development aid and to be able to renegotiate their debt repayment schedules.

Structural Adjustment and Food Crop Intensification

The following analysis refers to the relatively brief period under which SAP policies were implemented in sub-Saharan Africa. Being variously introduced some time during the first half of the 1980s, it is generally considered that the SAP period had ended by the mid-1990s (except for Ethiopia, where SAP was introduced in 1993) (see Fig. 5.1, Holmén, Chapter 5, this volume). Invariably, SAP was meant to result in a complete turnaround of the economy away from state-led development to a market-based economy. The new role of the government was to become an enabler rather than a manager. This meant, on the one hand, that macro-economic stability was imperative (balanced budget and devaluation of overvalued currencies to facilitate exports and curb imports, thereby creating incentives for producers for the home market). It also meant deregulation of markets and the liquidation or transfer of parastatals to the private sector. Moreover, subsidies and price guarantees were to be abolished, since these gave the wrong signals to traders and producers and would lead to misallocation of (scarce) resources. In short, the state should step out of agriculture and limit itself to strengthening the infrastructural and institutional framework within which markets operate.

Implementation of SAP, in many cases, meant a renewed priority to agriculture (Akande and Kormawa, 2003; Mulat and Teketel, 2003; Seini and Nyanteng, 2003). However, in most cases, emphasis was not on staple food production. Although agriculture's share of governments' budgets often decreased (Isinika *et al.*, 2003; Milner *et al.*, 2003; Akande, Chapter 9, this volume), farmers initially responded favourably to the policy changes (Isinika *et al.*, 2003) and production increases were sometimes substantial (Oluoch-Kosura, 2003; Saasa, 2003). To begin with, reforms emphasized currency devaluations and macro-economic reform, liberalization of grain markets and removal of price controls on agricultural commodities. Elimination of subsidies usually followed at a later stage.

That agriculture initially responded positively, most likely, is a consequence of: (i) growth in cash crops exports; and (ii) expanding market opportunities for staple crops, especially maize, when markets were deregulated and more traders appeared. In Malawi, smallholders began to substitute burley tobacco for maize (Milner *et al.*, 2003). In Zambia adoption of high yielding varieties of maize increased from 30% of smallholder farmers in 1985 to 57% in 1990 (Saasa, 2003). In Ghana high-yielding varieties of maize, rice and cassava were 'adopted widely by farmers' (Seini and Nyanteng, 2003). In Nigeria, deregulation led to soaring inflation, especially rural inflation, and 'food crop production responded favourably to increasing output prices' (Akande and Kormawa, 2003) as 'more people took to farming on part-time or full-time basis' (Akande and Kormawa, 2003:57). As indicated above, the reported positive impact of SAP may be explained by several factors. One, obviously, is that during the 1980s, high-yielding varieties of maize were released in several countries studied (Evenson and Gollin, 2003). This could easily be attributed to SAP itself. But it should also be remembered that the research leading to these releases had been initiated in the pre-SAP period (see Smale, 1995; Byerlee and Eicher, 1997; Smith *et al.*, 1997; Mareida *et al.*, 2000).

However, in the food crop sector, this positive response appears to have been a temporary improvement. Reporting from Kenya, Hassan and Karanja (1997) report that growth in maize production could not be sustained in the 1980s and 1990s. FAO data for food-staples reveal stagnation of both area harvested and yield for maize in most countries studied (FAOSTAT data, 2004). Where production did increase (Ghana, Nigeria, Uganda) this was due principally to area expansion. Rice production generally did increase, but from low levels and, in most cases, this was due to increased acreage. Yields only grew in Ghana and Kenya. Production of sorghum increased in most countries but these increases were almost invariably due to extended area with yield improvements found only in Ghana. Cassava production also stagnated in most cases but made substantial increases in Ghana and Nigeria; in the latter case due to increased acreage, in the former due to yield improvements.

With Ghana being the outstanding exception, it appears that SAP was no panacea for food self-sufficiency in SSA. While national food security theoretically could be attained through export of cash crops in exchange for cheap food from outside, this would not guarantee food security for smallholders or for the poor in the cities. This has disrupted the reform programme and played havoc with the legitimacy of governments meant to implement them. Moreover, the abovementioned pre-SAP tendency towards polarization appears to have been accentuated during SAP. In Zambia, large farmers abandoned maize for more lucrative export crops while many smallholders were either stuck with unprofitable maize or turned to low-yielding sorghum instead (Saasa, 2003). Also in Kenya, intensification tended to be a privilege for the wealthy. Hence, 'small-farm production increased mainly through area expansion . . . while large-farm output expanded mainly through increased yields' (Lele and Agarwal, 1989).

As mentioned (Holmén, Chapter 5, this volume), enhanced production levels of major food crops started to oscillate after SAP. No close relationship was found with weather adversities and years of low production. For example, Zambia had surplus maize production 'only four times over the 1986–1996 period and [a] deficit six times (only three of these by reason of drought)' (Saasa, 2003:14).

It should be noted that elimination of subsidies appeared at a later stage of reform implementation. More important, this part of the SAP reforms were much more sensitive and difficult for governments to carry out. Initial improvements of SAP were to a not insignificant degree a consequence of retained subsidies. Whereas Zambia's SAP started in 1983, it was temporarily abandoned in 1987. The reason was food riots in 1986 sparked by an increase in maize meal prices (Saasa, 2003). Renewed food riots in 1990, again over price increases for maize meal, led to the fall of the government (Pletcher, 2000). In Nigeria, SAP was followed by runaway inflation, a slow-down in agricultural growth rates, escalating unemployment and social dislocations (Akande and Kormawa, 2003), factors which no doubt contributed to the collapse of the democratization experiment in 1993. This is a threat facing all governments in the region. Hence, a number of countries, notably Kenya, Malawi, Nigeria and Zambia, have had 'stop-and-go' implementation of SAPs, in no small degree due to the political dangers involved. Agricultural input subsidies were either abolished but reintroduced, as in Malawi (Harrigan, 2003) and Tanzania (Isinika *et al.*, Chapter 11, this volume), or subsidies have been retained (Nigeria, Zambia) although the levels of subsidization have been repeatedly raised and reduced, as in Nigeria (Akande and Kormawa, 2003).

A number of conclusions seem to be implied by the above circumstances. First, although unstable climate is a problem for agriculture in most countries, ups and downs in food production appear to be more closely related to ups and downs in levels of input subsidies than to changing weather conditions. Second, the initial positive impact of SAP can equally well be related to the fact that it was only partially implemented as to SAP *per se*. More exactly, it can have been the combination of retained subsidization and deregulated markets that gave the positive effect. Third, due to the on-and-off patterns of state involvement in agricultural markets, it is not always so easy to make a clear distinction between SAP and post-SAP periods. By the mid-1990s, however, agricultural subsidies had been withdrawn in all countries concerned and this could be regarded as a convenient start of post-SAP (but see below).

Post-SAP Food Crop Performance

The generally positive effects on staple food production that followed immediately after SAP have, in most cases, not been sustained post-SAP.[3] In Ethiopia, Ghana, Kenya, Nigeria and Malawi yield growth of the major cereals has been marginal or stagnating, often less than 1% per annum post-SAP (Akande and Kormawa, 2003; Milner *et al.*, 2003; Mulat and Teketel, 2003; Oluoch-Kosura, 2003; Seini and Nyanteng, 2003). In Tanzania, yields of maize and rice declined post-SAP (Isinika *et al.*, 2003) and in Nigeria yields decreased for rice, cassava and yam (Akande and Kormawa, 2003). Only Uganda and Zambia have experienced significant growth in agriculture post-SAP but this growth is a consequence of a reorientation away from food-staples towards higher value export crops (Bazaara and Muhereza, 2003; Saasa, 2003).[4] In Uganda, it is also a consequence of the (almost) ending of civil war.

Hence, despite an average agricultural growth rate of 4.4% per annum in Zambia in the 1990s, agriculture's performance has only been 'moderately satisfactory' (Saasa, 2003). This is because diversification has occurred among large, commercial farmers, who abandoned maize during SAP (resulting in a decrease of total maize area by at least 30% over the 1989 to 1996 period), whereas smallholders remain 'stuck with maize' and have been 'progressively marginalized from, rather than being pulled closer into, the liberalized market' (Saasa, 2003:53). Also in Malawi, the little growth in agriculture that has occurred post-SAP has been mainly in the estate (export) sector while – except for cassava in the last few years – smallholder (food) production either stagnated or declined (Milner *et al.*, 2003).

Use of modern inputs

The decline or stagnation in staple food production post-SAP is directly related to reduced use of modern inputs. This is despite the fact that a range of new technologies (improved varieties) have been released and made available to farmers in many countries (Mareida *et al.*, 2000; Evenson and Gollin, 2003). Moreover, despite the frequently aired allegations that Green Revolution technologies are unsuitable for Africa (Platteau, 2000; Madeley, 2002; de Grassi and Rosset, 2003), they have been found to 'adapt favourably to local environmental conditions and show remarkable resistance to diseases' (Akande and Kormawa, 2003:31; see also Oluoch-Kosura, 2003). However, except seed, modern inputs 'are not being used' (Bazaara and Muhereza, 2003:18; see also Evenson and Gollin, 2003; Kelly, 2003b).

In most countries investigated, smallholders have faced major problems in accessing modern inputs during the 1990s (Akande and Kormawa, 2003; Bazaara and Muhereza, 2003; Milner *et al.*, 2003; Mulat and Teketel, 2003; Saasa, 2003). Not only are adoption levels generally low, but de-adoption of hybrids and fertilizers has occurred in recent years (Kelly, 2003b; Oluoch-Kosura, 2003). Nevertheless, in some cases, e.g. Zambia, the use of HYVs is high, reaching 57% of farmers in the mid-1990s (Saasa, 2003). Likewise, in Ghana about 60% of farmers use improved varieties of maize, rice and cassava (Seini and Nyanteng, 2003). The recent dramatic growth in cassava yields in Malawi (see Fig. 5.8, Holmén, Chapter 5, this volume) must also be a consequence of a widespread adoption of new, high-yielding varieties. These seem to be exceptional cases, however, and generally the use of modern inputs is much lower. In Tanzania, 27% of farmers used improved seeds in the mid-1990s (Isinika *et al.*, 2003) and in Ethiopia, in 2000/01, only 5.4% of the cereal area was planted with improved seeds (Mulat and Teketel, 2003). For maize the corresponding figure in 2003 was 12% (FAO/WFP, 2004).

Fertilizer use post-SAP

It has recently been claimed that fertilizer use is increasing in sub-Saharan Africa (Crawford *et al.*, 2003) and that 'fertilizer consumption has increased substantially in recent years in Kenya' (Jayne *et al.*, 2003b:301). However, 'use rates vary considerably throughout the country, ranging from less than 10% of households surveyed in the dryer lowland areas to over 90% in . . . high potential areas' (Jayne *et al.*, 2003b:301). On average, however, the application of fertilizer is inadequate in Kenya (Oluoch-Kosura, 2003), as it is in most of SSA (Isinika *et al.*, 2003; Milner *et al.*, 2003; Seini and Nyanteng, 2003). With agriculture, in large parts of the subcontinent, being constrained by leached and nutrition-poor soils, the need for fertilizer is great. Moreover, due to cattle diseases and shrinking farm sizes, which in many cases limit the number of cattle a smallholder can keep, access to manure is restricted (Government of Malawi 2002; Saasa, 2003). (See Figs 6.1 and 6.2.)

With low average yields and despite the need for fertilizer, decline of fertilizer use on food crops appears to be the most dramatic effect of the SAP-reforms. In aggregate terms (not confined to food crops), fertilizer

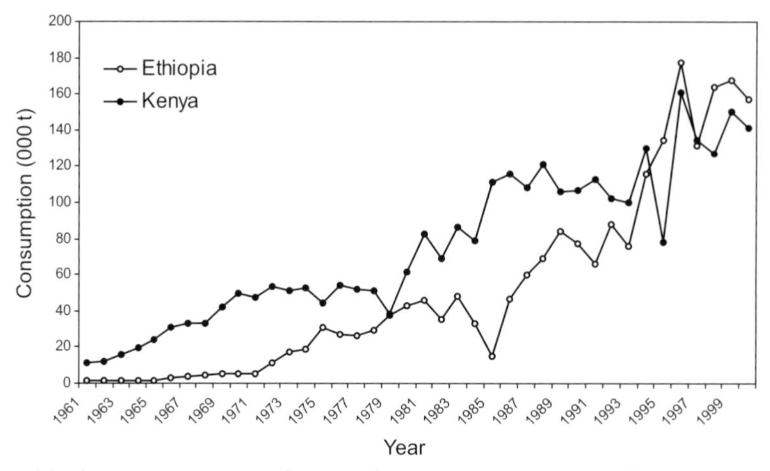

Fig. 6.1. Total fertilizer consumption in Ethiopia and Kenya. (Source: FAOSTAT data, 2004.)

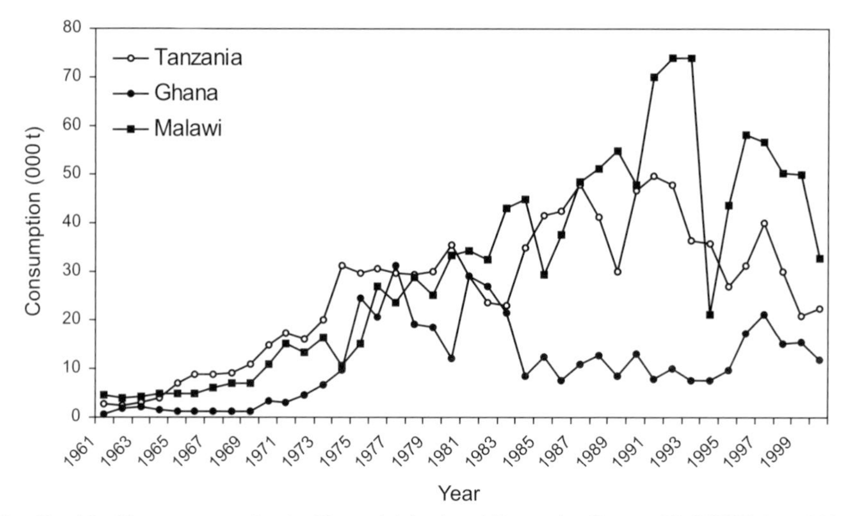

Fig. 6.2. Total fertilizer consumption in Ghana, Malawi and Tanzania. (Source: FAOSTAT data, 2004.)

consumption increased in all countries investigated except Uganda prior to the introduction of SAP (FAOSTAT data, 2004). (In Uganda's case, this was in large part due to civil war and general insecurity.) Only in Kenya and Uganda has growth been sustained through SAP, although in Kenya the trend has become more erratic post-SAP (FAOSTAT data, 2004). In the other countries, fertilizer consumption has either stagnated (Ethiopia) or declined (Malawi, Nigeria, Tanzania, Zambia) (FAOSTAT data, 2004). These trends of stagnating or declining fertilizer consumption post-SAP are also confirmed by the *Afrint* case studies (Bazaara and Muhereza, 2003; Isinika *et al.*, 2003; Oluoch-Kosura, 2003; Milner *et al.*, 2003; Mulat and Teketel, 2003; Seini and Nyanteng, 2003). It is one of the most deleterious consequences of SAP and poses a threat to the long-term ability of the subcontinent to feed its growing population.

These findings seem to contradict recent reports, e.g. Jayne *et al.* (2003a:1), who state: 'it is not the case that fertilizer use in SSA has declined [since SAP]'. Instead Jayne *et al.* found that mean fertilizer use per hectare rose by 5% between pre-SAP and post-SAP. Excluding Nigeria (which was a major consumer before fertilizer subsidies were eliminated) (Fig. 6.3), they find that fertilizer use increased by 15% over this period (Jayne *et al.*, 2003a:1). A few comments seem warranted. First, Jayne *et al.* include all land registered under arable and permanent crops, which includes non-edibles. Second, their study contains 26 countries, 10 of which are in West Africa and not included in this study. Third, most of these West African countries also represent the highest levels of increase in fertilizer consumption since 1980. When these countries are excluded, increases are less impressive and the apparent contradiction is dramatically reduced.

Actually, among the countries included in the *Afrint* study, Jayne *et al.* found significant increases in fertilizer consumption only in Ethiopia, Kenya and Uganda. In the case of Uganda it is to be noted that use levels remain very low (Fig. 6.4). Actually, at less than 0.5 kg/ha (Jayne *et al.*, 2003a:1), they remain among the lowest in SSA. On the other hand, Jayne *et al.* found that fertilizer consumption per cultivated hectare declined or stagnated since the introduction of SAP in Ghana, Malawi, Nigeria, Tanzania and Zambia. Also of interest is the observation that 'most of the increase in fertilizer consumption . . . occurred in the first half of the 1990s' (Crawford *et al.*, 2003:279), confirming our observation that after an initial positive response directly after SAP (when, in many

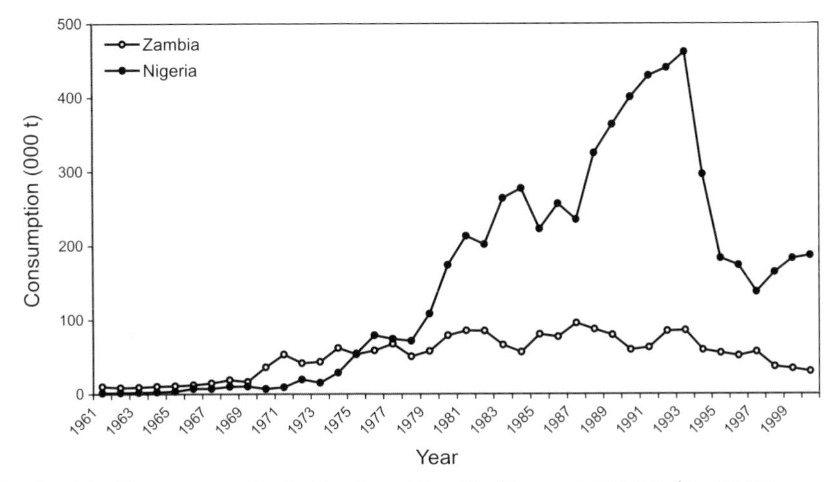

Fig. 6.3. Total fertilizer consumption in Nigeria and Zambia. (Source: FAOSTAT data, 2004.)

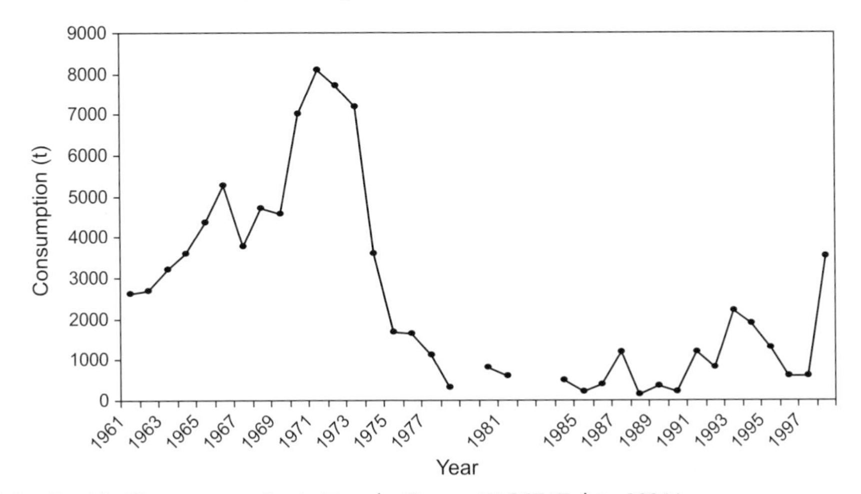

Fig. 6.4. Total fertilizer consumption in Uganda. (Source: FAOSTAT data, 2004.)

cases subsidies were retained), decline or stagnation set in post-SAP. In short, there seems to be no disagreement.

Leaving the aggregate level aside, a more interesting observation is that 'most of the increase in fertilizer use has been on [cash crops], with minor increases on maize' (Jayne *et al.*, 2003a:2). Also, Seini and Nyanteng (2003) report that fertilizers are allocated to rice and cash crops rather than to staple food crops such as maize. Fertilizer prices have increased post-SAP while grain prices have declined. Sometimes price increases on fertilizer have been extremely high – 150% in Meru, Tanzania (Larsson, 2001), some 200–300% in Ghana (Seini and Nyanteng, 2003), and 200–500% in Nigeria (Akande and Kormawa, 2003) – depending on the type of fertilizer. Fertilizer prices are presently reported to be 4.5 times as high in western Kenya as in Europe (Conway and Toennissen, 2003).

The pattern is not clear-cut, however. Jayne *et al.* (2003a:2) write: 'only in Ethiopia did food grains receive the lion's share of the increase in fertilizer use'. This is mainly because president Zenawi 'took a personal interest in . . . [the] new extension programme (NEP) to promote high external input technologies for maize, tef, sorghum and pulses' (Howard *et al.*, 2003:337), resulting in 'aggressive efforts' to increase the use of fertilizer (Mulat and Teketel, 2003). The campaign appears to have gained some success since 30% of Ethiopian farmers are reported to have used fertilizer (Howard *et al.*, 2003) and bumper crops were attained in 2000 and 2001. The Ethiopian government has increased its control over fertilizer pricing and distribution (Jayne *et al.*, 2003b), which, perhaps, explains this success. However, this has also resulted in 'high levels of rent seeking associated with a tendency to favour politically well-placed suppliers' (Kelly *et al.*, 2003a). Even though fertilizer use on food staples has increased in recent years in Ethiopia, total use is roughly equal to that in Kenya (Fig. 6.1) although Ethiopia's population is more than twice that of Kenya's. Moreover, fertilizer use is limited to accessible, high potential areas (Mulat and Teketel, 2003). In 2001, bumper harvests led to an 80% drop in maize prices in 'favoured' areas and threatened to ruin poor smallholders who bought inputs on credit. At the same time, drought hit other parts of the country but lack of transport infrastructure meant that surplus grain could not be moved to deficit areas, which instead had to be supplied by food aid from abroad. Moreover, in Ethiopia, fertilizer remains indirectly subsidized (Howard *et al.*, 2003) and it is increasingly dependent on donor assistance programmes (Jayne *et al.*, 2003b; Mulat and Teketel, 2003). Hence, one may raise questions about the system's sustainability.

Subsidies and agricultural credit

In general, a major problem facing African small-scale producers is not how to use inorganic fertilizer or high yielding varieties but, rather, how to afford them. For liquidity-constrained smallholders, credit is essential. With the dismantling of parastatals during SAP, if they ever did so, most governments have now ceased providing agricultural credits to smallholders. Instead, commercial banks are relied upon to extend loans but smallholders, generally, do not have the necessary collateral. And even if they do, interest-rates – 46% in Ghana and 48% in Malawi – are prohibitively high. Hence, small-scale peasants in sub-Saharan Africa, despite some exceptional cases, no longer have access to agricultural credit (Akande and Kormawa, 2003; Bazaara and Muhereza, 2003; Isinika *et al.*, 2003; Milner *et al.*, 2003; Oluoch-Kosura, 2003; Saasa, 2003; Seini and Nyanteng, 2003). For female-headed peasant households, access to credit is even more restricted (Milner *et al.*, 2003; Mulat and Teketel, 2003; Oluoch-Kosura, 2003).

There are a few exceptions, though. In Ethiopia, the government still extends credit to smallholders in the form of fertilizer loans. With unstable prices on maize (see above) and because farmers are obliged to pay their fertilizer loans immediately after harvest (when prices are low) (Mulat and Teketel, 2003), borrowing for cereal production is a risky undertaking. The government of Ethiopia 'has taken a very strong stand on repayment, with arrests or confiscation of assets where necessary' (Kelly *et al.*, 2003a:393). The Malawi government has temporarily supplied smallholders with subsidized small-scale packages of seed and fertilizer (Starter Pack, Targeted Input Programme). Having first been seen as a temporary relief measure, the government has sought to make permanent the programme under strong opposition from the World Bank (Harrigan, 2003) and donors are no longer prepared to support it.[5] In Zambia, the government in recent years distributed large quantities of fertilizer on credit through well-connected 'fertilizer-agents' – influential local elites or their proxies (Jayne *et al.*, 2003b:312) – while at the same time accepting loan recovery rates as low as 30–40% (Kelly *et al.*, 2003a). Zambia's fertilizer/credit programme not only reaches less than 10% of the smallholders (Saasa, 2003), it has been wide open to rent-seeking activities (Jayne *et al.*, 2003b) and it 'has become a virtual give-away' (Kelly *et al.*, 2003a:392), although apparently functional in fulfilling patronage objectives.

Infrastructure

Africa is a large continent with, in vast areas, low population densities. Consequently, provision of transport infrastructure is a costly undertaking. Nevertheless, it is essential for integration and development. Whereas, since SAP, governments should concentrate on a supportive role and provide necessary infrastructure, roads, storage and transport infrastructure are everywhere poorly developed. While generally expanding pre-SAP, maintenance tended to be inadequate and the standard of transport infrastructure deteriorated. Expansion of transport systems take time but only in Ghana and Tanzania have governments invested heavily to expand and upgrade roads and communications infrastructure post-SAP (Isinika *et al.*, 2003; Seini and Nyanteng, 2003). In contrast, in countries such as Kenya, Malawi, Uganda and Zambia, such investments have decreased in recent years (Akande and Kormawa, 2003; Bazaara and Muhereza, 2003; Milner *et al.*, 2003; Mulat and Teketel, 2003; Saasa, 2003). In Kenya, the state of infrastructure 'has deteriorated to the extent that it has become a hindrance to growth' (Oluoch-Kosura, 2003). Even where infrastructure has improved, the costs of transport have increased when governments have withdrawn (subsidized) transportation of produce (Isinika *et al.*, 2003). Hence, transport costs are high everywhere, posing disincentives for peasants to commercialize and for private traders to extend their activities beyond more densely populated and easily accessible areas.

Extension

In most countries, a similar effect can be seen in extension services. Whereas there appears to be no consistent trend post-SAP in terms of increasing or decreasing government budget allocations to extension (Akande and Kormawa, 2003; Mulat and Teketel, 2003; Oluoch-Kosura, 2003), these are in most cases being transformed into: (i) market-oriented and (ii) participatory undertakings, namely, at the same time as the concept of the 'progressive' farmer is revived, extension is now to be demand-driven.[6] In, e.g. Malawi and Uganda, the stated objective is to transform agriculture from subsistence orientation to a commercial enterprise via a stepwise development from 'subsistence-peasant' through 'emerging (progressive) farmer' to 'commercial farmer' (Government of Uganda, 2000; Government of Malawi, 2002; Bazaara and Muhereza, 2003; Milner *et al.*, 2003).

This reorientation towards participation and partnership seems, however, to contain a fair amount of lip-service. The Malawi Ministry of Agriculture's (Government of Malawi, 2002) suggestion that 'not many people take farming seriously' (p.35) and that smallholders 'should look at farming as a business and not just a hobby' (p.39) is a case in point. As this study shows, structural and institutional factors explain more about low adoption rates than do peasants' alleged 'improper attitudes'.

Top-down extension was not uncommon in Asia during its Green Revolutions. Arguably, this did not hinder their successes. Hence, the circumstance that African extension has been less effective than its Asian counterparts ought to be explained by other factors. Most likely, profitability explains the different outcome. Even if Asian smallholders have been taxed as well – actually at a higher rate than in Africa (Gunnarsson and Rojas, 1995) – their margins between input and output prices have been much greater. In Asia, price policies made it worthwhile for smallholders to invest and adopt new technology; in sub-Saharan Africa they do not. Without similar incentives, it is 'questionable whether [African] subsistence farmers want to become commercial producers' (Bazaara and Muhereza, 2003:32).

With the contemporary low profitability of staple food production, it is likely that a demand-driven extension service will concentrate on more profitable cash crops. This may perpetuate a big and 'progressive' farmer bias, especially as demand-led implies cost-sharing (Oluoch-Kosura, 2003). In, for example, Zambia, the government has handed over much of input provision and extension to privately operated out-grower schemes. Such schemes may be effective distributors of selected inputs. Nevertheless, they tend to favour the organizers (big farmers and agribusiness), and the technology provided is 'imposed upon smallholder farmers on a "take-it-or-leave-it" basis' without much reciprocity (Saasa, 2003:44). Considering the previous bias towards big farmers as well as the above mentioned trends towards spatial polarization, it is likely that demand for extension will be similarly geographically and socially uneven – strong from large farms and in already favoured areas, and weak among food-producing smallholders and in more peripheral locations. With, on average, half the population eking out a living under the poverty level, one may have doubts about how widespread effective demand will be.

Entrepreneurs and agricultural markets

Since the implementation of SAP, state monopolies on input provision and trade in agricultural produce have been eliminated. Likewise, the previously common prohibitions of trans-province trade and shipment of agricultural products have been abolished. Consequently, private traders and entrepreneurs have found some new room for manoeuvre. Crawford *et al.* (2003) conclude that reforms – when implemented as intended – in general have enhanced the private sector's capacity to serve the agricultural sector. They further found that progress has been concentrated in areas where agriculture is most profitable and export driven. Moreover, whereas also poor farmers located in these favoured areas have benefited from privatization, access to credit remains a problem for farmers with limited income from cash crops or non-farm activities.

Woodhouse (2003:1709), likewise, argues 'there seems little doubt that increased agricultural output has been largely driven by producers' response to market opportunities'. Also, Jayne *et al.* (2003b:297) find that (in Kenya) 'there has been an impressive private sector response to fertilizer market reform and that the market is generally competitive, particularly at the retail level'. These quotations may, however, give too rosy a picture of market development post-SAP, especially since reforms, generally, have not been implemented as intended. Spatial disparities are often reported to be on the increase as traders prioritize easily accessible and more productive areas, more or less abandoning peripheral regions. However, even if SAP has meant enhanced spatial polarization, the assumed break with the past should not be exaggerated. Pan-territorial pricing did not always mean pan-territorial supply.

Other studies present a quite opposite picture of the current state of affairs. Generally, they found, markets are undeveloped with a limited number of traders mostly engaged in produce trade but not in input supply (Milner *et al.*, 2003; Mulat and Teketel, 2003; Oluoch-Kosura, 2003; Saasa, 2003), the latter circumstance reflecting that whereas output markets may have been fully liberalized, input markets sometimes remain controlled (Pletcher, 2000). Liberalization has been half-hearted and patronage policies have been found to be compatible with processes of political and economic liberalization (Tangri, 1999; Rakner, 2003). In Ethiopia, *all* private fertilizer traders withdrew in 2001/02 due to unfair bureaucratic treatment (Mulat and Teketel, 2003). Instead monopolistic holding companies have been created with 'strong ties' to regional governments (Jayne *et al.*, 2003b). In other cases, privatized parastatals have been sold to top politicians' kin and proxies, sometimes at extremely low prices, resulting in private monopolies replacing state monopolies (Tangri, 1999; Craig, 2003). Cronyism and rent-seeking have been integral parts of liberalization (Pletcher, 2000; Cooksey, 2003). Hence, Kelly *et al.* (2003a: 393) stress the 'high levels of rent-seeking associated with a tendency to favour well-placed suppliers . . ., thereby constraining the development of lower cost, truly commercial input supply networks'.

This, however, is not to say that governments are indifferent to market development, only that opportunities to enter are artificially skewed. Moreover, emerging markets are increasingly threatened by external influences. Among other things, SAP aimed at the elimination of external trade-barriers. This has led to increased competition from imported food grains (at artificially low prices as rich countries dump their subsidized surpluses in poor countries). The profitability of staple food production is undermined and therefore the prospects of attaining food self-sufficiency look bleak. Governments have pursued different policies to come to terms with this situation. Uganda appears to have fully embraced SAP and aims for food-security through export of high-value crops, the foreign exchange earnings of which will permit imports of food grains if necessary. In Zambia, flight from food-crop production has occurred among larger farmers at the same time as fertilizer subsidies have been reintroduced (although distribution remains skewed). It is questionable how much of these fertilizers is allocated to food crops. In Tanzania, the revival of subsidized agricultural input supply in the Southern Highlands was announced in June 2003 (Isinika *et al.*, Chapter 11, this volume). To what extent these fertilizers will go to food crops remains to be seen. In Malawi, smallholders have to some extent abandoned unprofitable maize production for exportable cash crops (peppers, tobacco) while at the same time the state has tried to make permanent subsidized starter pack programmes.

Other routes have been followed as well. In Kenya, depending on the availability of domestically produced staple-food, the government raises import duties on food imports to restrict them when domestic supplies are high (to increase producer prices) and lowers import duties when there is a domestic deficit (to lower domestic prices) (Oluoch-Kosura, 2003).[7] In Nigeria, with the aim of attaining national food self-sufficiency (Shaib *et al.*, 1997), a retreat from a full-fledged SAP in 1993 inaugurated a policy of 'guided deregulation', namely 'in cases where domestic [food] production was adjudged adequate for the internal market at non-inflationary prices, imports were discouraged' (Akande and Kormawa, 2003:58). Initially, import of maize, sorghum and wheat was banned (Shaib *et al.*, 1997). The 'import prohibition list' was later expanded also to include millet, wheat flour, vegetable oils and (for health reasons) all types of meat (WTO, 1998).

It should be remembered that most potential customers of inputs and producers of output are small-scale and liquidity constrained. In the mid-1990s, 72% of Malawi smallholders each farmed less than 1 ha, and 41% less than 0.5 ha (Milner *et al.*, 2003). In Ethiopia, nearly 40% of peasant households each cultivates less than 0.5 ha (Milner *et al.*, 2003). Hence, even when smallholders respond positively to market signals, their supplies and effective demands are limited. Together with frequently dispersed settlement patterns and inadequate road infrastructure, this renders trade rather costly. Most traders also are small-scale with limited working capital and limited capacities both to store and to transport agricultural commodities. They are hardly ever capable of extending credit, especially not for food crops. They face difficulties in reaching economies of scale and most traders operate at very small margins. Jayne *et al.* (2003b:313) found transport and handling costs account 'for 50% or more of total domestic marketing margins. The sum of importer, wholesaler and retailer profit margins generally account for less than 10%' (see also Saasa, 2003).

Agricultural markets tend to be non-competitive (Kelly *et al.*, 2003a; Milner *et al.*, 2003). Traders are sometimes reported to establish 'associations' in various markets, covering specific commodities. Thereby they are said to be able to collude fixed prices, to control the flow of commodities, and to erect barriers against others entering the market (Seini and Nyanteng, 2003). Hence, complaints abound about peasants being cheated by unscrupulous traders using false weights and measures and/or selling false seeds (Bazaara and Muhereza, 2003; Milner *et al.*, 2003). Sometimes, it is the traders being cheated by peasants (Kelly *et al.*, 2003a). It is, of course, difficult to assert how much is true, how much is suspicion and how much is rumour with this type of allegation. Whatever the case, trust appears not (yet) to characterize these newly emerging markets.

For a number of reasons, then, markets remain poorly developed. Rural markets are characterized by high seasonality, strong price variations, large information asymmetries and high transaction costs. Small-scale traders do not possess the capital, managerial skills and business experience to scale up activities. Hence, African smallholders are facing serious marketing and price uncertainties, which have contributed to diminishing producer confidence in newly liberalized markets. This, in turn, has encouraged governments increasingly to intervene in order to correct 'market failures'.[8] Writing about Zambia, but widely applicable, Saasa (2003:40) argues that governments tend to under-estimate the capacity of the private sector and that 'the government's lack of confidence in the capacity of the private sector has become a self-fulfilling prophecy in the sense that its continuing involvement in the market despite market liberalization has fuelled the private sector's loss of confidence in this activity and, hence, a good justification for further government entry'.

There can be good grounds to doubt the motives for government's market interventions. But the imposed 'one-message-only' policies to substitute market for government, as the 'trigger' of development, has been preoccupied with hypothetical gains in ideal situations far removed from real world possibilities and constraints. Proponents of SAP have had the ill-founded expectation that market reform would be quick and that around the corner there were willing and able entrepreneurs just waiting for the 'go-ahead'. This turned out to be wishful thinking. For one thing, 'the argument for statism was normally not that state-organized economic activities were inherently superior, but that they were a substitute for a defective market. Now instead, a defective market is seen as a substitute for a defective state' (Hettne, 1992:5). Clearly, the World Bank's and many donors' diagnosis and proposed cure miss essential points. As noted by Simon (1999:29), 'SAPs . . . are directed at the symptoms rather than the underlying causes'.

The Role of NGOs and Peasant Associations

In the Asian leg of this study, the conclusion was reached that Asian Green Revolutions were based not only on state and market but that they also had a strong smallholder orientation. Before SAP smallholders in sub-Saharan Africa were quite neglected and productivity remained at a low level. True, African smallholders were often made to join cooperatives, which could have been a means for articulation of their needs and aspirations. More often than not, however, rural cooperatives were either established or hijacked by governments and the influence of smallholders was often negligible (Holmén, 1990). With the changed economic and political landscape post-SAP, it has been assumed that now opportunities will emerge for peasants to form their own organizations for economic and developmental purposes.

Compared with, for example, Asia, where non-governmental organizations (NGOs) have a long history, non-governmental organization activity in sub-Saharan Africa is of a fairly recent date (Farrington and Bebbington, 1993). During the last two decades there has been a formidable explosion in numbers as well as types of NGO. In, for example, Uganda, the number of NGOs started to increase in the late 1980s. By 2002 there were 2900 NGOs registered in the country, 400 of which were international (NGO Task Force, 2002). It has often been assumed that the NGOs of today represent a better organizational form than yesterday's parastatals. While sometimes correct, this is nevertheless a dubious assumption. NGOs are not one of a kind. They span a wide gap in terms of origin, size, resources, purpose, ideology and scope of activities.

Commonly, a distinction is made between indigenous and foreign or international NGOs. Both are intermediaries, supporting (or creating) organizations at 'lower levels', ultimately working with grassroot groups and community-based organizations (CBOs). 'Higher order' NGOs also often act as proxies for Western national as well as private aid agencies. Some foreign NGOs have budgets in parity with official donors. Indigenous NGOs at national and regional levels fulfil similar intermediary functions and often act as support organizations for local groups. NGO activities include, for example, extension, agricultural input provision, the organization of seed banks, agro-processing, micro-credits and small-scale enterprise, but also advocacy for human rights, women's emancipation and empowerment.

Even among those NGOs orientated primarily or exclusively towards agricultural development, their ideas about what African agriculture needs vary greatly and are sometimes conflicting. Some big, international NGOs such as Sasakawa Global 2000 (active in a dozen countries in SSA) work directly with governments, extend a 'High External Input' type of agriculture, and introduce 'classic' packages of grain seed and fertilizer. Others promote diversification and the growing of export crops (Bazaara and Muhereza, 2003; Milner *et al.*, 2003). Others, yet, are orientated more towards small-scale seed and fertilizer provision for food crops among poor peasant households (Seward and Okello, 2003). In contrast, quite a few NGOs are sceptical about Green Revolution technologies (and market orientation) and embrace a 'small is beautiful' concept emphasizing no or few external inputs, recycled seeds, grassroots empowerment and development from below (if and as defined by grassroots). While there are examples of successful empowerment (e.g Krishna *et al.*, 1997), it is not uncommon that 'empowerment' means that villagers shall learn to see things the outsider's way (Holmén and Jirström, 1996).

In rural Africa, grassroots associations and CBOs tend either to be non-existent or weak (Howard *et al.*, 2003; Saasa, 2003). NGOs, domestic as well as foreign, have stepped in and tried to form local groups, sometimes with good results (Ellis *et al.*, 2003). However, for all their virtues, an unease has evolved over the nature of the NGO case (Tripp, 2000:1). Reporting from Nigeria, Akande and Kormawa (2003:67) found that, in most cases, farmer groups are not only organized but also managed by NGOs. Sometimes NGOs are successful, e.g. in raising yields, only because, due to preferential treatment, they have been able to circumvent those obstacles that stand in the way of ordinary people and the market (Reardon *et al.*, 1999). For example, some NGO projects show higher adoption rates than official extension services and are more popular among smallholders because they provide subsidized inputs in a supposedly post-subsidization era (Holmén and Jirström, 1996; Kelly *et al.*, 2003a).

Whereas rural farmer associations tend to be few and weak, and therefore seem to need external support, such support tends to create dependencies. The gap between supporters and receivers is often great and the agenda of the supporter need not overlap with that of the supported. Cleary (1997:228) frankly states: 'all developing country NGOs are part of their societies' elite'. Seini and Nyanteng (2003:21) found NGOs in Ghana to be 'of a top-down nature. They are urban based and have little or no contact at all with grassroot farmers'. Like fortune hunters, some NGOs have been established primarily to tap the aid-flows now bypassing governments (Holmén, 2002; Bingen *et al.*, 2003; Seini and Nyanteng, 2003). In Uganda, Dicklich (1998) found that quite a few were one-man enterprises. Not uncommonly, NGOs are used as springboards for political careers and tend to 'withdraw from grassroots activities in favour of higher level policy processes' (Ellis and Bahiigwa, 2003:997).

Not surprisingly, governments have displayed quite ambiguous attitudes towards NGOs. At first there was a natural suspicion on the part of many governments. NGOs were seen as competitors and were expected to fill various gaps left by a retreating state. They were also increasingly competing with governments for aid money. Most of all, NGOs were expected to act as watchdogs and keep reluctant governments on the 'narrow road'. This, naturally, fostered a hostile attitude in many governments towards NGOs. Over time the parties have tended to accommodate and it is now emphasized that states and NGOs have complementary rather than competing roles. In quite a few sub-Saharan countries the legal status, and hence their formal recognition, has been asserted through NGO Bills. Nevertheless, some still work under 'a cumbersome legal and institutional framework' (Seini and Nyanteng, 2003) and in, for example, Ethiopia, the present government is 'unwilling to tolerate independent unions or associations' (Mulat and Teketel, 2003:29).

Post-SAP

In what sense is it meaningful to talk of a post-SAP period in contemporary Africa south of the Sahara? To begin with, that is a matter of definition. Post-SAP could mean that the liberalization policies have been successfully carried out and that, in a real sense, a new economy has been established. If that were the case, SAP would stand for a sustainable shift to a different 'mode of production'. This is implied in statements that the new policy agenda had been largely implemented by the mid-1990s (Friis-Hansen, 2000). Another possible interpretation is that SAP is still going on – muddling through in the right direction but at a slower pace than originally anticipated – and that post-SAP has not yet been reached. Then, of course, there is the possibility that governments in sub-Saharan Africa have turned away from SAP and instead are trying to resume a more active involvement in economic affairs. In that case, SAP merely stands for brief and not sustained 'spurts in reform'. Indications of all three possibilities are observable.

Governments in sub-Saharan Africa implemented SAP more or less reluctantly and in a number of cases reforms (especially in the case of fertilizer subsidies) were introduced, delayed and/or withdrawn in a stop-and-go manner. Since then, a growing number of sub-Saharan governments 'have all to varying degrees turned away from market-based policies, and are steadily "bringing the state back in"' (Cooksey, 2003:68). At the time of writing, the state maintains a trade monopoly on grain in Ethiopia and it has resumed the role of 'buyer of last resort' of food staples in e.g. Kenya, Nigeria and Zambia (Akande and Kormawa, 2003; Oluoch-Kosura, 2003; Saasa, 2003). The Zambian government 'has announced that it plans to implement a floor price policy [on maize] for the whole country' (FAO/WFP, 2003:3). In Kenya, 'almost all the marketing boards – there is at least one board for each major crop – are still in operation, albeit with relatively limited powers' (WTO, 2000). The Kenyan state further protects domestic staple food production by means of a flexible customs policy. For similar reasons, Nigeria has placed a number of domestically produced food crops on the Import Prohibition List. In 2003, the Tanzanian government resumed subsidizing fertilizer in order to ensure national food security.

Since the introduction of SAP, the World Bank and the donor community have been pushing hard for African governments (and peasants) to abandon the goal of food self-reliance and instead opt for food security by means of diversification, prioritizing export crops and investment in non-farm sources of livelihood. To some extent such reorientations have been brought about in all countries studied, but most prominently in Uganda. In Zambia, large farmers abandoned maize during SAP in favour of export crops and, on a smaller scale, many smallholders in Malawi replaced maize with tobacco (previously a privilege for estate owners) in the 1990s. Non-farm activities are growing everywhere (except, perhaps, in Ethiopia). In, for example, Malawi 'in 1992 an estimated one million people (out of a total labour force of about four million) were engaged in micro-enterprises and small and medium enterprises . . . overwhelmingly (90%) in rural areas' (Milner *et al.*, 2003:28f). On the other hand, national food self-sufficiency is still a declared objective in Ethiopia, Kenya, Malawi and Nigeria (WTO, 1998, 2000; Harrigan, 2003; Milner *et al.*, 2003; Mulat and Teketel, 2003).

The Model Revisited

As shown in Chapter 5 (Holmén, this volume), sub-Saharan Africa has not been without growth in staple-food production. Average annual maize production, for example, has, over a 40-year period, quadrupled in Ghana, Nigeria, Tanzania and Uganda. Population growth has, however, outpaced food production and, in most cases, both food security and national food self-sufficiency have deteriorated. Intensification of food production has taken place, albeit in a spatially and temporarily uneven manner. In general, however, intensification has been of minor importance, especially among the smallholders, and extensification has been the major strategy to raise food production. This is not because there have been no attempts to implement Green Revolutions on the subcontinent – on the contrary. However, such efforts have invariably taken the form of temporary 'spurts of production' rather than led to a sustainable increase in productivity (see, e.g. Haggblade, Chapter 8, this volume).

Comparing with the Asian experiences, which were broad-based and pulled the smallholders along in the development process, African attempts at Green Revolutions have, largely by default rather than by design, been exclusive and largely bypassed smallholder peasants. Hence, African Green Revolutions were never realized. This is partly because in the 1960s and 1970s national food self-sufficiency was generally not perceived as a problem. Partly it has, for most of the post-independence period, been easier to expand agriculture horizontally than to embark on the much more demanding road of intensification. Moreover, compared with their Asian counterparts, African governments have had weaker roots in society and, hence, a more restricted ability to impose their will and implement policy. At the same time, African states have faced fewer external threats, which could have driven them to take an effective lead in development.

Although national food self-sufficiency has officially been a prioritized objective in most countries, both the overall food situation and economies in general deteriorated in the 1970s. This coincided with external shocks such as oil-price increases and falling world-market prices on major export goods. The imposition of SAP in the 1980s and the subsequently reduced levels of foreign aid, especially for agriculture, were intended as a remedy and were expected to finally make development take off. As demonstrated above, SAP only had short-lived positive effects on staple-food production and this, moreover, appears to have been because SAP in many cases was not fully implemented. Polarization, however, has been accentuated with large, sometimes absent, farmers reaping the benefits of the policy shift whereas smallholders in general have been progressively marginalized.

After some 20 years of reform, about half the population ekes out a living below the poverty line. Markets remain undeveloped and peasant associations are either weak or non-existent. For the vast majority of farmers, yields of major food crops remain low – much under their potential. Most smallholders use simple tools, few have access to credit, there has been widespread de-agrarianization and de-adoption of modern inputs, and only a declining minority of smallholders can afford fertilizer. Many are not food-secure throughout the year. In 2000, *all* countries included in this study received food aid from abroad (FAOSTAT data, 2004).

In Asia, the Green Revolutions were successful because they were state-driven, market-mediated and smallholder-based. Also, in sub-Saharan Africa governments have assigned to themselves a leading role in similar endeavours. However (with the exception of Ethiopia since 1974 and, perhaps, Tanzania during Ujamaa), it would be an exaggeration to say that these attempted Green Revolutions were actually *led* by governments. Neither have they been market-mediated since markets in various ways have been undermined, pre-SAP by design and post-SAP, perhaps, more by default. Nor have African Green Revolution attempts been smallholder-based. Patronage policies and 'indirect rule' in combination with a disregard for peasants and low expectations about their ability to raise productivity and respond to market signals, led to a neglect of smallholders and, hence, to their increasing impoverishment.

Which, then, are the prospects for successful Green Revolutions in years to come? In order to answer that question, we need to look closer at the geopolitical situation prevailing in the early years of the third millennium.

The Contemporary Geopolitical Situation

What brought forth the Green Revolution in Asia in the 1960s and 1970s – and what made these Green Revolutions become inclusive (i.e. both market-mediated and small farmer-based) – was a series of circumstances which in large part had to do with the then prevailing geopolitical situation. The perceived severity of manifest or potential external threats, food shortages and fear of being cut off from external food-supply, and a high price on staple grains on the world market, contributed not only to putting food-crop intensification and national food self-sufficiency high on the political agenda, but also to the firmness with which the new policies were carried out. Arguably, the U-turns taken by different regimes were not taken without great agony (Djurfeldt and Jirström, Chapter 4, this volume). They were the result of a strongly felt pressure for change. Is there today a corresponding combination of circumstances which could induce governments in sub-Saharan Africa to make similar U-turns?

On the positive side we find that there is today available a range of technologies which are more Africa-friendly than they were, say, 30 years ago. Hence, nature is no obstacle and a number of studies stress that sub-Saharan Africa possesses a still untapped agricultural potential (Alexandratos, 1995; Eicher, 2001; Fischer *et al.*, 2002; Evenson and Gollin, 2003; Ortiz and Hartmann, 2003), which is also indicated by contemporary yield-gaps in *Afrint* research findings at micro level (Larsson, Chapter 7, this volume).

The prevailing food shortages in many countries and the fact that the land-frontier, in large parts of the subcontinent, is or is about to be reached represents a 'classical' situation where intensification ought to take place (Boserup, 1965). Insofar as development aid has had the negative side-effect of serving as an artificial life-line for governments and, thus, easing the pressure on the state to develop internal resources, the fact that aid has diminished during the last decades could paradoxically be a positive circumstance as well. Seen from this perspective, diminishing aid could have the same function as high population pressure as a 'trigger' for intensification and development.

On the negative side, the state has been weakened since the introduction of SAP and it is an open question whether contemporary African states can take on such a strong leading role, as did the Asian states when they implemented their Green Revolutions. This question is all the more relevant since African governments with few exceptions do not face external threats on a similar magnitude as did many Asian states in the 1960s and 1970s. If such threats were present, they could have forced states to become stronger. Also, declining aid is not *only* a positive factor, especially not since foreign aid directed at agriculture – i.e. at where development must start – has been even more reduced than aid in general (Von Braun, 2003).

This has also been reflected in declining financial contributions from governments in the rich countries to public and international agricultural research institutes (Eicher, 2001; Greenpeace, 2002; World Food Price Foundation, 2002; IFPRI, 2003). These institutes provide public goods and, during the time of the Asian Green Revolutions, they shared without cost their knowledge and technologies with governments and national agricultural research centres in the Third World. This is one important reason why the Asian Green Revolutions came about at all (see Otsuka and Yamano, Chapter 13, this volume). Today, rich-country governments are increasingly withdrawing from such 'wasteful' behaviour and instead increasingly leaving germplasm research etc. to private, in most cases transnational, agribusiness companies.

The once guiding lodestar, to indigenize knowledge and technology, is presently being turned into its opposite. Akande and Kormawa (2003:75) conclude 'the [World] Bank succeeded in locating the base of the technological practices in [African] agriculture in foreign countries and foreign technology'. Moreover, since the big agribusiness companies patent their (and others') crops, seeds, etc., the costs in adopting new technology have become higher for today's African farmers and governments than they were when the Asian Green Revolutions were launched. This makes an African Green Revolution less likely to happen.

Whereas high world market prices on food grains made it economically defensible to subsidize agricultural inputs such as seed and fertilizer in the 1960s and 1970s (and thereby guarantee high adoption rates), cereal prices are now less than half the level that prevailed then (ODI, 2003). Today it would not make economic sense for governments to subsidize inputs when food can be bought much cheaper on the world market. Hence, the contemporary orthodoxy in development thinking says: (i) that Africa should diversify and opt for extra-agricultural income generating activities rather than agriculture (the 'sustainable livelihood' school); and (ii) that Africa should produce export crops (fruit, vegetables, spices, etc.) and import food from abroad. The World Bank (2003:43) thus states: 'The narrow agricultural focus [and the] . . . previous focus on productivity should tilt [be reversed] . . . High value crops should be a priority, not staple crops like hitherto.' An amazingly large number of aid-agencies and writers on the subject (Diao *et al.*, 2003; ODI, 2003; Pingali, 2003) take the present world-market price levels as given, even unchangeable, and therefore say the same as the World Bank.

The prevailing food-grain prices, however, are not 'given' in the sense that they reflect a free market – and, therefore, represent in some sense the best solution. Instead, they are manipulated. Today's world-market prices of agriculture products are a consequence of subsidized over-production in the rich countries, which is then dumped in poor countries at prices below production cost. Whereas the rich countries impose SAPs and force poor countries to eliminate subsidies, it has been estimated that the 'OECD member countries spend about US$75 billion annually on subsidies to their own farmers and agricultural industries . . . [which is] about six times more than these same developed countries provide to the developing world in official development assistance' (Von Braun, 2003:2). Not only do rich-country governments find it tremendously difficult to practise what they preach, this biased trade-policy effectively undermines any effort to invest and raise productivity in African staple-food agriculture.

Moreover, the World Bank and donor community give the same message to all less developed countries. They are all expected to export out of poverty. But the market for their products is limited due to the elasticity of demand for the kinds of produce that they can export. The logic of supply and demand further says that prices will fall when a large number of suppliers compete on the same market. The rich countries also 'protect' themselves behind tariff-walls. These trade-barriers are not only much higher for products that the poor countries can export than for goods produced in other rich countries (De Vylder *et al.*, 2001), they are also progressively raised if poor-country exporters try to add value by postharvest processing (De Vylder *et al.*, 2001). There are thus few niches open for would-be exporters in Africa. As long as these biased structures remain, a widespread export drive would therefore mostly benefit customers in already-rich countries, with few financial benefits for the producers in poor countries.

The above identified, positive pressures for intensification are thus 'balanced' by external forces, eroding the economic rationale for intensifying. If the rich countries stopped undermining internal markets in poor countries by dumping subsidized staplefood, there would be an incentive for peasants to invest and raise productivity, and use could finally be made of Africa's as yet largely untapped agricultural potential. If the rich countries stopped denying would-be exporters in poor countries access to their markets, they would create incentives not only for peasants to intensify and for entrepreneurs to create a market, they would also provide incentives for African governments to take a more active role in the development of 'their' countries' internal resources. Eventually, they might even launch a broad-based African Green Revolution.

In a certain way it seems fair to say that Africa's problem is not so much SAP as the fact that SAP has not been fully implemented. Reluctant governments, in the rich countries, while propagating the 'market' as a solution to Africa's problem, do their best to hinder the development of markets. As long as truly liberating steps are not taken by rich-country governments, it is not only cynical but outrageous to propagate 'good governance' and to blame poverty and declining food security in Africa on the subcontinent's insatiable 'crooks and cleptocrats'.

Notes

[1] For detailed information on the *Afrint* project and studies, see Djurfeldt (Chapter 1, this volume) and www.soc.lu.se/Afrint

[2] Even if agricultural modernization was an important part of the Ujamaa project, this is not to say that the primary purpose of the Ujamaa policy was to increase production. Other objectives may have been at least as important (see below).

[3] As mentioned, it is not always easy to distinguish between SAP and post-SAP periods. The following sections therefore contain some unavoidable overlaps. The main focus, however, is on the most recent period.

[4] This trend may again be altered. It has recently been reported that, in Zambia, 'big farmers [are] turning away from soybeans towards increased maize acreage following the entry of cheap soya products for poultry feed from Zimbabwe onto the Zambian market' (FAO/WFP, 2003).

[5] The Starter Pack programme ran from 1998/99 to 1999/2000 and meant free distribution of seed and fertilizer in small quantities, which initially permitted all farmers to plant 0.1 ha of maize/legume intercrops. The purpose was twofold: (i) to enhance household food security by higher yields; and (ii) as indicated by the name, demonstration-effects, which would create demand for modern inputs. In 2000/01 the programme was replaced by a scaled-down 'targeted' input programme (TIP) officially directed at the most needy third of the smallholder population (H. Potter, Lilongwe, 2001, personal communication). The TIP, however, encountered numerous difficulties in directing the inputs to those smallholders most in need, partly because it was open to manipulation and partly because it is difficult to determine who needs it most when virtually all farm households are in need.

[6] The exception being Ethiopia, which is presently focusing on food crops and mustering more than 15,000 extension agents (Mulat and Teketel, 2003). The authorities aggressively promote an 'intensified package approach' (FAO, 2003), which, however, is characterized by top-down blanket recommendations and leaves little room to articulate peasants' interests or active participation (Mulat and Teketel, 2003).

[7] This protectionism of domestic producers benefits all categories of farmers but possibly smallholders mostly. Hence, this is a rather egalitarian policy. However, large-scale farmers are also selectively protected as 'heavy machinery is [encouraged] through zero rated custom duties and value added tax. Hand and animal drawn equipment however, have custom duties and value added tax charged [rendering] small-scale farmers disadvantaged in the use of machinery' (Oluoch-Kosura, 2003:23). This, together with the elite's practice of land-grabbing (Klopp, 2000) indicates that Kenya's post-SAP trade policies are not so inclusive after all.

[8] Jayne *et al.* (2003b:295) make the point that the common notions of 'market failures' in SSA often miss the point. True, subsidies and monopolies distort markets, but rather than malfunctioning, it is often

instead a question of missing markets, i.e. markets are not only thin and volatile, they often 'do not arise at all'.

References

Akande, T. and Kormawa, P. (2003) *Afrint* case study of Nigeria: Macro Study Report (final) (mimeo.). International Institute of Tropical Agriculture, Ibadan, Nigeria.

Alexandratos, N. (1995) *World Agriculture: Towards 2010. An FAO Study*. FAO/John Wiley & Sons, Chichester, UK.

Andræ, G. and Beckman, B. (1985) *The Wheat Trap*. ZED, London/The Scandinavian Institute for African Studies, Uppsala, The Bath Press, Avon, UK.

Bates, R. (1983) Governments and agricultural markets in Africa. In: Johnson, D.G. and Shuh, G.E. (eds) *The Role of Markets in the World Food Economy*. Westview Press Inc., Boulder, Colorado, pp. 153–183.

Bayart, J.-F., Ellis, S. and Hibou, B. (1999) *The Criminalization of the State in Africa*. Indiana University Press: James Curry, Bloomington, Indiana and Oxford, UK.

Bazaara, N. and Muhereza, F. (2003) *Afrint* Macro Study Report – agricultural intensification and food security issues in Uganda (draft) (mimeo.). Centre for Basic Research, Kampala.

Bingen, J., Serrano, A. and Howard, J. (2003) Linking farmers to markets: different approaches to human capital development. *Food Policy* 28, 405–419.

Boserup, E. (1965) *The Conditions of Agricultural Growth*. Aldine, Chicago, Illinois.

Byerlee, D. and Eicher, C.K. (eds) (1997) *Africa's Emerging Maize Revolution*. Lynne Rienner Publishers, Boulder, Colorado and London.

Cleary, S. (1997) *The Role of NGOs under Authoritarian Systems*. Macmillan Press, London.

Conway, G. and Toennissen, G. (2003) Science for African food security. *Science* 299, 1187–1188.

Cooksey, B. (2003) Marketing reform? The rise and fall of agricultural liberalization in Tanzania. *Development Policy Review* 21, 67–91.

Craig, J. (2003) Privatisation in Africa: promoting local enterprise? ID21 – Communicating Development Research. http://www.id21.org/getweb/s7ajc1g1.html (16 May, 2003).

Crawford, E., Kelly, V., Jayne, T.S. and Howard, J. (2003) Input use and market development in sub-Saharan Africa: an overview. *Food Policy* 28, 277–292.

de Grassi, A. and Rosset, P. (2003) A Green Revolution for Africa? Myths and realities of agriculture, technology, and development. http://www.ocf.berkeley.edu/~degrassi/gra.htm

De Vylder, S., Axelsson-Nycander, G. and Laanatza, M. (2001) *The Least Developed Countries and World Trade*. Swedish International Development Cooperation Agency (Sida), Stockholm.

Diao, X., Dorosh, P. and Rahman, M. (2003) Market opportunities for African agriculture: an examination of demand-side constraints on agricultural growth, IFPRI. http://www.ifpri.org/divs/dsgd/dp/dsgdp01.htm

Dicklich, S. (1998) Indigenous NGOs and political participation. In: Hansen, H.B. and Twaddle, M. (eds) *Developing Uganda*. James Curry Ltd, Oxford, Kampala, Athens, Nairobi, pp. 145–158.

Economist (2004) Opportunities, mostly missed. *The Economist* 370, 17 January.

Eicher, C.K. (2001) Institutions and the African farmer (draft). http://us.f113mail.yahoo.com/ym/ShowLetter?box=inbox&MsgId=490_2380199_13379 (1 July, 2001).

Ellis, F. and Bahiigwa, G. (2003) Livelihoods and rural poverty reduction in Uganda. *World Development* 31, 997–1013.

Ellis, F., Kutengule, M. and Nyasulu, A. (2003) Livelihoods and rural poverty reduction in Malawi. *World Development* 31, 1495–1510.

Evenson, R.E. and Gollin, D. (2003) Assessing the impact of the Green Revolution, 1960 to 2000. *Science* 300, 758–762.

FAO (2003) Integrated natural resources management to enable food security: the case for community-based approaches in Ethiopia. Environment and Natural Resources Working Paper, No. 16. http://www.fao.org/sd/2003/EN11013_en.htm

FAO/WFP (2003) *Special Report: Crop and Food Supply Assessment Mission to Zambia. 10 June 2003*. FAO/WFP, Rome.

FAO/WFP (2004) *Special Report: Crop and Food Supply Assessment Mission to Ethiopia, 12 January 2004*. FAO/WFP, Rome.

FAOSTAT data (2004) Food and Agriculture Organisation of United Nations. http://www.fao.org/waicent/portal/statistics_en.asp

Farrington, J. and Bebbington, A. (1993) *Reluctant Partners? Non-governmental Organizations, the State and Sustainable Agricultural Development*. Routledge, London.

Fischer, G., v. Velthuizen, H.S.M., and Nachtergaele, F. (2002) *Global Agro-ecological Assessment for Agriculture in the 21st Century: Methodology and Results*. IIASA/FAO, Vienna.

Friis-Hansen, E. (1994) Hybrid maize production and food security in Tanzania. *Biotechnology and Development Monitor.* June, 12–13.
Friis-Hansen, E. (ed.) (2000) *Agricultural Policy in Africa after Adjustment.* CDR Policy, Copenhagen.
Government of Malawi (2002) *Agricultural Sector Priority Constraints, Policies and Strategies Framework for Malawi.* Ministry of Agriculture and Irrigation. Malawi Sector Investment Programme.
Government of Uganda (2000) *Plan for Modernization of Agriculture.* Ministry of Agriculture/ Ministry of Finance, Kampala.
Greenpeace (2002) Empty promises: The 'Rome Declaration on World Food Security' in 1966 and today's realities. International Genetic Engineering Campaign, Background Information 06/02. http://www.greenpeace.org/-geneng
Gunnarsson, C. and Rojas, M. (1995) *Tillväxt, stagnation, kaos.* SNS, Stockholm.
Harrigan, J. (2003) U-turns and full circles: two decades of agricultural reform in Malawi 1981–2000. *World Development* 31, 847–863.
Hassan, R.M. and Karanja, D.D. (1997) Increasing maize production in Kenya: technology, institutions and policy. In: Byerlee, D. and Eicher, C.K. (eds) *Africa's Emerging Maize Revolution.* Lynne Rienner Publishers, London.
Hettne, B. (1973) *Utvecklingsstrategier i Kina och Indien.* Internationella folkhögskolan/ Studentlitteratur, Lund, Sweden.
Hettne, B. (1992) The future of development studies. (Mimeo.) Paper presented at the 40th anniversary of ISS. Institute of Social Studies, den Haag.
Holmén, H. (1990) State, cooperatives and development in Africa. Research Report No. 86. The Scandinavian Institute of African Studies, Uppsala, Sweden.
Holmén, H. (2002) NGOs, networking, and problems of representation. ICER Working Paper No. 33/2002. International Centre for Economic Research, Turin, Italy.
Holmén, H. and Jirström, M. (1996) *No Organizational Fixes: NGOs, Institutions and Prequisites for Development.* Publications on Agriculture and Rural Development: No. 4. Swedish International Development Cooperation Agency (Sida), Stockholm.
Howard, J., Crawford, E., Kelly, V., Demeke, M. and Jeje, J.J. (2003) Promoting high-input maize technologies in Africa: the Sasakawa-Global 2000 experience in Ethiopia and Mozambique. *Food Policy* 335–348.
Hyden, G. (1983) *No Shortcuts to Progress – African Development Management in Perspective.* Heinemann, London/Ibadan/Nairobi.
IFPRI (2003) *Trade Policies and Food Security.* IFPRI, Washington, DC.
Isinika, A., Ashimogo, G. and Mlangwa, J. (2003) *Afrint* country report – Africa in transition: Macro Study Tanzania (final). Department of Agricultural Economics and Agribusiness, Sokoine University of Agriculture, Morogoro, Tanzania.
Jayne, T.S., Kelly, V. and Crawford, E. (2003a) Fertilizer consumption trends in sub-Saharan Africa. FS II Policy Synthesis No. 69. July. http://www.aec.msu.edu/agecon/fs2/psynindx.htm
Jayne, T.S., Govereh, J., Wanzala, M. and Demeke, M. (2003b) Fertilizer market development: a comparative analysis of Ethiopia, Kenya, and Zambia. *Food Policy* 293–316.
Kelly, V., Adesina, A.A. and Gordon, A. (2003a) Expanding access to agricultural inputs in Africa: a review of recent market development experience. *Food Policy* 379–404.
Kelly, V., Crawford, E. and Jayne, J. (2003b) Agricultural input use and market development in Africa: recent perspectives and insights. FS II Policy Synthesis No. 70. July. http://www.aec.msu.edu/agecon/fs2/psynindx.htm
Klopp, J. (2000) Pilfering the public: the problem of land grabbing in contemporary Kenya. *Africa Today* 47, 7–26.
Krishna, A., Uphoff, N. and Esman, M.J. (1997) *Reasons for Hope. Instructive Experiences in Rural Development.* Kumarian Press, West Hartford, Connecticut.
Larsson, R. (2001) Between crisis and opportunity: livelihoods, diversification, and inequality among the Meru of Tanzania. Lund Dissertations in Sociology 41. Department of Sociology. Lund University, Lund, Sweden.
Lawson, C. and Kaluwa, B. (1996) The efficiency and effectiveness of Malawian parastatals. *Journal of International Development* 8, 747–765.
Lele, U. and Agarwal, M. (1989) Smallholder and large-scale agriculture in Africa. Are there trade-offs between growth and equity? MADIA Discussion Paper 6. The World Bank, Washington, DC.
Mabogunje, A.L. (1980) *The Development Process: a Spatial Perspective.* Hutchinson, London.
Madeley, J. (2002) *Food for All – the Need for a New Agriculture.* Zed, London.
Mareida, M.K., Byerlee, D. and Pee, P. (2000) Impacts of food crop improvement research: evidence from sub-Saharan Africa. *Food Policy* 531–559.

Milner, J., Tsoka, M. and Kadzandira, J. (2003) *Afrint* country report: Malawi Macro Study (draft) (mimeo.). Centre for Social Research, University of Malawi, Zomba, Malawi.

Mulat, D. and Teketel, A. (2003) Ethiopian agriculture: macro and micro perspective. *Afrint* Macro Study (final) (mimeo.). Department of Business Administration, Addis Ababa.

Naustdalslid, M. (2001) Kenya. Afrika.no, Fellesrådet for Afrika. http://www.afrika.no/norsk/Land/_st-Afrika/Kenya/ (16 December, 2003).

Netting, R.M. (1993) *Smallholders, Householders. Farm Families and the Ecology of Intensive, Sustainable Agriculture*. Stanford University Press, Stanford, California.

NGO Task Force (2002) *The Proposed Response of NGOs to the Amendments of the 1989 NGO Statute*. (Mimeo.) Kampala.

ODI (2003) Rethinking rural development, ODI Briefing Paper. *Currents* 31/32, 28–32.

Olayide, S.O. and Idachaba, F.S. (1987) Input and output marketing systems: a Nigerian case. In: Mellor, J.W., Delgado, C.L. and Blackie, M.J. (eds) *Accelerating Food Production in Sub-Saharan Africa*. The Johns Hopkins University Press, Baltimore, Maryland.

Oluoch-Kosura, W. (2003) *Afrint* country report: Kenya Macro Study. College of Agriculture and Veterinary Sciences, Department of Agricultural Economics, Nairobi.

Orr, A. (2000) Green gold? Burley tobacco, smallholder agriculture, and poverty alleviation in Malawi. *World Development* 28, 347–363.

Ortiz, R. and Hartmann, P. (2003) Beyond crop technology: the challenge for African rural development. Paper presented at the workshop 'CGIAR Challenge Program for Sub-Saharan Africa' (mimeo) Accra, 9–15 March.

Pingali, P. (2003) Sustaining food security in the developing world: the top five policy challenges. *Quarterly Journal of International Agriculture* 42, 259–270.

Platteau, J.P. (2000) *Institutions, Social Norms and Economic Development*. Harwood Academic Publishers, Amsterdam.

Pletcher, J. (2000) The politics of liberalizing Zambia's maize markets. *World Development* 28, 129–142.

Rakner, L. (2003) *Political and Economic Liberalization in Zambia*. The Nordic Africa Institute, Uppsala, Sweden.

Reardon, T., Barrett, C., Kelly, V. and Savadogo, K. (1999) Policy reforms and sustainable agricultural intensification in Africa. *Development Policy Review* 17.

Saasa, O. (2003) *Afrint* country study: agricultural intensification in Zambia. Macro Study (final) (mimeo.). Institute of Economic and Social Research, University of Zambia, Lusaka.

Scott, J. (1998) *Seeing Like a State. How Certain Schemes to Improve the Human Condition Have Failed*. Yale University Press, New Haven, Connecticut.

Seini, W. and Nyanteng, V. (2003) *Afrint* Macro Study: Ghana Macro Report (final) (mimeo.). Institute of Statistics, Social & Economic Research, ISSER, Legon, Ghana.

Seward, P. and Okello, D. (2003) *SCODP's Mini-pack Method – the First Step Towards Improved Food Security for Small Farmers in Western Kenya*. SCODP, Sega.

Shaib, B., Aliyu, A. and Bakshi, J.S. (1997) Nigeria: national agricultural research strategy plan, 1996–2010. Federal Ministry of Agriculture and Natural Resources, Department of Agricultural Sciences, Abuja, Nigeria.

Simon, D. (1999) Development revisited: thinking about, practising and teaching development after the Cold War. In: Simon, D. and Närman, A. (eds) *Development as Theory and Practice*. Longman, Harlow, UK.

Smale, M. (1995) Maize is life: Malawi's delayed Green Revolution. *World Development* 23, 819–831.

Smith, J., Weber, G., Manyong, M.V. and Fakorede, M.A.B. (1997) Fostering sustainable increases in maize productivity in Nigeria. In: Byerlee, D. and Eicher, C.K. (eds) *Africa's Emerging Maize Revolution*. Lynne Rienner, Boulder, Colorado, pp. 107–124.

Tangri, R. (1999) *Parastatals, Privatization & Private Enterprise – the Politics of Patronage in Africa*. James Curry/Africa World Press/Fountain Publ., Oxford, Trenton, Kampala.

Teketel, A. (1998) *Tenants of the State. The Limitations of Revolutionary Agrarian Transformation in Ethiopia, 1974–1991*. Department of Sociology, Lund University, Lund, Sweden.

Tomich, T.P., Kilby, P. and Johnston, B. (1995) *Transforming Agrarian Economies. Opportunities Seized, Opportunities Missed*. Cornell University Press, Ithaca, New York.

Tripp, R. (2000) GMOs and NGOs: biotechnology, the policy process, and the presentation of evidence. Natural Resources Perspectives No. 60, September. http://www.odi.org.uk/nrp/60.html

Von Braun, J. (2003) *Overview of the World Food Situation. Food Security: New Risks and New Opportunities*. IFPRI, Washington, DC.

Wohlgemut, L. (1994) *Bistånd på Utvecklingens Villkor*. Scandinavian Institute of African Studies, Uppsala, Sweden.

Woodhouse, P. (2003) African enclosures: a default mode of development. *World Development* 31, 1705–1720.

World Bank (2003) The World Bank: reaching the rural poor – a renewed strategy for rural development. (Summary condensed by *Currents*.) *Currents*: 42–45.

World Food Price Foundation (2002) The World Food Price Laureates' statement on world hunger. http://www.worldfoodprize.org/statement.htm (20 June, 2002).

WTO (1998) Trade Policy Reviews. Nigeria: June 1998. http://www.wto.org/english/tratop_e/tpr_e/tp75_e.htm (17 December, 2003).

WTO (2000) Trade Policy Reviews. Kenya: January 2000. http://www.wto.org/english/tratop_e/tpr_e/tp124_e.htm (19 December, 2003).

7 Crisis and Potential in Smallholder Food Production – Evidence from Micro Level

Rolf Larsson
Department of Sociology, University of Lund, Lund, Sweden

This chapter brings out some of the farm-level factors that are crucial for intensified food crop production in sub-Saharan Africa (SSA). Although not ruling out the role of 'traditional' farm inputs, the chapter argues that farmers' access to 'Green Revolution' technologies and to viable and stable markets is among the most effective means for achieving higher agricultural productivity. It further argues that the current food crop crisis is policy related in that the use of industrial inputs, especially fertilizer, is marginal for most farmers, who also consider production for the market too risky or uneconomical. Currently, only a minority of farmers can manage commercial agriculture with purchased inputs. Their level of production, however, points to the large potential of food crop agriculture in SSA should more favourable market conditions be created. The analysis is based on the *Afrint* survey of more than 3000 farming households in eight countries, focusing on four major staples (maize, cassava, sorghum and rice) in areas thought of as having a potential for increased production (for methodology, see Djurfeldt *et al.*, Chapter 1, this volume).

In the first part of the chapter, we will give an account of the current trends and indicators of intensification, including yield, total production per farm and consumption unit, and sale of crops. Intensification is commonly understood as a process of increasing yields and returns to labour, often but not exclusively based on an increased use of industrial inputs and/or driven by a growing market demand. Our approach here is slightly different in that we discuss intensification from data that is cross-sectional rather than longitudinal. We argue that the smallholder food crop agriculture in most of SSA currently is in crisis, indicated by, *inter alia*, (i) generally low levels of production per farm, yields and return to labour; (ii) small quantities of food crops marketed and hence low incomes derived from food crop agriculture; and (iii) a large gap between a highly productive minority of commercially orientated farmers and the majority.

The generally low level of production per farm as well the low average and median yields recorded in the survey indicates the problematic food production situation in SSA as demonstrated earlier in this volume. The low levels of production and yields recorded can be assumed to negatively affect farmers' market integration and contribute to the persistence of poverty in rural SSA. While leaving the policy implications of this issue aside, we will later in the chapter discuss the nature of households' income composition and the importance of other types of household incomes, especially those stemming from non-farm or off-farm sources. In the final part of the chapter we will bring these aspects together in a concluding discussion of some of the factors that condition farm-level developments in SSA.

The Farm Households

The sampled households are typical of the family farms, which constitute the backbone of agriculture in most of SSA. Typical is the generally small area under cultivation (Table 7.1). Production is partly for subsistence, partly for sale. Fields are worked by family members mainly, with women performing the bulk of farm labour using simple hand tools. Locally, fields are prepared by the use of ox-drawn ploughs and, occasionally, by tractors.

Although monetary income was not measured in the *Afrint* survey, other data bear evidence of the insecurity and poverty that many of the interviewed households are facing. The large subsistence element, the low and irregular incomes earned, the low productivity and the constant risk for crop failure are among the indicators. The common supplementary reliance on low-income jobs outside farming also bears evidence of widespread vulnerability.

Production and Yields

Maize

Both average production and yields of maize over the period 2000–2002 are generally low with an overall mean yield of 1.3 t/ha and year (Table 7.2). This is at level with the FAO estimate of 1.2 t/ha for SSA as a whole for the same period (FAOSTAT data, 2004).

As indicated in the table for maize (and for other crops as well), there is a variation in country means but also between farmers within the same region and village. It is worth noting, however, that a small number of farmers (the 5% best performing) manage to realize yields that are substantially higher. We will come back to the issue of farm-level yield gaps shortly.

The large differences observed in regional and village yield aggregates bear evidence of a considerable diversity in the conditions facing farmers. Variation in macro-economic conditions, agroecological factors, proximity to

Table 7.1. Land under cultivation (total and per crop in ha) and percentage of households cultivating by type of crop.

	Total	Maize	Cassava	Sorghum	Rice	Other food crops	Non-food crops
Mean farm size (ha)	2.6	1.0	0.6	1.0	0.8	0.7	1.0
Median farm size (ha)	1.8	0.7	0.4	0.8	0.6	0.5	0.5
Households cultivating (%)	100	85	40	23	25	81	37

Table 7.2. Maize production (t/farm) and yields (t/ha) seasons 2000–2002.*

Country	Proportion of sampled farmers growing maize (%)	3-seasons mean production (t/farm)	3-seasons mean yield (t/ha)	3-seasons median yield (t/ha)	5% best performing farmers' yield (t/ha)**
Ethiopia	52	1.2	1.2	1.0	4.0
Ghana	49	0.8	1.2	0.8	5.2
Kenya	100	0.9	1.6	1.1	4.7
Malawi	99	0.6	0.9	0.8	2.2
Nigeria	98	3.7	1.8	1.4	3.4
Tanzania	89	0.9	1.0	0.8	2.6
Uganda	91	1.1	1.5	1.2	4.4
Zambia	100	1.5	1.1	0.9	2.8
Total	85	1.5	1.3	1.0	3.4

*Yields above 10 t/ha at farm level have been excluded.
**Based on village aggregates.

markets, quality of road networks, price margins and incentives, availability of land and labour, access to high yielding and labour saving technologies, etc., are among the factors which influence yields. The higher production per farm and cultivated area in Nigeria, for example, probably partly reflects the more favourable macro conditions facing farmers there (see Akande, Chapter 9, this volume).

Cassava

A similar pattern as for maize emerges for the other crops (Tables 7.3 and 7.4). The 3-year average yield of cassava at 5.4 t/ha, which for most farmers reflects harvesting during the 2000–2002 period, is considerably less than the corresponding FAO estimate for SSA as a whole, which for the same period stands at 8.9 t/ha (FAOSTAT data, 2004). A major difficulty here has been the problem of accurately estimating what is the seasonal or annual cassava production.[1] Although we attempted to obtain such estimations in interviews with farmers, we found the reporting to be too unreliable to be used for most of our statistical analysis. Where we do cite production figures, these should be read with caution and as indicators of trends or differences rather than as absolute figures.

Higher cassava yields were reported from countries and by farmers who had been exposed to the new high yielding and virus resistant varieties developed by the International Institute of Tropical Agriculture (IITA) in Nigeria (see Haggblade, Chapter 8 and Akande, Chapter 9 in this volume). This was, for example, the case in Uganda and Ghana. The highest yields were reported by Nigerian farmers and may reflect an ongoing cassava revolution there.

Sorghum

For sorghum, average yields based on the survey stand at 0.9 t/ha, which is also the FAO estimate for SSA as a whole for the period 2000–2002 (Table 7.3) (FAOSTAT data, 2004). Also in this case, there is a pronounced variation between countries, regions, villages and farm households.

Paddy/rice

Overall average yield of paddy stands at 1.4 t/ha, which is somewhat less than the FAO estimate for SSA as a whole (1.6 t/ha) for the 2000–2002 period (FAOSTAT data, 2004). Rice is grown by nearly half the sampled farmers in Ghana, Tanzania and Uganda. Although urban demand for rice is rapidly picking up in eastern and southern Africa (it has for long been demanded in West Africa), average yields continue to lag behind those obtained elsewhere in the world. Some individual farmers, however, experience yields comparable to those recorded in Asia. At farm level, the highest yields observed derive from farmers in Nigeria and Tanzania (7.4 and 7.7 t/ha, respectively). A quarter of the rice farmers (24.5%) have all or part of their land under

Table 7.3. Sorghum production (t/farm) and yields (t/ha).*

Country	Proportion of sampled farmers growing sorghum (%)	3-seasons mean production (t/farm)	3-seasons mean yield (t/ha)	3-seasons median yield (t/ha)	5% best performing farmers' yield (t/ha)
Ethiopia	55	1.2	1.1	1.0	2.4
Ghana	50	0.4	0.5	0.4	1.3
Nigeria	42	2.0	1.2	0.9	2.0
Tanzania	< 1	–	–	–	–
Zambia	19	0.4	0.6	0.5	1.7
Total	23	1.1	0.9	0.7	1.8

*Yields above 3 t/ha at farm level have been excluded.

irrigation. They have higher yields than farmers without irrigation (Table 7.4).

Yield gaps

The above description shows that in what can be assumed to be potentially dynamic areas of SSA, the majority of farmers get yields far below those possible to obtain under present agroecological conditions.

Yield potential was defined as the mean yield of the 5% best performing farmers per crop and village (outliers excluded). The country-level aggregates based on such village means are given in Tables 7.2–7.4. Table 7.5 gives the summary yield gaps for all four staple crops, which point to a general gap of about 60% between the majority and the best performing farmers within the same village.

A crucial question is, why is the yield gap so large? Part of the explanation may be intra-village differences in agroecological conditions (soils, slope, etc.) favouring some farmers while handicapping others. The main reasons for the yield gap, however, as we will discuss below, are economic and political. The majority of smallholders either lack the resources for purchasing yield-improving inputs and/or consider marketing of food staples to be too risky. In the present situation, only a minority of the farming population has sufficient capital for purchasing farm inputs or managing market risks. As we will demonstrate later on, it is the largely suboptimal application of industrial technology inputs, notably fertilizer, which explains the poor yields of the farmer majority and the gap they experience versus the more resourceful farmers.

Technology Adoption, Land Use and Labour

There is a heavy reliance on manual labour in farm operations (land preparation, weeding, harvesting, transportation, processing etc.). Most of the interviewed households do hire labour in order to cope with peak periods in farming, but this is complementary to and does not substitute for the family labour on which the entire farm depends.

Hoe cultivation is by far the dominating kind of land preparation among the farm households interviewed, as well as in SSA generally (Binswanger and Pingali, 1988), a circumstance that sets definite limits to farm size and total output. In tsetse-free areas where it has been possible to raise cattle, the use of oxen for ploughing and transportation has increased the productivity of labour and

Table 7.5. Summary of yield gaps, all crops.

	Mean yield (t/ha)	Potential yield (t/ha)	Yield gap (%)
Maize	1.3	3.4	–60.3
Cassava	5.4	14.0	–57.6
Sorghum	0.9	1.8	–53.5
Rice	1.4	3.6	–58.9

Table 7.4. Paddy production (t/farm) and yields (t/ha).*

Country	Proportion of sampled farmers growing rice (%)	3-seasons mean production (t/farm)	3-seasons mean yield (t/ha)	3-seasons median yield (t/ha)	5% best performing farmers' yield (t/ha)
Ghana	40	0.5	1.0	0.8	2.8
Malawi	24	0.7	0.7	0.6	1.9
Nigeria	21	2.0	2.2	1.8	3.4
Tanzania	43	1.6	1.6	1.4	4.2
Uganda	51	0.5	1.5	1.4	4.5
Total	25	1.0	1.4	1.1	3.5
Partly or fully irrigated rice	–	0.8	1.7	1.5	4.4

*Yields above 8 t/ha at farm level have been excluded.

hence the area that can be cultivated. Ethiopia is one example in which 95% of the sampled farmers plough their land with their own or hired oxen but where land shortage and other factors constrain total output per household (Mulat and Teketel, 2003). In Kenya, Uganda, Tanzania and Zambia between a quarter and a half of the respondents similarly plough their maize fields with oxen, however, in situations where land or other production factors often remain limiting (Ashimogo *et al.*, 2003; Karugia, 2003; Wamulume, 2003).

Only a minor percentage of the farmers, mainly the wealthiest strata, have access to or can afford to hire tractors for ploughing and other tasks. Also, tractor ploughing is typically concentrated in areas adjacent to large-scale or estate farms, and/or where there is a clear market demand, as in the case of Tanzania's Kilombero District (see Isinika *et al.*, Chapter 11, this volume and Ashimogo *et al.*, 2003). Here more than a fifth of the farmers hire tractors for the ploughing of rice fields. Another example is found in northern Ghana, where two thirds of the farm households use either tractors (35%) or oxen (28%) in preparing their rice fields (see Seini and Nyanteng, Chapter 12, this volume).

As will be further discussed below, the use of oxen and tractors in the preparation of maize and rice fields is positively associated not only with higher production and marketing of crops, but also with considerably higher yields. This bears evidence of the larger resources, including yield-raising inputs that can be afforded by those farmers who have access to traction.

Although land under irrigation may be more common in SSA than is officially recognized, it still constitutes only a small fraction of all land cultivated. Of the land cultivated by the sampled farmers, about 7% is under some kind of irrigation. Virtually all irrigation systems recorded are small-scale and managed by individual farmers or groups of households. Irrigation systems mostly concern the cultivation of vegetables and to some extent rice.

Three quarters of all rice cultivated is rainfed and lowland, often under conditions of natural flooding in which case the need for irrigation is limited. Irrigation is positively associated with higher yields and production for the market (Table 7.4). A larger proportion of farmers who irrigate also sell the crop (87%) as compared with those who grow non-irrigated paddy (69%).

Industrial inputs – seeds and fertilizers

We found adoption rates of high yielding seed varieties to be quite high, notably in the case of maize. In fact, adoption rates of high yielding varieties are higher in Africa today than was the situation in South Asia in the 1970s (Evenson and Gollin, 2003), suggesting that this aspect of technology is not as constraining as may be popularly assumed. Table 7.6 depicts the application percentages of high yielding varieties (HYVs) for the different crops.

The relatively high percentage of farmers using maize hybrids and open pollinated varieties (OPVs) is probably due to the long history of maize breeding in SSA, especially in southern and eastern Africa (see Haggblade, Chapter 8, this volume). It should be noted, however, that although farmers may report use of hybrid seeds, such statements sometimes refer to *recirculated* hybrid seeds with a poorer production potential than hybrids proper. In this sense, the figures in the table may give a somewhat exaggerated view of seed adoption. The highest adoption rates of improved maize varieties were recorded for Kenya, Zambia and Nigeria with 75–80% of the sampled farmers using either hybrids or composite varieties. In Tanzania, the situation was the opposite with 80% of the farmers using traditional varieties as their main type of seed.

For cassava, the majority of farmers (60%) use traditional varieties. The clear

Table 7.6. Main type of seed. Percentage of farmers using.

	Maize	Cassava	Sorghum	Rice
Traditional	43	61	86	71
Improved/OPV	24	39	13	29
Hybrid	33	–	1	–
Total	100	100	100	100

exception is Nigeria, which accounts for most records of improved cassava in the sample. In Nigeria, 82% of the sampled cassava growers use improved planting material. Adoption rates of improved virus resistant varieties, released in the mid-1990s by the IITA in Nigeria, are relatively high also in Uganda, Ghana and Malawi.[2]

In the case of sorghum and rice, improved varieties seem to have partly replaced traditional ones. For sorghum, adoption rate is highest in Kenya, where a third of the sampled farmers (33%) use improved seed. For rice, about 60% of the farmers in Nigeria and 30% in Ghana and Uganda use improved varieties.

The use of pesticides is modest compared with other continents. Less than a fifth (17%) of the rice farmers applied pesticides during the most recent cropping season. On maize and sorghum, about 12% and 10% did the same while on cassava the use of pesticides was marginal (3%).

The marginal use of fertilizer

In contrast to the relatively high adoption rate of improved seeds, the use of chemical fertilizer is extremely low. With due consideration taken of the fact that fertilizer recommendations differ with crop type, agro-ecological characteristics, type of fertilizer used etc., our data point to generally very low application rates on staple food crops. This finding is in line with observations elsewhere. In Tanzania, for example, data from a plant nutrition project in the early 1990s showed that the amount of nutrients per hectare of cultivable land was 3.3 kg of nitrogen, 1.9 kg of phosphate and 1.1 kg of potassium (Quinones *et al.*, 1992). Similarly, a World Bank estimate for 1993–1996 revealed that farmers in SSA used a mere 15 kg of fertilizer nutrients per year and hectare of arable land for all crops, compared with 180 kg and 75 kg in Asia and Latin America respectively (World Bank, 2000).[3]

In the case of maize, more than half of the sampled farmers (53%) did not apply any chemical fertilizer at all during the 2002 season and most of those who did used very small quantities. The average application rate was 14 kg/ha.[4] There is, however, considerable variation in the national and regional average application rates on maize, as well as in the amounts reported by individual farmers. For example, Kenyan and Zambian average rates on maize reach 31 and 37 kg/ha while in Uganda and Ghana rates are negligible.

For the other crops, the amounts applied are even smaller. Chemical fertilizer is nearly non-existent on cassava for which only 6% of the farmers applied some chemical fertilizer during the most recent season. On rice, one third of the farmers (30%) applied some fertilizer. On sorghum, 45% of the farmers used fertilizers, albeit in very small quantities. A summary of fertilizer use is given in Table 7.7.

Non-industrial or organic inputs

The limited use of chemical fertilizer is to some extent compensated for by fallowing, crop rotation, intercropping and by the incorporation of organic matter in cultivated fields. In the case of maize, crop rotation and intercropping (most often with legumes) are practised by almost half of the farmers and another third use fallowing for restoring soil fertility. About as many apply compost material, most often in the form of crop residues. A quarter of the farmers use animal manure, a practice that is associated with ownership of or access to livestock (three quarters of the households do not have cattle and more than half of the farm households do not have goats or sheep). About a third of the sampled farmers have taken on additional conservation and investment measures on their maize

Table 7.7. Fertilizer adoption rate (per cent farmers using) and average amount (kg) applied per crop and hectare.

	Fertilizer use (%)	Amount (kg/ha)
Maize	47	14
Cassava	6	1.6
Sorghum	45	6.6
Rice	30	7.4

farms in the form of planting trees and grass strips, constructing levelling bunds and, in some cases, have built terraces on sloping land.

The picture is similar for the other staple crops. Crop rotation, intercropping, application of animal manure and compost material, as well as fallowing, are major means for replenishing soil fertility. In the case of rice, many farmers rely on nutrients swept into the fields during annual flooding, which in, for example, Tanzania Kilombero District forms a crucial part of the seasonal crop cycle (Ashimogo *et al.*, 2003; see also Isinika *et al.*, Chapter 11, this volume).

In some locations (for example reflected in the Zambian subsample) there is a widespread adoption of conservation farming, here referring to particular farming practices such as 'pot holing' or 'rip ploughing'. In some environments, these techniques have been found to improve the water retention of the plants and hence their ability to withstand drought (RELMA/Sida, 2001; see also Akande, Chapter 9, this volume).[5]

With the exception of fallowing, most of the measures mentioned are well established practices for maintaining soil fertility in situations of permanent cultivation. The thorough land preparation required in such circumstances, including the collecting and spreading of animal manure and of other organic matter, as well as the need for repeated weeding, conservation measures etc., comprises labour demanding tasks. We may include here the building and maintenance of irrigation structures, and the mentioned practice of conservation farming, notably the 'pot holing' technique. These measures have the potential of increasing overall yields but may not necessarily improve labour productivity, i.e. increase production per labour input unit, and this may be an important constraint to their diffusion.

Returns per consumption unit

Not only is the overall production of staple crops per farm small, returns per labour or consumption unit within the farm households also bear evidence of an agricultural crisis in SSA. If one assumes that 220 kg of grain equivalents per person (consumption unit) and year is roughly what is required to be food secure from own production, one can infer that below this amount, households are net buyers of staples, and above this amount they are surplus producers.[6] Table 7.8 outlines the distribution of grain equivalents per consumption unit or labour unit. Due to large measuring errors in estimating cassava production, figures are presented both including and excluding cassava. What is evident from the table is not only a remarkably low production per consumption unit but also a highly variable and skewed distribution. Production for the lowest 10% is negligible while the highest 10% produce a surplus exceeding two to three times their own consumption needs. More than half the households (55%) fail to produce above 220 kg of grain equivalents per consumption unit and year and consequently are net buyers of basic food items. When cassava is included, this figure is 62%.

Households may secure their food or incomes to buy food from sources other than staple crop production, including cultivation of other food crops (e.g. vegetables, root crops etc.) and sale of so called export or cash crops (e.g. cotton, coffee, tea, cocoa, etc.). They may also work off-farm for cash. Next, we will more closely look into these aspects.

Commercialization and Market Integration

The multifaceted side of rural livelihoods can be illustrated by the manner in which households divide their production between own consumption and sale, and share their labour time between farm and non-farm or off-farm activities. The balance between these livelihood components depends on a number of factors, including the general conditions of agricultural production and marketing of food crops in SSA.

In the case of maize, cassava and sorghum, only about half the growers had anything to sell after their last harvest in 2002.

Table 7.8. Production of grain equivalents (kg) per year and consumption unit per household; mean, median, percentiles and standard deviation (SD).

	Mean	Median	10%	90%	SD	220 kg pcu	Total no. of cases
kg grain eq. pcu	270	156	34	542	566	55%	2707
kg grain eq. incl. cassava pcu	332	196	39	677	607	62%	2728

pcu, per consumption unit.

Table 7.9. Percentage selling and amount marketed by type of crop.

	Maize	Cassava	Sorghum	Rice	Other food crops	Non-food crops	Any type of crop
Proportion of growers who sell the crop (%)	48	57	49	74	70	100	87
Average (and median) amount sold, all farmers (t)	0.6 (0.0)	1.4 (0.3)	0.3 (0.0)	0.5 (0.2)	– –	– –	– –
Average proportion of total production sold, all farmers (%)	24	31	18	33			

Furthermore, the amounts marketed were modest. In the case of maize, the average amount sold per household was 0.6 t, all growers considered (Table 7.9). Looking at sellers only, the average amount sold was 1.2 t/household; however, there was a large variation between households.

For maize, as well as for the other staple crops, the bulk of marketed production comes from a commercially orientated minority of the farmer population. The skewed distribution of marketed production is evident when average and median sales are compared (Table 7.9). As far as staple crops are concerned, marketed production is marginal for the vast majority of farmers. Although most of them (87%) did sell at least something of at least one crop in the year preceding the survey, the amounts sold and the incomes generated are small.

A few further comments may be in place. The widespread view of cassava as a subsistence and security crop to which households may turn in times of drought is true, but only partly. The relatively high marketing rate and sales of cassava that can be seen in the table are influenced by the situation in Nigeria and Ghana, where about 90% of the interviewed cassava growers produce for the market.

The highest marketing rates are for rice and other food crops (vegetables, beans, potatoes, etc.), both of which may reflect the economic liberalization that swept across the continent in the 1990s. Production of both rice and vegetables is market driven, and in the latter case, is an important and immediate source of cash income that can be regularly tapped throughout the year by households in need of cash (Ponte, 1998).

Off-farm activities provide another, and to judge from the literature, increasing source of household incomes in SSA (Bryceson, 1997a). The most common types of activity are various micro-businesses and employment. Our definition of 'employment' is crude but most often refers to low-income jobs of a casual labour type found in both the farm and the non-farm sectors. Micro-business includes various self-employment activities such as brewing, petty trade and retailing, crafts, etc. Incomes earned in micro-business can be assumed to be higher than from employment. Large-scale business, finally, refers to self-employment activities that in terms of scale, investments and returns surpass those of micro-business. Various kinds of transportation, construction, manufacturing and trade belong to this category. Only a few households are involved in these kinds of wealth generating activities (Table 7.10).

About half the households (53%) have at least one adult member who regularly earns

Table 7.10. Type of off-farm activities, all households (hh) (per cent).

Activity	No off-farm income	Employment only	Micro-business only	Employment and micro-business combined	Large-scale business solely or combined
Per cent hh	47	19	22	11	2

income from activities outside the farm. And, on average, more than a quarter of all adult members (28%) are regularly involved in some kind of income earning activity outside farming. This pattern is fairly uniform in all countries except for Ethiopia and Nigeria. In the former both the proportion of households (21%) and the proportion of members (11%) involved in off-farm activities are remarkably low, a circumstance that reflects the low level of urbanization and diversification of the rural economy. In Nigeria, on the other hand, more than three quarters of the households (77%) obtain incomes from outside their farms.

'Deagrarianization'

What are the driving forces and implications of households' (increasing) dependence on off-farm activities in SSA? The process has been coined 'deagrarianization' by Bryceson, who claims that households' increasing resort to off-farm income sources has been reinforced by the recent decades' erosion of the public sector and by the liberalization of global agricultural markets, both of which have negatively affected smallholders' ability to earn a living from farming (Bryceson, 1997ab, 1999). In contrast to the Asian situation, in which agricultural growth contributed to the diversification of the rural economy through positive intersectoral linkages, diversification in the African case seems to have come about foremost as a result of a crisis in smallholder agriculture. It is the young households in particular who resort to off-farm incomes as a risk-coping strategy in a situation where farming opportunities represent too costly or too risky alternatives. Here, Nigeria is a possible exception. The higher yields and returns per consumption unit there, combined with a considerably higher proportion of household members drawing incomes from outside farming than in the rest of Africa, suggest the presence of positive intersectoral synergy effects of a kind that characterized much of the Asian development.

Despite the Nigerian exception, however, the general picture conveyed in the preceding section is one of a problematic livelihood situation for the majority of farming households. Although income was not measured directly in the *Afrint* survey, the proxy indicators we have been discussing, including production per farm unit and per consumption unit, as well as yields and marketing of food crops, all bear evidence of an agricultural crisis that manifests itself in persistent poverty for the rural population. For most farmers, non-farm and off-farm work generate only small incomes, which, although important, are inadequate to alter the poverty conditions they are experiencing. Similarly, incomes from sale of other food crops may provide important supplements to their household budgets but, with few exceptions, take place on a limited scale, involve high labour and transport costs and are prone to large price fluctuations. For what could be termed traditional export crops or non-food cash crops the situation is also problematic. About a third of the households produce non-food cash crops, however, often under conditions that have deteriorated due to increased competition by new players on the world market (cocoa, coffee etc.) or due to price dumping by Western producers and governments (e.g. cotton).

The gap between the minority of highly productive and market-orientated farmers and the majority of low producing, low income earning households is a clear illustration of the crisis. Why is the gap so large? Why are the majority of farmers marginalized from the food crop market? In addressing these crucial issues, we will pick up the questions asked initially and attempt an answer by statistical regression analysis. While the ultimate solution to the crisis necessarily has to do with conditions and policies at macro level, our analysis will be indicative of those problems and bottlenecks faced by farmers and which need to be addressed at macro and policy level.

Determinants of Yields – Agroecology, Technology and Markets

In this section, we will look at farmers' yields against a number of causal and conditioning factors at meso and household level. In the text the general findings will be summarized and discussed while in an appendix to the chapter some of the regression tables are presented for the interested reader.

Factors referring to the various agroecological settings of the surveyed villages include rainfall pattern, soil quality, slope and proportion of irrigated land. These factors were assessed in participatory rural appraisal (PRA) interviews with village leaders and resulted in a subjective ranking of the villages with respect to their agricultural and marketing potential. In the same way, villages were assessed with respect to their market potential, taking into account factors such as access to road infrastructure, distance to towns and permanent crop outlets, access to electricity and telecommunication etc.

This crude classification of the surveyed villages into low/good potential with respect to agroecology and market factors (agricultural dynamism) is validated by the fact that yields, production and marketing are higher in villages with both good agroecological and market conditions than where such preconditions are missing. The results on yields are summarized in Table 7.11.

While it can be concluded from Table 7.11 that both markets and agroecology matter, it is also evident that average yields and production remain low also in areas where relatively favourable conditions are present. This circumstance points to the existence of a number of general and macro based mechanisms that constrain small farmer activity (see Holmén, Chapters 5 and 6, this volume). In the following, we will analyse how these macro conditions affect farmers' activities through a number of intermediate factors at meso and household level. The following independent variables are included in the analysis.

Apart from agricultural dynamism, as described above, we have entered into the model two binary variables assumed to influence production conditions at village level, and which have been further discussed in Chapter 7. Also, these variables derive from interviews with village leaders. They are: (i) state/NGO, a binary variable indicating the presence in the village of an external project or support (e.g. extension, credit, input supply etc.) related to food crop agriculture; and (ii) the presence in the village of any kind of small farmer organization. Neither of these variables contains specific information as to the kind and extent of the support rendered. Hence external support and farmer organizations may imply quite different things in different villages. Despite the shortcomings in the operationalization of these variables, however, it was hoped that they would give some indication of the potential role of these actors for staving off some of the insecurity that is presently associated with smallholder based commercial production. As a general observation, state intervention and farmer organization are not very prominent features in the post-SAP situation of SSA.

In the model we have also entered a number of binary variables related to the kind of technological inputs (both industrial and non-industrial) and land use practised by the

Table 7.11. Yield (t/ha) by village type of agricultural dynamism (potential).

	Maize	Cassava	Sorghum	Rice
Low agri/market potential	1.2	4.9	0.7	1.0
Mixed agri/market potential	1.2	3.9	0.9	1.5
Good agri/market potential	1.7	9.6	1.0	1.5

individual farmers, and which were discussed earlier. These include the use/non-use of chemical fertilizer, improved seeds and pesticides (i.e. industrial inputs), as well as the use/non-use of various non-industrial or 'traditional' inputs and practices, such as animal and green manure, crop rotation and fallowing. Here we have also noted if land is ploughed with oxen or tractors, and if irrigation is used for the production of staple crops.

Controlling for country-specific factors

As seen earlier (Tables 7.2–7.4), there are considerable differences in average yields between the individual countries in the sample. In fact, a large part of the overall variation observed in yields and production refers to differences between the countries. Since the country samples are not statistically representative, observed differences to some extent reflect difficulties of applying identical regional and village sampling criteria in all countries. This being said, however, we are also inclined to believe that observed differences can be traced back to differences in government policies and other macro conditions affecting agriculture, as we have discussed in more detail in Chapter 6.

We have consequently used Nigeria as a reference category in the regression analysis. This is on the basis of the generally higher yields and production in Nigeria as compared with the other countries, this in turn probably reflecting the generally more favourable conditions for market-orientated food production there (see Holmén, Chapter 6 and Akande, Chapter 9, this volume). In the regression tables presented in the Appendix, the beta values indicate how much lower or higher the yields or production would be in a country other than Nigeria, other factors held constant. In contrast to the rather dynamic and promising situation in Nigeria, it is hard to describe the Malawian situation in any other way than as a crisis. The bimodal agrarian structure, the declining land:man ratio, the AIDS pandemic and the widespread poverty are key elements, made worse by a severe drought affecting the country at the time of the *Afrint* survey (Holmén, Chapter 6, this volume).

Determinants of yields and production

For all crops the findings remain fairly robust when it comes to agroecological and market conditions. Farmers tend either to score higher on the dependent variables where conditions are favourable or to score lower where they are unfavourable, this underlining the importance of both good agroecological conditions (especially access to water and adequate rainfall) for obtaining higher than average yields, as well as access to market outlets for surplus production.

The variables on farmer organization and state support provide a mixed picture, however. Neither has any significant impact on yields when it comes to maize. However, for maize there is a weak but positive relationship between marketing and the presence of state/ NGO support. In rice cultivation, higher yields tend to be associated with the simultaneous presence of farmer organizations, while for sorghum higher yields are associated with state or NGO projects. However, in the present SSA situation where the support provided to farmers by the state is generally weak and where viable farmer organizations are rare or missing altogether, the crude character of these independent variables makes it tricky to analyse their precise impact and to identify what aspect of state intervention is responsible for higher yields. A cautious interpretation is to say that the results do not contradict the Asian model wherein external (state) intervention or the presence of local farmer organizations may have a positive impact on yields and hence on farmers' ability to produce for the market (for regression details, see Table 7A.1 in the Appendix).

In the regression model, the impact of 'modern' technology on crop yields is the perhaps most apparent finding. Chemical fertilizer, improved seeds and pesticides, as well as mechanical means of land preparation (tractor and oxen ploughing) are generally associated with higher yields (in the case of maize each factor adds about 500–700 kg/ha).

It is worth noting that we observe this effect on the basis of a rather crude division of households into users/non-users of industrial inputs that gives little consideration to the variation in actual application rates. In respect of the generally low quantities applied of, for example, fertilizers, as we demonstrated in Table 7.7, the findings point to a large potential for obtaining higher yields in SSA, *should farmers' adoption and use of industrial inputs increase*.

Leaving the modern inputs out of the model, it is interesting to note that both fallowing and intercropping are associated with *lower* yields. The former is the conventional way of restoring soil fertility where land is abundant and where incentives for using purchased inputs are lacking and hence production is for subsistence, circumstances that are rather common in the sample. The negative association between yields and intercropping, on the other hand, is probably due to problems of measuring yield under intercropping conditions. Although the overall volume of production may be higher, yields for an individual crop in an intercropped field may well be lower than when the same crop is grown in a single stand.

Application of animal and green manure both have a significant impact on maize and sorghum yields (each increasing, for example, maize yields by about 250 kg/ha). These are typical measures for intensifying production where land is short and/or where farmers find industrial inputs too costly. They have some apparent drawbacks, however. Access to animal manure is a problem for households lacking livestock. More important, both are labour demanding and set definite limits to the farm area that can be supplied by these inputs given the farm labour available.

When industrial inputs and land use practices (tractor and oxen ploughing) are entered into the model for maize, the significant effects of both fallowing and animal manure disappear. Farmers that score high on the use of chemical fertilizer also seem to use manure to a greater extent, with the result that the effect of the latter tends to be downplayed by the greater impact of chemical fertilizer.[7] However, it is when the scale of farm operations and labour concerns are taken into account that the limited potential of using animal manure for raising productivity becomes obvious. Manure increases yields where farm size is small and land can be operated manually. When oxen or tractors permit the cultivation of a larger area, the effect of manure disappears while that of fertilizer remains. The highest yields and returns to labour are found among farmers who combine oxen/tractor ploughing with industrial inputs, notably fertilizer.

Also, the negative association between fallowing and maize yields disappears when chemical fertilizers are introduced, indicating a potential of obtaining higher yields and production in areas with land abundance should market conditions facilitate the adoption of fertilizers there. The effect of green manure (e.g. crop residues left on the land after harvest or compost material applied to small areas) seems independent of industrial inputs or type of land use, at least in the case of maize where it contributes positively to higher yields. Crop rotation, on the other hand, seems to have no traceable impact on yields. While, generally, it can be observed that fertilizer use seems to imply that crops are rotated less frequently, it should be noted that combining fertilizer use with crop rotation in the case of rice results in higher yields than with either of the two measures used alone. This interaction effect, where fertilizer reinforces the effect of crop rotation and vice versa, is a likely reason for the apparently low impact of crop rotation seen in the regression model (Appendix, Table 7A.1).

Finally, the positive impact on rice yields of irrigation technology is noteworthy. Irrigation is not commonly adopted but where it occurs, almost exclusively for rice and vegetables, it has a positive impact on yields, as can be seen in the model (Table 7A.1).

The cassava revolution?

According to some studies, the release of virus resistant and high yielding varieties of cassava by IITA in the 1990s has had a revolutionary impact on the production of this crop in various countries, notably in Nigeria where it was first released (Nweke *et al.*, 2002; see also Akande, Chapter 9, this volume). Regrettably, our data on cassava yields are not sufficiently robust to confirm this picture or to allow a meaningful analysis of the impact of technology and other factors involved in this process. It is noteworthy that in Nigeria, where adoption of the new varieties is widespread, and which is confirmed by the *Afrint* study, national data do not indicate any yield increase but point to a rise in production associated with a corresponding increase in cultivated area. Although circumstantial evidence seems to point in the direction of some kind of a revolution in the case of cassava in Nigeria, and perhaps in other countries as well, there is a conspicuous absence of production and yield data that could substantiate this contention.

The key role of fertilizer

In summarizing the findings under this section, the reader's attention should be drawn to the documented positive impact of industrialized inputs on yields, notably chemical fertilizer, and, at the same time, the strikingly low adoption and application rates of these inputs by farmers in general. The findings indicate that markets can be a driver of higher yields and production in places where agroecological conditions are favourable. However, the low average yields obtained by the majority of farmers in such settings indicate that essential elements for realizing a surplus production (i.e. market incentives) are lacking. In the final section, we will look into some of these aspects by examining the how the socio-economic characteristics of the farming population in SSA relate to yields, production and marketing patterns.

The Role of Household Characteristics, Gender and Wealth

How do the sex, age and educational level of the farm manager influence production, yields and the propensity to sell any of the food crops? Similarly, what do household and farm size, use of hired labour and, not least, the poverty/wealth situation of the households mean for their propensity to produce for the market?

In the preceding sections we have indicated that yield, production and marketing of food crops contain dimensions related to gender and poverty/wealth. We have argued that certain groups (i.e. poor farmers and female-headed households) are at a disadvantage in terms of risk management and access to the assets and inputs required to increase their production for the market.

Along the same line, we have indicated that the bulk of marketed production comes from a rather small group of wealthy male farmers, who, relatively speaking, are both high yielders and large producers. Through a combination of farm and off-farm incomes, and probably also due to a better access to state and NGO support, they are able to access a range of crucial assets and farm inputs through which both labour and land productivity can be raised but which, in the present state, are beyond the reach of the ordinary farmer.

The wealth and gender distribution of the household survey sample is given in Table 7.12. The sex of the farm manager as stated by the respondents is in the vast majority of cases equivalent to that of the household head.

Table 7.12. Wealth and gender (farm manager) distribution of sampled households.

Wealth group	Per cent	Cumulative per cent
Very poor (1)	26.0	26.0
Below average (2)	34.9	60.9
Average (3)	29.1	89.9
Above average (4)	7.7	97.7
Very wealthy (5)	2.3	100.0
Total	100.0	

Sex of farm manager	Per cent
Male	78.5
Female	21.5
Total	100.0

Production and yields

Using maize as an example, Figs 7.1–7.3 illustrate fundamental differences between households in farm performance and market integration. These differences relate to the wealth of the households and to the sex of the farm managers. The classification of households into wealth groups is based on a subjective judgement made by the interviewers. Against this method can, of course, be raised a number of methodological objections, especially since we have lumped together respondents from quite different socio-economic contexts. However, if we accept the wealth groups as a valid, yet crude way of classifying African smallholders, the resulting analysis is quite revealing about the conditions under which the majority of rural households in Africa are farming.

Almost 90% of the households are found in the first three wealth groups (Table 7.12) for which average maize yield does not exceed 1.4 t/ha and average production per farm is not higher than 1.8 t/year (Figs 7.1–7.3). In terms of average production per consumption unit, it is only the two wealthiest groups that are net surplus producers. The wealthiest group has, on average, more than twice as high a yield as the poorest group, and in terms of total production and production per farm and consumption unit, scores seven to 10 times higher (Figs 7.1–7.3). The higher production for the wealthy groups is a result not only of their higher yields but also of their larger farm size and the labour saving assets they have access to in the form of oxen and tractors to work the land.

The next set of graphs (Figs 7.4–7.6) show the gender dimension of farm performance in maize. The difference in yield is to the advantage of male-headed households but is not very large, and at country level is statistically significant only in the case of Malawi, Kenya and Tanzania. When it comes to production per consumption unit, the difference is statistically significant but not exceptionally large (females: 184 kg; males: 228 kg). Total production per farm, on the other hand, differs substantially in that male-headed households

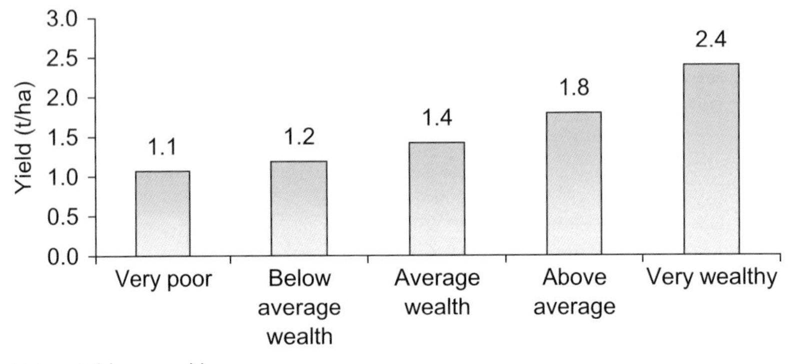

Fig. 7.1. Maize yield per wealth group.

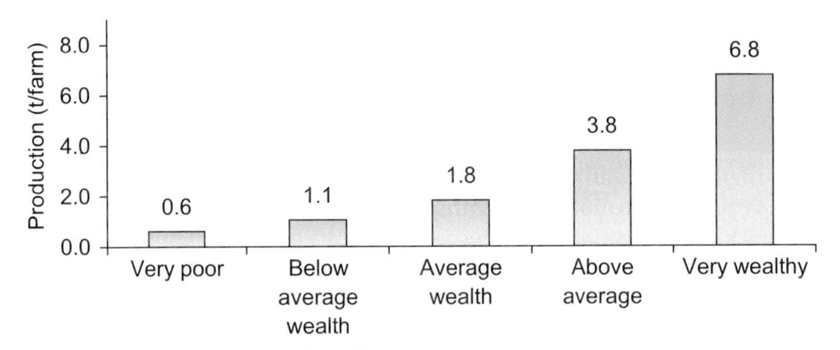

Fig. 7.2. Maize production per farm and wealth group.

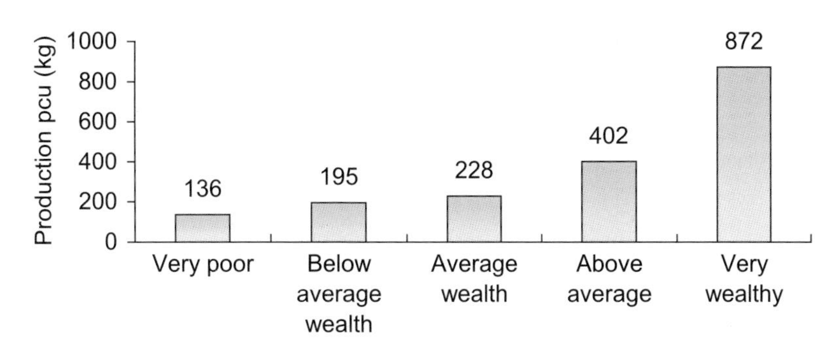

Fig. 7.3. Maize production per consumption unit and wealth group.

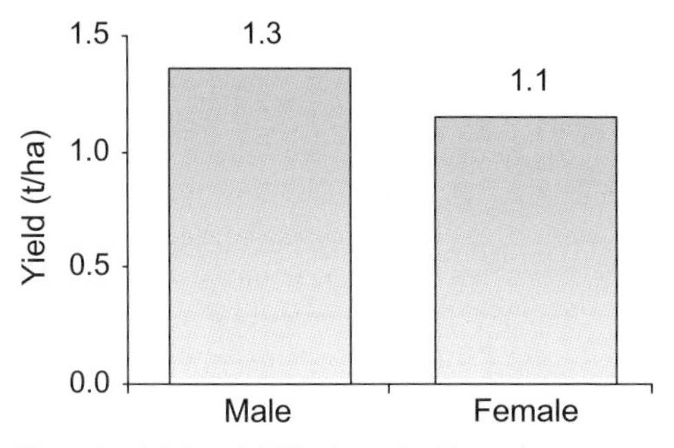

Fig. 7.4. Maize yield by household gender.

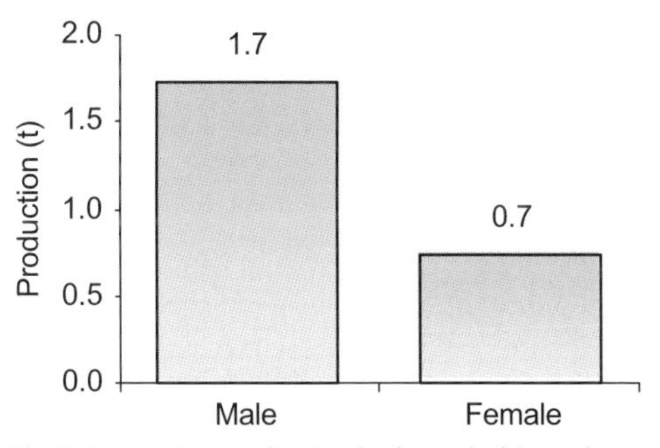

Fig. 7.5. Maize production by household gender.

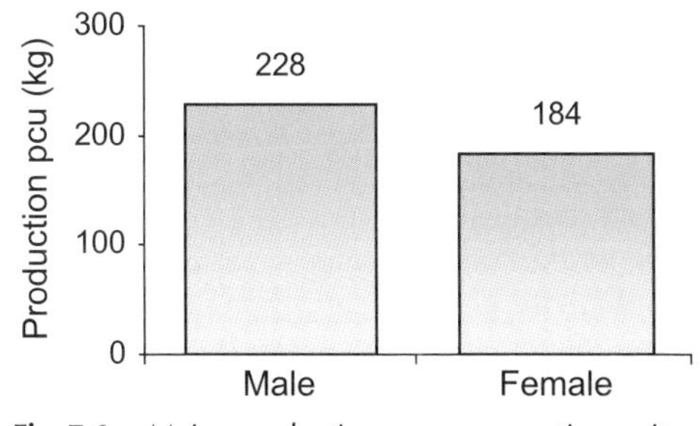

Fig. 7.6. Maize production per consumption unit by household gender.

on average produce more than twice as much as female households on a yearly basis. This suggests that gender differences in farming primarily refer to the scale of farm operations, not to the level of intensification. As we will illustrate shortly, female-headed households more often tend to lack the resources necessary (e.g. land, labour etc.) for producing a surplus. Compared with male-headed households, their farming is to a greater extent subsistence than market orientated.

The differences described for maize pertain to cassava,[8] sorghum and rice as well, as can be seen in Table 7.13 below summarizing production and yield for all staple crops and for the different wealth and gender categories of household. A couple of interesting observations can be made in relation to the wealth groups. First, for all crops, there seems to be a marked leap in both level of production and yield between the households of the two or three lowest wealth groups and the rest. The lowest two groups encompass about 60% of the households, this indicating the approximate proportion of households that are trapped in a situation of low production, low yields and low income.

Second, there are differences in the pattern for maize on the one hand, and cassava, sorghum and rice on the other. For cassava and rice, both average production and yield is lower for the wealthiest group than for the groups in the middle (groups 3 and 4). For sorghum, total production is highest for the wealthiest group while yield for this group is considerably lower. In terms of production per consumption unit, the pattern is similar, with group 3 (average wealth) having the highest production of cassava and sorghum

Table 7.13. Summary of production per farm and consumption unit (within brackets) and yield of staple crops by wealth group and the sex of the farm manager.

	Production (t) and pcu (kg)				Yield (t/ha)			
Wealth and gender	Maize	Cassava	Sorghum	Rice	Maize	Cassava	Sorghum	Rice
Very poor (1)	0.6	1.5	0.4					
	(136)	(381)	(62)	0.5	1.1	3.4	0.6	1.1
Below average (2)	1.1	2.1	0.7	0.8				
	(195)	(452)	(87)	(112)	1.2	3.9	1.2	1.4
Average (3)	1.8	5.5	1.8	1.4				
	(228)	(866)	(204)	(176)	1.4	7.6	1.5	1.6
Above average (4)	3.8	9.2	2.3	3.3				
	(401)	(1,298)	(167)	(243)	1.8	8.2	2.8	2.2
Very wealthy (5)	6.8	8.3	3.9	1.5				
	(871)	(594)	(128)	(446)	2.4	3.1	1.0	1.2
Male	1.7	4.5	1.2	1.1				
	(228)	(758)	(153)	(190)	1.3	5.8	1.3	1.4
Female	0.7	2.0	0.4	0.7				
	(184)	(426)	(75)	(187)	1.1	3.9	0.6	1.2
Total	1.5	4.0	1.1	1.0	1.3	5.4	1.2	1.4

while in the case of maize and rice, the highest production is by the wealthiest group. This suggests that the wealthiest strata concentrate on maize as a cash crop, where they are able to take advantage of their larger farm size and better access to labour saving technology in the form of tractors etc. Apart from commercial maize production, the wealth of this group probably derives from a number of other sources, i.e. various kinds of profitable off-farm enterprises.

Market integration

When it comes to marketing, gender and wealth differences are as striking as for production. Table 7.14 gives the percentages of households producing for the market and the average share of the harvest sold by the different wealth groups and sex of the farm manager. The table reveals a pattern resembling that of Table 7.13. It shows the significance of maize as a market crop in that commercial production to a greater extent than for the other staples involves the wealthiest group. For cassava, sorghum and rice, the share of farmers producing for the market and the proportion of harvest sold is highest for the middle groups. Overall, however, most of the marketing is done by groups 3–5, while groups 1–2 show a considerably lower level of market integration. As demonstrated earlier, however, quantities sold are very modest (see also Table 7.9).

Not only do poorer groups market their crops less often and in smaller quantities, the price they obtain is generally lower than for the wealthier groups. Crop sale by poorer groups often takes place as distress sale at a time when prices are at the bottom. Wealthier farmers, on the other hand, can afford to store part of their harvest until demand and prices rise. They are also generally in a better position to negotiate sale prices, transport costs etc. (Table 7.15). Male and female farm managers, however, obtain about the same farm-gate price for maize. In terms of gender, the major differences pertain to the scale of farm operations and to the level of market orientation and income rather than to productivity and unit price for sales. By and large, food production by female-headed households is to a much larger extent than for male headed ones aimed at home consumption.

Table 7.14. Marketing by wealth group and household gender.

Wealth groups and gender	Share of farmers selling (%)				Share of harvest sold (%)			
	Maize	Cassava	Sorghum	Rice	Maize	Cassava	Sorghum	Rice
Very poor (1)	33	49	19	64	17	29	6	27
Below average (2)	45	57	45	74	21	29	18	31
Average wealth (3)	55	73	62	79	28	35	28	39
Above average (4)	67	71	66	92	29	27	29	49
Very wealthy (5)	75	62	44	82	37	38	15	42
Male	53	67	54	75	27	34	20	34
Female	31	50	32	64	13	26	14	30
Total	48	63	51	74	24	33	19	33

Table 7.15. Lowest and highest seasonal farm-gate price by wealth group (US$ per 100 kg bag).

Wealth group	Lowest price	Highest price
Very poor (1)	10	15
Below average (2)	12	19
Average (3)	14	22
Above average (4)	15	21
Very wealthy (5)	16	23
Total	13	20

Off-farm linkages

A significant part of households' market integration is their involvement in off-farm activities, as described earlier. What then are the links between off-farm activities/incomes and farm performance?

For all crops there is a relationship between the type of off-farm incomes on the one hand, and farm performance (yields) and commercialization of farm production on the other. This means that incomes earned from certain off-farm activities are likely to enhance yields and the propensity to produce for the market. In Table 7.16, displaying the figures for maize, the listed types of activity can be assumed to represent different income levels, with the highest incomes earned by households involved in large-scale business. Households with no off-farm income are included as a reference category. Households relying on employment as their sole source of off-farm income have, regardless of crop, low yields and seem less likely than the other household categories to market their produce. The low rate of crop sales for this group suggests that the income generated from employment is too small to be reinvested in agriculture. For this category of households, farm production appears to be predominantly subsistence orientated. In contrast, income from micro-business implies a much higher degree of commercialization measured as the share of households producing for the market and the share of production sold. The most pronounced commercialization is found among households drawing on large-scale business incomes, a finding indicating that this group of households produce at a larger scale and can afford labour saving and yield improving inputs (Table 7.16).

Moreover, looking at producer price level, a similarly distinct pattern is revealed.[9] Higher incomes from certain kinds of off-farm activity seem to imply a higher farm-gate price. This suggests that access to sufficiently large off-farm incomes render the households better equipped to obtain a better price, for instance by storing a part of the harvest until the demand is higher. This capacity is obtained when incomes derive from micro-business and large-scale business activities. Off-farm income in the form of employment, however, has the opposite effect. Employment in the form of low-income jobs underlines the poverty condition of affected households and indicates that crop sales for this group are distress sales. This is a condition they share with households without access to off-farm incomes. (Table 7.17).

Summarizing the results given above, we have found substantial differences between

Table 7.16. Maize yield (t/ha) and marketing of maize by type of off-farm income.

Type of off-farm income source	Mean yield	Share of crop sold (%)	Share of hh selling (%)
No off-farm income	1.2	20	43
Employment	1.3	18	35*
Micro-business	1.3	27*	53*
Large-scale business	1.9*	48*	83*
All households	1.3	22	46

*Statistically significant at 0.05 level (Anova). 'No off-farm income' is reference category.

Table 7.17. Lowest and highest mean farm-gate price per season and by the households' kind of off-farm activity (US$/100 kg bag).

Type of off-farm income source	Lowest price	Highest price
No off-farm income	12	17
Employment solely	10	15
Micro-business	14*	21*
Large-scale business	17*	27*
Total	13	19

*Statistically significant at 0.05 level (Anova). 'No off-farm income' is reference category.

households in terms of production (per farm and consumption unit), yields and crop sales. A majority of the rural farm population, including a disproportionate number of female-headed households, is economically marginalized in that their level of production is too low to permit more than irregular sales of very small quantities. Off-farm activities provide an important complementary income source. When such incomes are high, they seem to go along with higher yields and better farm-gate prices.

Regression analysis of socio-economic factors

Lastly, we will in a stepwise manner examine the above factors and a number of related variables in a couple of regression models in which the dependent variables are: (i) production per farm and consumption unit; (ii) yield; and (iii) crop marketing. We will concentrate our analysis on maize on the basis of its significance as a market crop and widespread occurrence in all country samples. It should be noted, however, that the evolving pattern is fairly homogenous as regards the other staple crops and the investigated household variables.

In order to reduce the number of variables in the regressions, the technological variables presented earlier (Appendix, Table 7A.1) have been grouped into indices, following their positive or negative correlation with maize yields. The variables 'improved seeds', 'fertilizers' and 'pesticides' have been transformed into an index called 'industrial inputs'. Similarly 'green manure' and 'animal manure' have been indexed 'preind 1', and the farming practices of 'crop rotation', 'intercropping' and 'fallowing' 'preind 2'. The variables 'oxen ploughing' and 'tractor ploughing' have been recoded into a binary variable called 'traction'. The effects of these variables/factors on production per farm and consumption unit, yields and marketing of maize are summarized in the Appendix, Table 7A.2.

Most of the institutional and technological factors discussed earlier in relation to yield remain important also for production per farm and consumption unit, and for marketing (percentage of harvest sold). The presence at village level of a small farmer organization seems to imply both a higher production and a larger share of the harvest that goes to the market. Similarly, favourable market and agroecological conditions positively influence production. So does the use of industrial inputs and access to animal or tractor draught power, and to a lesser extent the use of animal and green manure. As in the case of yields, there are large differences in production and marketing between the countries in the sample, with most countries scoring below Nigeria.

Adding household factors to the models does not, on the whole, alter the picture as regards the role of technological factors (Appendix, Table 7A.3). It is hardly surprising but deserves to be emphasized that farmers who 'sell maize' have higher yields and production (per farm and consumption unit) than farmers who are subsistence orientated.

Despite the institutional and structural constraints circumscribing commercial production of food crops, the data clearly point to the crucial role of the market as a driver of increased production, yields and income earning. Given the low market integration generally, the findings indicate a large production potential that so far is only realized by a minority of farmers.

Among the other socio-economic factors examined, 'total farm size' and 'hired labour' are positively associated both with production per farm and per consumption unit. Household 'wealth group' influences production in that poverty implies a lower production and wealth a higher production compared with the average household. So far, the results are consistent with the bivariate analysis presented above.

In addition, higher 'age' and 'education' seem to imply a somewhat higher production per farm but not per consumption unit. Similarly, 'household labour' (i.e. number of adults in the household) is positively correlated with production per farm but negatively correlated with production per consumption unit, a somewhat puzzling finding at first sight. Although production per consumption unit tends to increase with farm size, other factors held constant, it may decrease if the number of adult household members increases with farm size held constant. Under such circumstances, a crucial factor for improving production per consumption unit is access to new technology.[10]

It is noteworthy that the pronounced difference in production per farm and consumption unit between male- and female-headed households found in the bivariate presentation disappears in the regression models. This is probably so because 'gender' (sex of farm manager) conditions the scale of farm operations through intervening variables such as farm size and the use of tractor and oxen in land preparation. When these scale-orientated variables are excluded from the analysis, the sex of the farm manager is significantly and negatively correlated with production, implying that female-headed households score significantly below male-headed ones. When farm size is controlled for, however, the association disappears. This suggests that observed gender differences in production are largely a result of differences of scale, i.e. women's lesser access to capital assets, not lack of technology as such. Modern inputs are positively associated with higher production, regardless of the sex of the farm manager. Similarly, the association between gender and production per farm weakens (but is still significant) as 'household labour' and 'hired labour' are brought into the model, indicating that female-headed households' lesser access to labour may limit their ability to produce a surplus for the market.

Factors accounting for the variation in maize yield are similar to those found for production (Appendix, Tables 7A.2 and 7A.3). Of the socio-economic variables, higher than average wealth, hiring of farm labour and market orientation ('sell maize') are positively correlated with higher yields, the result of which is telling evidence of the yield gaps presented earlier. It should be noted that what constitutes 'wealth' is not only a larger than average farm size but also access to a range of other resources and social networks, including those from outside agriculture. In this sense, wealth is as much a precondition for higher yields and production as it is the result of commercial agriculture and other incomes. Farm size has no independent effect on yields but indirectly contributes to higher production and marketing (see Production and Marketing columns, Table 7A.3). Interestingly, yields increase with higher education, possibly because the more educated farmers are better informed about new technologies and market options, have access to a larger social network and may gain higher incomes from off-farm activities.

In line with our discussion on gender above, yield data do not support the view that female-headed households are discriminated against when it comes to accessing and using modern or other kinds of yield raising technology *as such*. Their lower production follows from their smaller capital and labour assets. When it comes to market integration, in the regression defined as the proportion sold of the maize harvest, both female-headed households and poor households score significantly lower, which is expected following their low production.

Finally, a note of caution should be raised. We have demonstrated clear differences between households in terms of yields and production on the basis of their use or non-use of specific inputs and presence/non-presence of a number of social conditions. It is reasonable to assume, however, that households differ not only in their use of farm inputs *per se* but also in the *quantities* and *rates* of such inputs that they apply. Most likely, both female-headed and poor households face constraints in affording sufficient amounts of yield-improving inputs. And, the higher yields found for wealthy strata are likely to derive not only from the fact that they tend to use industrial inputs but also from their more frequent and higher application of such inputs, particularly fertilizer. Although these are important aspects, the survey was not primarily designed to uncover the fine-tuned aspects of various farm technologies, mainly because of problems of measuring and quantifying inputs. As has been demonstrated here, the agricultural crisis is apparent enough to be revealed even without such precise measuring.

Summary Conclusion

Our data support the view that smallholders in SSA face a prolonged and multidimensional crisis, i.e. a high degree of subsistence farming, low productivity, low and uncertain incomes, a high risk exposure to market failures and climatic adversaries, and an increasing resort to multiple sources of off-farm income.

While the adoption rates of high yielding seeds are relatively high, the use of chemical fertilizer is marginal for most farmers and for most crops. Chemical fertilizer is the one input that for all crops and in all analyses has the strongest and most consistent effect on yield and level of production (per farm and consumption unit). Yet few farmers see it as worthwhile or can afford to use fertilizer.

Our argument is that the performance of African smallholders is held back by a number of economic, political and institutional factors at regional, national and international level. Under present conditions, only a small number of wealthy households have access to the resources and financial security that make it possible for them to improve yields, raise production and market anything more than a marginal surplus. The performance of these farmers and the gap between them and the majority clearly shows that the African food crisis is policy related. It also points to a large but so far untapped production potential. Where markets and agroecological conditions are favourable, and production is orientated towards the market, households respond with higher production and higher yields. On the whole, however, the majority of the farm population, including most female-headed households, are trapped in a situation of low and uncertain incomes, financial and institutional insecurity, inadequate on-farm resources, and low labour and area productivity. In the present situation, few households can afford or see it worthwhile to invest in productivity-raising and labour-saving technologies in order to produce a marketable surplus of staple food crops. As a result, they are poorly integrated into domestic markets for food staples.

Notes

[1] The problems of yield estimation are several. As described in a recent report from FAO, cassava is planted throughout the year and a single plot will contain plants of different ages, new plants replacing old ones as they are consumed. Furthermore, during its life span of up to 2 years, cassava may be harvested at any time (FAO/WFP, 2003).

[2] The survey question on 'improved' cassava does not adequately distinguish between different types of improved variety. Some improved varieties may simply be existing traditional varieties that are recommended by extension staff, others may be of the high yielding and virus resistant TMS (Tropical Manioc Selection) type developed by IITA.

[3] The Asian estimate may include more than one crop per year.

[4] Application rates above 100 kg/ha for maize are treated as reporting errors.

[5] We were unable to document any significant effects in the form of higher yields of maize in areas

where this particular farming practice was present, possibly because of spurious data following the frequent misunderstanding by farmers and enumerators of this particular question. Other measures that can contribute to improved soil fertility through non-industrial inputs include the improved fallowing technique as reported in Haggblade, Chapter 8, this volume.

[6] In calculating grain equivalents, the following weights have been used: paddy 0.8 and cassava tubers 0.3. Consumption units: adults (15–60 years) 1.0; children (< 15 years) 0.5; old (> 60 years) 0.75. The figure 220 kg grain per consumption unit and year is taken from Sukhatme (1970) and indicates the approximate minimum food and calorie intake required to keep a person alive, corresponding to 2200 kc or 600 grams of grain per day.

[7] In order to check for the interaction effect between fertilizer and animal manure, a new variable 'fertilizer–manure' was constructed. When entered into the model, the effect of both fertilizers and manure is reduced, as expected. However, while fertilizer remains statistically significant, confirming its greater and independent impact on yields, the manure and the interaction variable suffer from problems of multicollinearity. The interaction variable is not included in the model of Table 7A.1.

[8] We have included cassava here with a cautious note regarding the validity of the figures presented.

[9] There is a large variation in reported farm-gate prices depending on season, village location, country for production, type of maize sold (e.g. green or mature maize) and, not least, due to reporting errors stemming from definition and measurement problems.

[10] Although not in the model in Table 7A.3, it should be noted also that the consumer:worker ratio is significantly and negatively correlated with production per consumption unit, implying that the larger the proportion of dependents in a household, the lower the production per consumption unit. One factor that can maintain production per consumption unit in a situation of an increasing consumer:worker-ratio is improved technology. Exploration of off-farm income opportunities is another activity that can affect both production per consumption unit and the overall income situation of a household, either positively or negatively, depending on type of income source and the overall market situation for agriculture. In the regression model, off-farm work is negatively correlated with production per consumption unit, suggesting it generally functions as an income substitute in the absence of viable food crop markets. However, the association is too weak to be statistically significant (albeit barely so = 0.07).

References

Ashimogo, G.C., Isinika, A.C. and Mlangwa, J.E.D. (2003) Africa in transition. Micro Study Tanzania. Sokoine Agricultural University, Morogoro, Tanzania.

Binswanger, H.P. and Pingali, P. (1988) Technological priorities for farming in sub-Saharan Africa. *World Bank Research Observer* 3, 81–98.

Bryceson, D.F. (1997a) De-agrarianisation in sub-Saharan Africa: acknowledging the inevitable. In: Bryceson, D.F. and Jamal, V. (eds) *Farewell to Farms. De-agrarianisation and Employment in Africa*. Ashgate Publishing Ltd, Aldershot, UK.

Bryceson, D.F. (1997b) De-agrarianisation: blessing or blight? In: Bryceson, D.F. and Jamal, V. (eds) *Farewell to Farms. De-agrarianisation and Employment in Africa*. Ashgate Publishing Ltd, Aldershot, UK.

Bryceson, D.F. (1999) *Sub-Saharan Africa Betwixt and Between: Rural Livelihood Practices and Policies*. Afrika Studie Centrum, Leiden, The Netherlands.

Evenson, R.E. and Gollin, D. (2003) Assessing the impact of the Green Revolution, 1960 to 2000. *Science* 300, 758–762.

FAO/WFP (2003) *Special Report. Crop and Food Supply Assessment Mission to Malawi*. Food and Agriculture Organisation of UN, Rome.

FAOSTAT data (2004) Food and Agriculture Organisation of United Nations. http://www.fao.org/waicent/portal/statistics_en.asp

Karugia, J.T. (2003) A micro level analysis of agricultural intensification in Kenya: the case of food staples. Department of Agriculture Economics, University of Nairobi, Nairobi.

Mulat, D. and Teketel, A. (2003) Ethiopian agriculture: macro and micro perspective. University of Addis Ababa, Addis Ababa.

Nweke, F.I., Spencer, D.S.C. and Lynam, J.K. (2002). *The Cassava Transformation – Africa's Best-kept Secret*. Michigan State University Press, East Lansing, Michigan.

Ponte, S. (1998) Fast crops, fast cash: market liberalization and rural livelihoods in Songea and Morogoro rural districts, Tanzania. *Canadian Journal of African Studies* 33, 316–348.

Quinones, M.A., Foster, M.A. and Sicilima, B. (1992) The Kilimo/Sasakawa-Global 2000 Agricultural Project in Tanzania. In: Russell, N.C. and Dowswell, C.R. (eds) *Africa's Agricultural Development. Can it be sustained?* Sasakawa Africa Association, Mexico City, pp. 26–43.

RELMA/Sida (2001) Annual Report 2000. Regional Land Management Unit, RELMA, Nairobi.
Sukhatme, P.V. (1970) Incidence of protein deficiency in relation to different diets in India. *British Journal of Nutrition* 24 (June), 477–487.
Wamulume, M. (2003) Zambia *Afrint* Micro Report. Institute of Economic and Social Research, University of Zambia, Lusaka.
World Bank (2000) *Can Africa Claim the 21st Century?* World Bank, Washington, DC.

Appendix: Regression Tables

Table 7A.1. Village conditions and technology impact on crop yield, controlling for country-specific factors. (Significant variables shown in bold.)

	Maize yield (kg)		Sorghum yield (kg)		Rice yield (kg)	
	Beta	Sign	Beta	Sign	Beta	Sign
(Constant)	1297.8	0.000	770.3	0.000	1366.6	0.000
Farmers' organization	–50.1	0.354	–108.8	0.102	**495.8**	0.000
State/NGO	62.3	0.285	**237.8**	0.003	–77.6	0.488
Low dynamism	**–389.0**	0.007	181.2	0.162	–499.1	0.069
Good dynamism	**331.9**	0.000	**243.3**	0.000	63.7	0.627
Hybrid seeds	**120.6**	0.047				
Improved seeds	88.9	0.185	14.1	0.844	–82.9	0.446
Fertilizer	**470.2**	0.000	121.1	0.088	**369.3**	0.008
Pesticides	136.3	0.077	**257.6**	0.001	112.1	0.367
Oxen plough	**456.5**	0.000	22.3	0.696	237.8	0.128
Tractor plough	**654.4**	0.000			256.5	0.078
Crop rotation	–77.8	0.121	51.1	0.309	247.4	0.071
Intercropping	**–128.5**	0.016	–5.2	0.925	–460.5	0.086
Fallowing	–59.5	0.267	–29.3	0.659	–25.1	0.832
Animal manure	51.1	0.400	**190.5**	0.002	8.0	0.967
Green manure	**240.6**	0.000	10.5	0.852	62.6	0.607
Irrigation					**547.1**	0.000
Ethiopia	**–960.7**	0.000	98.3	0.258		
Ghana	**–352.3**	0.001	**–685.8**	0.000	**–716.3**	0.000
Kenya	**–300.8**	0.005	–170.0	0.377		
Malawi	**–992.2**	0.000	**–531.4**	0.008	**–631.4**	0.002
Tanzania	**–816.0**	0.000	–518.1	0.154	–100.7	0.590
Uganda	–26.2	0.799	266.4	0.156	–342.0	0.155
Zambia	**–921.7**	0.000	**–338.1**	0.004		
R^2	0.21		0.37		0.26	
Total no. of cases	2263		611		575	

Table 7A.2. Village conditions and technology impact on maize production (total and per consumption unit), yield and marketing, controlling for country-specific factors. (Significant variables shown in bold.)

	Total prod. (kg)		Prod. pcu (kg)		Yield (kg/ha)		Marketing (%)	
	Beta	Sign	Beta	Sign	Beta	Sign	Beta	Sign
(Constant)	**1058**	0.000	**88.4**	0.000	**1233.8**	0.000	**36.5**	0.000
Small farmers' organization	**255.9**	0.000	**22.6**	0.004	55.0	0.157	**2.6**	0.018
State/NGO	**–178.4**	0.016	–8.8	0.285	**80.1**	0.050	1.0	0.378
Good agri. potential	**611.4**	0.000	**41.7**	0.000	**295.4**	0.000	2.2	0.130
Low agri. potential	–95.7	0.213	0.3	0.975	**–240.4**	0.000	1.3	0.294
Industrial inputs	**417.2**	0.000	**46.0**	0.000	**225.8**	0.000	**1.9**	0.001
Traction (oxen and tractor)	**1142.0**	0.000	**105.6**	0.000	**371.4**	0.000	**3.6**	0.006
Preind1 (manure)	**285.9**	0.000	**27.4**	0.000	**110.3**	0.000	–0.5	0.505
Preind2 (fallowing etc.)	**–172.9**	0.000	–3.4	0.428	**–105.3**	0.000	–0.6	0.323
Ethiopia	**–1448.0**	0.000	**–74.0**	0.000	**–593.2**	0.000	**–23.5**	0.000
Ghana	**–383.0**	0.006	**32.9**	0.040	**–197.3**	0.012	**16.6**	0.000
Kenya	**–1688.0**	0.000	**–124.7**	0.000	**–221.4**	0.001	**–37.3**	0.000
Malawi	**–1065.0**	0.000	16.1	0.224	**–710.1**	0.000	**–38.5**	0.000
Tanzania	**–854.9**	0.000	**32.3**	0.011	**–466.0**	0.000	**–20.2**	0.000
Uganda	**–965.4**	0.000	**–43.1**	0.003	58.3	0.409	**–6.2**	0.002
Zambia	**–860.6**	0.000	**–44.7**	0.002	**–647.9**	0.000	–36.9	0.000
R^2	0.285		0.191		0.234		0.449	
Total cases	2496		2387		2406		1999	

Table 7A.3. The impact of household factors on farm performance.

	Total prod. (kg)		Prod. pcu (kg)		Yield (kg)		Marketing (%)	
	Beta	Sign	Beta	Sign	Beta	Sign	Beta	Sign
(Constant)	–574.2	0.000	2.7	0.932	841.8	0.000	45.7	0.000
Farmers' organization	38.5	0.283	8.4	0.410	12.2	0.774	1.3	0.247
State/NGO	65.8	0.074	9.7	0.358	**165.4**	0.000	2.2	0.063
Good agri. potential	**340.7**	0.000	**40.1**	0.002	**268.2**	0.000	1.1	0.481
Low agri. potential	–31.3	0.420	–0.1	0.991	**–239.9**	0.000	1.2	0.352
Industrial input index	**225.7**	0.000	**45.4**	0.000	**179.0**	0.000	**1.5**	0.017
Traction (oxen and tractor)	**519.1**	0.000	**105.8**	0.000	**306.7**	0.000	2.0	0.128
Preind 1 (manure)	**65.2**	0.011	**26.1**	0.000	**95.3**	0.002	–0.3	0.683
Preind 2 (fallowing etc.)	–19.3	0.309	–11.1	0.041	**–112.1**	0.000	–0.2	0.694
Sell maize	**362.8**	0.000	**115.0**	0.000	**391.5**	0.000		
Sex, farm manager	–42.5	0.234	–3.9	0.703	–17.7	0.679	**–3.2**	0.008
Age, farm manager	**2.9**	0.008	0.4	0.196	–0.2	0.903	**–0.1**	0.001
Education, farm manager	**7.9**	0.054	–0.3	0.787	**12.6**	0.008	0.2	0.205
Adults in hh	**58.5**	0.000	**–18.5**	0.000	3.8	0.619	–0.3	0.174
Hired labour	**193.9**	0.000	**42.5**	0.000	**100.3**	0.014	0.3	0.780
Total farm size	**136.5**	0.000	**26.0**	0.000	–8.6	0.219	**0.5**	0.004
Wealth group 1 (poorest)	–52.9	0.239	**–33.0**	0.012	0.4	0.994	**–8.3**	0.000
Wealth group 2	–24.9	0.501	**–26.1**	0.015	–15.9	0.717	**–4.6**	0.000
Wealth group 4	**172.0**	0.004	**43.4**	0.008	**258.4**	0.000	1.0	0.588
Wealth group 5 (wealthiest)	**654.3**	0.000	**102.1**	0.001	**385.6**	0.002	**8.4**	0.013
Ethiopia	108.9	0.266	5.6	0.849	**–337.6**	0.003	**–14.7**	0.000
Ghana	25.3	0.735	17.2	0.428	**–300.5**	0.001	**20.6**	0.000
Kenya	**–152.5**	0.032	–19.9	0.343	83.5	0.320	**–33.5**	0.000
Malawi	**145.6**	0.037	**72.5**	0.000	**–308.5**	0.000	**–36.1**	0.000
Tanzania	**180.7**	0.002	**61.3**	0.000	**–405.8**	0.000	**–16.5**	0.000
Uganda	–57.2	0.398	–15.7	0.434	27.7	0.731	–2.0	0.356
Zambia	119.8	0.076	22.2	0.276	**–348.7**	0.000	**–30.8**	0.000
R^2	0.522		0.348		0.283		0.430	
Total cases	2051		1948		2082		2076	

8 From Roller Coasters to Rocket Ships: the Role of Technology in African Agricultural Successes

Steven Haggblade
International Food Policy Research Institute, Lusaka, Zambia

Broad-based poverty reduction in Africa will require significant improvements in agricultural performance. Only growing agricultural productivity can simultaneously reduce food prices, which govern real incomes and poverty in urban areas, and increase incomes of the majority of Africa's poor, who currently work in agriculture. For this reason, agricultural growth provides a central thrust around which the battle against African poverty must be waged. Indeed, available empirical evidence suggests that gains in agricultural productivity and income translate directly into poverty reduction (Thirtle *et al.*, 2003).

The question is how. How can Africa accelerate its agricultural growth in the first place? The *Afrint* project has focused on this question by examining the stellar performance of Green Revolution Asia and exploring how Africa might learn from this experience in order to jump-start its own agricultural economy.

To complement *Afrint*'s cross-continental comparison, this chapter draws inspiration from home-grown agricultural successes within Africa. It summarizes findings from a recent review of African case studies conducted by the International Food Policy Research Institute (IFPRI) on 'Successes in African Agriculture'. Like the *Afrint* project, IFPRI's review aimed to study superior agricultural performance in order to gain insights into how future improvements might be introduced into African agricultural systems. While *Afrint* researchers turned to the Asian Green Revolution for inspiration, the IFPRI review focused instead on local success stories from within Africa. As with positive deviants studies in the field of nutrition, the IFPRI team aimed to identify superior performers operating within the same institutional, cultural, political and agroecological constraints. In doing so, it provides a useful contrast with *Afrint*'s Asia–Africa comparisons.

This chapter summarizes key lessons emerging from the IFPRI review of African agricultural successes. In order to confine this overview to manageable proportions, discussion focuses primarily on the role of technology in triggering advances in African agriculture.

Learning from Past Successes

Identifying successes

Though inadequate in number and scale to counter sub-Saharan Africa's daunting demographic bulge, African farmers and agricultural policy makers have achieved a series of temporally and regionally scattered successes in agricultural development (Wiggins, 2000; Carr, 2001; Gabre-Madhin and

(eds G. Djurfeldt, H. Holmén, M. Jirström and R. Larsson)

Haggblade, 2003). In order to identify priority areas for its future work in Africa, the IFPRI has undertaken a broad review of these 'Successes in African Agriculture' (see Haggblade, 2004). By examining a series of instances in which important advances have occurred in the past in African agriculture, this effort aims to identify promising avenues for achieving similar success in the future.[1]

The IFPRI review began by conducting a broad consultation with Africa-based policy makers, scientists and researchers. The analytical team asked each expert to identify the instances they considered most important in advancing the state of African agriculture. The roughly 250 nominations they received are summarized and described in detail by Gabre-Madhin and Haggblade (2003, 2004). In conducting this review, the IFPRI team defined 'success' in African agriculture as: *a significant, durable change in agriculture resulting in an increase in agriculturally derived aggregate income, together with reduced poverty and/or improved environmental quality.*

Case study investigations

In order to critically assess whether these nominations constitute actual 'successes', we have reviewed available secondary evidence for roughly a dozen of the most commonly cited cases (Gabre-Madhin and Haggblade, 2003). In addition, a series of case study teams has compiled primary as well as secondary data for a representative selection of identified cases.[2]

Together, these case studies provide a series of important contrasts – among private as opposed to public instigators of change, points of intervention, levels of subsidy involved, food and export crops, regional diversity, duration and scale of impact. The following thumbnail sketches offer quick highlights of these accomplishments, in roughly chronological order.

Bananas in the Central Highlands

Beginning about AD 500, for nearly 800 years, farmers in the Great Lakes Region experimented intensively with bananas, a new crop imported from South Asia by Arab traders. The new crop's lower labour requirements, high calorie yield per hectare and favourable effects on soil erosion attracted keen interest. Yet adaptation proved arduous because most edible bananas are seedless and must be reproduced by vegetative propagation. This severely limits the prospects for varietal development, since genetic variation in the clones emerges only irregularly through flaws in cell reproduction. In spite of these difficulties, through assiduous detection and selection of mutant cultivars, farmers succeeded in identifying and isolating a wide range of varieties suitable for human consumption. Led by inventive local farmers, these efforts launched an extraordinary agricultural and demographic revolution in the Central African Highlands beginning about AD 1300 and laid the foundation for the subsequent political rise of the Buganda kingdom (Schoenbrun, 1993; Reader, 1997). By mid-20th century, Ugandan farmers were cultivating 60 different cultivars, the largest pool of genetic diversity anywhere in the world (De Langhe *et al.*, 1996; Reader, 1997). Given the infrequency and irregularity of genetic mutation in plant reproduction, most experts marvel at the rapidity with which African farmers achieved such genetic diversity (Simmonds, 1959; McMaster, 1962). In doing so, they developed an important food security crop that currently accounts for over one-quarter of caloric consumption in the region (FAOSTAT data, 2004).

Cassava

The cassava breeding, pest and disease fighting efforts of the past three decades have improved the lives of probably a hundred million poor consumers and farm family members across West, Central and southern Africa. IITA and associated government researcher programmes have averted a series of devastating mealybug and mosaic virus attacks across the continent. They have simultaneously produced a series of improved cassava varieties that yield 40% more than traditional varieties, even without fertilizer (Haggblade and Zulu, 2003; Nweke,

2003). Without recourse to purchased inputs, farmers in many locations across Africa are adopting improved cassava varieties, thereby placing downward pressure on staple food prices and benefiting not only farm families but also the urban poor who consume cassava products. Dubbed Africa's 'best-kept secret' by Nweke *et al.* (2002), these efforts have arguably proven to be the continent's most powerful poverty fighter to date.

Maize in East and southern Africa

The development and diffusion of modern, high-yielding varieties of maize have transformed this imported cereal from a minor crop in the early 1900s into the continent's major source of calories today. Maize breeding in southern Rhodesia and Kenya launched the first major breakthroughs during the 1960s, though research efforts subsequently spread throughout the continent with strong support from international centres such as the International Maize and Wheat Improvement Center (CIMMYT) and IITA from the 1970s onwards. Although unsustainable financial subsidies artificially inflated production gains in many locations, the breeding breakthroughs have proven to be an undeniable success, with improved maize germplasm probably benefiting five to ten million small farms throughout Africa as well as tens of millions of its urban consumers (Smale and Jayne, 2003).

Cotton in West Africa

Since independence in the 1960s, West African cotton production and exports have both grown rapidly, at a compound annual rate of about 6.5% per year over the past 40 years. Francophone Africa's share in world exports has grown from near zero to 16%, making it the world's third largest cotton exporting block after the USA and former USSR. Roughly one million smallholder farm families produce cotton in francophone Africa. Their cotton profits have enabled them to build up agricultural assets, particularly oxen and ploughs, making them the region's most productive cereal producers as well (Gabre-Madhin and Haggblade, 2003; Tefft, 2004).

Dairy production in Kenya

Dairy production in Kenya has grown rapidly in recent decades resulting in per capita production double the levels found anywhere else on the continent. Smallholders have captured a steadily rising share of that market so that, today, some 600,000 small farmers operating from one to three dairy cows produce 80% of Kenya's milk. As a result, recent panel data indicate that by the year 2000 nearly 70% of Kenyan smallholders were producing milk and that it had become their fastest growing income source. Among the small farmers who produce milk, annual net earnings from milk average $370/year (Ahmed *et al.*, 2003; Gabre-Madhin and Haggblade, 2003; Ngigi, 2003).

Rice production in Mali

Policy reform in rice milling and marketing has radically altered opportunities and incentives for Mali's rice producers over the past decade and a half. Beginning in 1987, the Malian government initiated a gradual set of reforms. These included price deregulation together with the dismantling of the monopolies on paddy assembly, milling and rice marketing held by the Office du Niger (ON) and Office des Produits Alimentaires du Mali (OPAM). As a result, small private dehuller mills, operating at one-quarter milling cost of the cost of the large state mills, began to appear in the Malian delta region. And these private millers and retailers began to offer higher prices for preferred varieties and for more carefully processed grains. The subsequent 50% devaluation of the CFA franc, in January 1994, further boosted producer incentives. Import prices doubled overnight pulling up domestic rice prices sharply in their wake. Producers responded rapidly to these new options and incentives and Malian rice production has more than tripled since 1985, growing by 9% annually over the past 20 years (Diarra *et al.*, 2000).

Horticulture exports from East Africa

From the early 1970s onwards, Kenya's private traders have steadily expanded high-value exports of fruits and vegetables from Kenya. Smallholders supply about 75% of all vegetables and 60% of all fruits. By the mid-1990s, between 100,000 and 500,000 Kenyan farmers and distributors were earning income from this horticultural export trade. One of the country's fastest growing foreign exchange earners, horticultural exports have tripled in real terms over the past 30 years, growing to $175 million in 2000 (Minot and Ngigi, 2003).

Control of rinderpest disease in livestock

Since its accidental introduction from Asia to tropical Africa, rinderpest has remained the continent's most deadly threat to livestock and to many wild animals as well. The initial rinderpest epidemic of 1890 killed an estimated 95% of Africa's cattle (Mack, 1970; Reader, 1997) and has continually diverted veterinary resources from other animal health and improvement activities. To address this widespread threat, the Organization of African Unity established an Inter-African Bureau on Animal Resources (IBAR) to coordinate an all-out international effort to control rinderpest. Assembled in 1986, this alliance involved national governments, their veterinary services, international centres and donors as a coalition of 35 countries launched the Pan Africa Rinderpest Campaign (PARC). Their concerted efforts resulted in the development of a tissue culture attenuated vaccine for the control and eradication of rinderpest (Plowright and Ferris, 1962; Provost, 1982). Following development of the vaccine, government and private veterinary services across the continent distributed the vaccine (Scott, 1985; Wamwayi *et al.*, 1992). Recent assessments estimate total income gains in the order of $50 million for livestock producers in 10 of the 35 countries evaluated. The production gains have generated $1.80 in net income for every dollar invested in the vaccination programme (Tambi *et al.*, 1999).

Sustainable natural resource management

Old strategies for coping with new pressures on Africa's natural resource base are becoming increasingly infeasible. Classic systems of fertility replenishment via shifting cultivation and long-term fallows break down as population pressure reduces the interval between fallows as well as their duration. The withdrawal of fertilizer subsidies across much of Africa during the structural adjustment liberalizations of the 1990s and the collapse of rural credit systems has rendered reliance on chemical fertilizers increasingly less profitable for farmers. Consequently, Africa's farmers and researchers have developed new solutions to increasing pressure on its soil and water resources. Among many hundreds of innovative efforts across the continent, our analytical teams have reviewed two sets of responses that have emerged in different locations. First is the use of planting basins, which has emerged in recent decades in both the Sahel and Zambia (Haggblade and Tembo, 2003; Kaboré and Reij, 2003). The second strategy involves the use of improved fallows, introduced over the past decade in eastern Zambia and western Kenya (Kwesiga *et al.*, 2003; Place *et al.*, 2003). Though most efforts are still in their early years of development and dissemination, preliminary results suggest that yields and returns to labour increase substantially under these new management practices and that adoption rates are increasing.

Generalizing from past successes

Levers for initiating change

Since African governments have long since abandoned the era of state farms and direct government management of on-farm production, future improvements will depend on improved performance by millions of individual African farmers. Therefore, the IFPRI case study teams have adopted a dynamic analytical framework placing farmer decision-making at its core (Fig. 8.1). Agricultural systems evolve continuously as

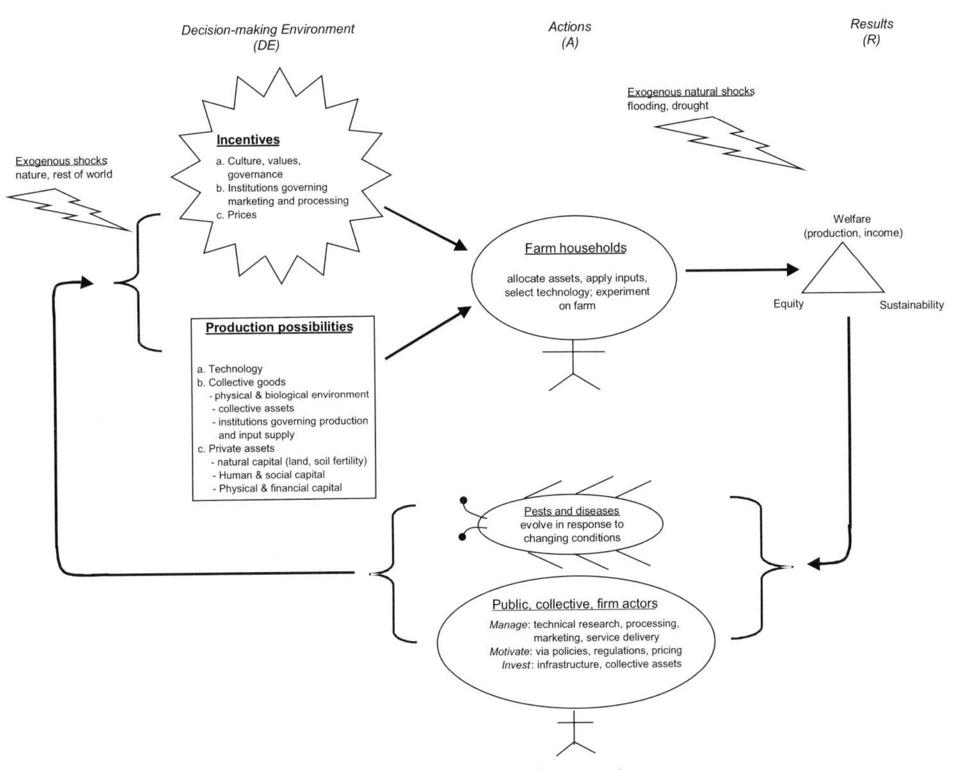

Fig. 8.1. The dynamics of agricultural change: the DE-A-R framework.

individual crops and their human managers respond to ever-adapting pests, diseases, weed species and environmental conditions. In this inherently dynamic system, two key structural features of the agricultural system govern human responses at any given point in time (IFPRI, 2003). First, production possibilities place initial bounds on the scope of action available to farmers. These opportunity sets depend on the stock of available biological and agronomic technology; on the available quantity, productivity and distribution of key productive assets such as land, labour, capital, and water; on the state of physical infrastructure; and on supporting institutions for resource management and input supply. Second, from within the available opportunity sets, prevailing incentive structures subsequently determine which of the many available options farmers will select. Incentives such as enhanced food security, social solidarity, or risk reduction influence individual and household decision-making, while market prices affect input supply as well as production, storage, processing and marketing of outputs.

Levers available for initiating change thus fall into these two categories:

1. Expanded production possibilities:
   - Technology.
   - Collective goods (physical environment, collective assets, institutions governing production and input supply).
   - Private assets (soil fertility, human capital, physical and financial capital).
2. Improved farmer incentives:
   - Governance, values, culture.
   - Institutions governing marketing and processing.
   - Prices (exchange rates, tariffs, taxes, market interventions).

Examining the case studies in depth has permitted the analytical teams to explore a series of important questions: which levers of change have proven most powerful? Who has taken the key initiatives? Which policy environments have most effectively facilitated scaling up to achieve large-scale impact? Which technologies, institutions and processes can be most readily replicated and upscaled?

Assessing replicablity

In some instances, technologies can directly transfer from one location to another. SR52, the breakthrough hybrid maize first released by the Zimbabwean agricultural service in 1961, spread rapidly in Zimbabwe and also to the surrounding countries of Malawi and Zambia where it remains important today in breeding lines (Smale and Jayne, 2003). Improved banana varieties developed in central Uganda spread rapidly throughout the Great Lakes Region of Central Africa (Schoenbrun, 1993; Reader, 1997). Similarly, a network of allied regional cotton research centres has developed a stream of improved plant genetic material used by farmers across the Sahel (Coton et Développement, 1998; Tefft, 2003).

Yet, in other instances, technologies prove location-specific. Direct import of IITA cassava varieties into Zambia, for example, has not fared well because of different altitude, temperature, soils and rainfall (Haggblade and Zulu, 2003). Many varieties of hybrid maize from temperate zones will not flower in equatorial regions because differences in daylight hours trigger tasselling. Pests, soils and policy environment vary across locations, making direct technology transplants uncertain. The International Center for Research in Agroforestry's (ICRAF) work with improved fallows demonstrates quite clearly the need for location-specific adaptive research (Kwesiga *et al.*, 2003; Place *et al.*, 2003). In these instances, the processes of change may prove more replicable than the individual technologies themselves.

Changing environment

In addition to questions of replicability, policy makers recognize that they will face a different environment in the future than they have in the past. The successes of the past arose in an era when world prices were buoyant, donor aid was plentiful and governments were heavily involved in agriculture. Yet many features of the agricultural landscape have changed over the past decades (Chigunta *et al.*, 2003; Eicher, 2003; Hazell *et al.*, 2003).

In the international arena, the composition of trade has changed substantially (Hazell *et al.*, 2003). In 1980, the value of bulk farm commodity exports worldwide stood at double the value of processed agricultural products. Yet by the year 2000 processed goods had surpassed the bulk commodities in value (Hazell *et al.*, 2003; Regmi and Gehlhar, 2003). The structure of trading systems has changed as well. Since 1990, relaxation of restrictions on international trade, foreign direct investment and foreign exchange markets has launched rapid consolidation in food retailing and marketing worldwide. In Kenya alone, vegetable and fruit markets now sell through more than 200 supermarkets, which account for up to 30% of food retailing in the country (Weatherspoon and Reardon, 2003). The rapid scaling up of procurement through these large retail outlets radically changes marketing requirements for African farmers. Likewise, the advent of biotechnology in the global environment has redefined technology options for African agriculture.

Aid flows for African agriculture have fallen by half – to $1 billion per year – from the late 1980s to the late 1990s (Eicher, 2003; Hazell *et al.*, 2003). Yet OECD farm subsidies cost farmers in sub-Saharan Africa nearly double that amount – $1.8 to $1.9 billion/year in lost agricultural income according to recent world trade simulations (Beghin *et al.*, 2002; Diao *et al.*, 2003). Net gains to African farmers from OECD country taxpayers have turned substantially negative.

Domestically, too, conditions have changed (Chigunta *et al.*, 2003). Growing population places increasing pressure on land resources across much of sub-Saharan Africa, narrowing the scope for further expansion of cropland. As a result, old strategies for coping with these new pressures on the natural resource base are becoming increasingly infeasible. Classic systems of replenishment via shifting cultivation and long-term fallows break down as population pressure reduces the interval between fallows as well as their duration. The withdrawal of fertilizer subsidies across much of Africa during the structural adjustment liberalizations of the 1990s and the collapse of rural credit systems has rendered reliance on chemical fertilizers increasingly less profitable for farmers. Farmers, therefore, are actively experimenting with new ways of maintaining soil fertility.

By the year 2020, HIV/AIDS may have reduced the agricultural labour force by as much as 26% in ten of the most affected African countries. AIDS afflicts agricultural scientists and professionals as well. In Kenya's Ministry of Agriculture, 58% of all staff deaths in the past 5 years have been AIDS-related, while in Malawi's Ministry of Agriculture and Irrigation at least 16% of the staff are HIV-infected (GTZ, 1999; Chigunta *et al.*, 2003; Topouzis, 2003).

Marketing systems, too, have changed substantially as governments have withdrawn support for parastatal marketing companies, dramatically reduced input and output marketing subsidies and relaxed regulatory restrictions on private trade. Uncertainties remain, however, making private traders nervous and slowing the development of efficient new post-reform marketing systems (Chigunta *et al.*, 2003).

Collective judgement

All of these changes imply that policy makers will have to apply the lessons from past successes in a very different environment going forward. To generalize and draw sensible inferences from past successes will require considerable experience, judgement and collateral knowledge.

For that reason, the IFPRI analytical team assembled a group of experienced agricultural specialists from government, the private sector and from across Africa to help with this synthesis effort. The conference organizers – IFPRI, NEPAD, Technical Centre for Agricultural and Rural Cooperation ACP-EU (CTA) and Internationale Weiterbildung und Entwicklung (InWEnt) – complemented this core of agriculturalists with representation from ministries of finance and trade, key governors of agricultural innovation and growth. During the first week of December 2003, 70 professionals from across Africa met to review the case studies, assess changes in the external environment and draw inferences on how to apply lessons from the past into the future. The results of these deliberations are summarized in Haggblade (2004). The ensuing discussion plumbs the case study material to focus on findings related to technological change as a motor of agricultural advance.

The Role of Technology

Technology as a driver of change

The IFPRI expert survey respondents cited improved technology as the key driver of change in over half of all cases cited (Table 8.1). This proved true regardless of disciplinary background of the respondents. Closer inspection by the case study teams suggests that improved technologies consistently proved instrumental in triggering sustained agricultural growth in the cases the analytical team studied in depth (Table 8.2). All the commodity-specific successes we investigated involved some sort of improved technology. The most common institutional successes cited were the build-up of African national agricultural research systems during the 1970s and 1980s (Gabre-Madhin and Haggblade, 2004). Similarly, the most commonly cited activity-specific successes – those involving new methods of soil fertility maintenance – revolved around new management practices. Even policy-induced changes in incentives frequently initiated more rapid expansion of improved technologies, lower-cost milling technologies in the cases of maize and rice market reforms; greater application of fertilizer and improved cotton seeds following the CFA devaluation of 1994 (see Gabre-Madhin and Haggblade, 2004).

Categories of technological change

In general, improved performance in agriculture requires some form of new technology –

Table 8.1. Motors of change in agriculture. (Source: IFPRI expert survey, Gabre-Madhin and Haggblade, 2004.)

Actors	Respondent categories*				
	Technical researchers	Social scientists	Implementors	Government/ donors	Total
A. Improving opportunities					
Increase farmer assets (%)					
Soil fertility	6	3	7	3	5
Irrigation	1	2	8	6	4
Farm and processing equipment	0	1	7	4	3
Draft power	0	0	2	2	1
Subtotal	7	7	***23***	15	12
Develop new technology (%)					
Higher productivity	30	27	22	19	25
Disease resistance	15	0	2	7	5
Introduce new species	2	0	2	0	1
Other	3	0	5	8	3
Subtotal	***50***	27	***31***	***35***	***34***
Improve access to superior technologies (%)					
Extension	13	15	18	8	14
Seeds	7	8	5	17	9
Fertilizer and pesticides	7	0	2	1	2
Credit	3	3	6	3	4
Subtotal	30	27	31	29	29
B. Improving incentives					
Macro policy (devaluation, trade liberalization) (%)	0	9	0	3	4
Agricultural policy (market reform, taxation) (%)	9	9	2	4	7
Private marketing (%)	2	7	4	6	5
Public marketing agencies (%)	0	7	5	6	5
Growing markets (%)	1	5	4	2	3
Land rights (%)	2	1	0	0	1
Subtotal (%)	14	***39***	15	21	25
Total interventions identified					
Per cent	22	39	20	19	100
Number**	111	202	103	98	514

*Differences among respondent categories are significant at the 1% level. ***Bold italics*** indicate above-average representation.

**Because many respondents cited multiple interventions, the totals here exceed the total number of cases proposed.

either more productive inputs or improved management practices. Though many classifications are possible, Table 8.3 categorizes observed technological changes into three broad categories: (i) development of improved genetic material; (ii) increased use of collateral modern inputs; and (iii) improved management practices. Improved genetic material proved central in the cases of maize, cotton, cassava and dairy. Greater use of purchased inputs proved essential in most cases: annual purchases of improved seeds in the cases of maize, cotton and horticulture; one-time procurement of improved planting material in the cases of cassava and improved fallows; and disease control inputs such as fungicides, pesticides and veterinary services in the cases of cotton and dairy.

New management practices offered significant gains in many instances: confinement

Table 8.2. Contrasting sources of change. (Source: Haggblade, 2004.)

	Bananas	Cassava	Maize	Cotton	Horticulture	Dairy	Sustainable natural res. mgmt	
1. Region, time period	Great Lakes Region AD 500 to 1300	West, Central & southern Africa 1900 on	East & southern Africa 1960 on	West Africa 1960 on	Kenya, Ivory Coast 1970 on	Kenya 1900 on	Planting basins Burkina, Zambia 1980 on	Improved fallows Kenya, Zambia 1990 on
2. Who initiated change?								
a. Key instigators?	• NARs	• Colonial research stations • Rural artisans • IITA • NARs	• Commercial farmers • Government breeders • Government policy makers • Parastatal marketing companies	• Donor and national governments • Parastatal marketing companies	• Private traders	• Commercial farmers • Government policy makers • Parastatals	• Private farmers	• ICRAF
b. Supporting actors?	• NGOs	• Private oil companies • NGOs	• Private seed companies				• Govt extension • NGOs	• Farmer researchers • NGOs • Govt extension
3. What interventions triggered change?								
a. Expanded production possibilities								
• Technology	•••	•••	•••	••	•	•••	•••	•••
• Input supply			•••	•••	••	•••	•	•
• Investments in asset base				•		•	•••	•••
b. Improved incentives								
• Political lobbying			•••	•		•••		
• Output marketing			•••	•••	•••	••		
4. Market outlet	• Domestic	• Domestic	• Domestic	• Export	• Export	• Domestic	• Domestic	• Domestic
5. Were large recurrent public subsidies involved in sustaining smallholder growth?	• No	• No	• Yes	• Yes	• No	• Yes	• No	• No

Table 8.3. Categories of technical change in the case studies. (Source: Haggblade, 2004.)

Case studies	Source of improved technology	Categories of technical change: a) improved genetic material	b) = a) plus purchased modern inputs	c) improved management practices
Bananas	Private: farmers	•••		
Cassava	Public: IITA, NARs	•••		
Maize	Public: NARs, IITA, CIMMYT	•••	•••	
Cotton	Public: IRAT, CIRAD	•••	•••	•
Horticulture	Private traders	•••	•••	•
Dairy	Public subsidy, control and monitoring of AI; private breeding	•••	•••	•
Planting basins	Private farmers			•••
Improved fallows	ICRAF			•••

••• critically important; • supplementary.
IRAT, Institut de Recherches Agronomiques Tropicales et des Cultures Vivrières.

and supplementary feeding of improved breeds among smallholder dairy farmers in Kenya; dry-season land preparation and early planting in the case of the planting basin systems in the Sahel and in Zambia; long-term management of plot rotations alternating sequences of leguminous fallows and alternating harvest years in the case of improved fallows. Farm trials with cotton and maize suggest that farm yields fall by 1% to 2% for each day farmers plant after the first rains (Howard, 1994; Arulussa Overseas Ltd., 1997; Elwell *et al.*, 1999). For this reason, management changes such as dry season minimum tillage offer significant potential for output gains even with existing plant genetic material.

Improved genetic material

Viewed as a group, the case study analyses suggest that these three categories of technical change can be usefully viewed as a progression representing increasing levels of difficulty in inducing farmer adoption. The banana varietal selection and development of improved TMS cassava varieties offer the simplest models of technical change. Without changing their cropping calendar or management practices, and without recourse to credit, new equipment or any recurrent purchased inputs, farmers can simply plant new varieties in the same way they have before. With TMS cassava varieties, their output increases by 40% on average but up to 100% in other cases and even more in distress cases such as Uganda where a virulent new form of mosaic virus caused the disappearance of 500 local varieties (Otim-Nape *et al.*, 2000). The ease of adoption explains the runaway success of the new cassava and banana varieties when introduced in a range of agroecologically appropriate producing zones across Africa.

Purchased inputs

When new technology requires annual purchases of modern inputs – hybrid seed, fertilizer, pesticides or herbicide – adoption becomes more difficult. The collapse of government input delivery and credit systems following the liberalization of the 1990s has led to retrenchment in maize production throughout East and southern Africa. In the wake of this withdrawal, only two systems of input provision and rural credit appear to have functioned effectively for African farmers. First is the vertically integrated contract farming model, in which private or parastatal firms supply inputs and recoup their cost at harvest time as in the cotton and horticulture case studies. Second is reliance on non-farm income which enables farm households to meet lumpy liquidity requirements from intra-household cash flows. This is the default system to which maize and dairy farmers have reverted and non-cash crop farmers have resorted throughout much of Africa. Since not all households have access to the necessary seasonal liquidity, high-input agriculture has slumped throughout much of Africa since the early 1990s.

Changing management practices

The most difficult changes to effect appear to be those requiring new management practices. The sustainable natural resource management systems offer the clearest example of this. In large part, they replace purchased inputs with household labour and new management systems. Improved fallows arose in eastern Zambia as a response to increasingly scarce and high-priced fertilizer. The conservation farming and zai planting basins likewise enable reduction in the use of purchased inputs, though because of their water harvesting properties they prove highly complementary with strategic doses of chemical fertilizer inputs. They offer important alternatives to African farmers in which improved agronomic practices can significantly increase output. In general, adoption numbers suggest that the new management systems are the slowest to take off and require the most extension support. Though many efforts are still in their early years, some evidence suggests that clustered extension efforts can produce critical masses of adopters most quickly and effectively (Kwesiga *et al.*, 2003; Haggblade and Tembo, 2003).

Sources of new technology

Public funding

Publicly funded research has proven responsible for the bulk of new technology development in the cases we studied. With maize, national agricultural research systems in Zimbabwe (then southern Rhodesia) and Kenya launched the first major maize breeding programmes in Africa, in the 1930s and 1950s respectively, in response to political pressure from large commercial farmers (Smale and Jayne, 2003). After decades of careful research, both countries released major breakthroughs in hybrid maize during the 1960s (Gerhart, 1975; Eicher, 1995). Maize breeding subsequently spread throughout the continent with strong support from international centres such as CIMMYT and IITA from the 1970s onwards (Byerlee, 1994; Manyong *et al.*, 2000a). Today in East and southern Africa, farmers plant about 58% of maize area in improved varieties, where they achieve yield gains of about 40% over local varieties (Morris, 2001). Though seed industries remain largely in private hands, the initial breeding work took place in government research programmes.

Similar lobbying by Kenyan dairy farmers resulted in long-standing public investments in dip tanks, quarantine laws, veterinary clinics and medicines, as well as support and monitoring for artificial insemination services since the 1930s (Ngigi, 2003). Together, this package of publicly subsidized inputs has led to rapid expansion of improved cattle breeds and per capita production levels double those found elsewhere in Africa (Staal *et al.*, 1997).

With West African cotton, a combination of French and African government funding has supported cotton research for over 50 years, generating a steady stream of improved varieties and management practices that have increased yields fivefold over the past four decades (Coton et Développement, 1998; Tefft, 2003). CIRAD, the French agricultural research institute, has coordinated a regional breeding programme that has enabled the Sahelian countries to economize on research overheads and share research and development costs. Mali, for example, has introduced six new cotton varieties to farmers over the past 40 years. Of these, only one was developed in Mali; the other five came from sister institutes across the Sahel (Dembélé, 1996).

Recent cassava breeding efforts in Africa have their epicentre at IITA in Ibadan, Nigeria. Beginning in 1971, the IITA team began a highly productive cassava breeding programme building on genetic material developed over 22 years of colonial research at the

Amani Research Station in Tanzania, from 1935 to 1957. Within 7 years the IITA team developed a series of mosaic virus resistant strains, the Tropical Manioc Selection (TMS) series, that has subsequently served as foundation stock in most national cassava breeding programmes (Nweke, 2003). Local adaptive breeding efforts by a series of National Agricultural Research Systems (NARs) across Africa, backstopped by a steady inflow of improved material from IITA, have resulted in a series of diffusions of improved cassava varieties since the mid-1980s (Manyong, 2000b; Nweke *et al.*, 2002). Time and again, this ongoing research capacity has proven crucial in confronting new threats, most recently in Uganda during the 1990s in turning back a virulent new strain of mosaic virus, which within 5 years had destroyed 80% of national cassava production. Four subsequent years of intense collaborative effort, led by Uganda's National Agricultural Research Organization (NARO) to distribute resistant material from IITA, succeeded in stemming this collapse and boosting production back to above pre-attack levels (Otim-Nape *et al.*, 1997; Legg, 1999; University of Greenwich, 2000).

Similar regional collaboration has proven instrumental in successfully combating the cassava mealybug, a devastating pest accidentally imported to Africa from South America in the 1970s. By the early 1980s the mealybug had eaten its way across most of Africa's cassava belt, resulting in crop losses of up to 80% and triggering an urgent regional response. By identifying a predator wasp in South America, international research centres, African NARs and donors launched a mass rearing and distribution programme that led to the biological control of the mealybug threat by 1988 (Norgaard, 1988; Herren and Neuenschwander, 1991).

Of more recent vintage, work on improved fallows by ICRAF researchers began in eastern Zambia in 1987 and in Kenya in 1992. Though driven by formal researcher systems, this work involved early on-farm trials and close collaboration with farmers from a very early stage. A series of farmer-designed and farmer-managed trials led to significant changes in recommended practices – including the use of intercropping in the early years of fallow establishment and the use of bare-root seedlings instead of nursery seedlings (Kwesiga *et al.*, 2003; Place *et al.*, 2003). National agricultural research programmes and a consortium of NGOs have participated closely in the extension of this new technology.

Private innovation

In several of the case studies, private innovation spearheaded the development of new technologies. The on-farm identification and selection of new banana varieties in the Great Lakes Region of Africa was a purely private initiative. Similarly, the planting basins developed independently in the Sahel and in southern Africa were both initiated by private farmers. Farmer-innovators in the Sahel developed the modified planting basin technology in response to several decades of crippling drought and land quality degradation. By digging 10,000 to 20,000 zai basins/ha during the dry season, then placing organic material in the basins and planting early, with the first rains, farmers in the Sahel achieved yield gains in sorghum of about 375 kg/ha (from 125 kg to 500 kg). On-farm trials using identical input packages on zai basins and conventionally ploughed fields in Mali suggest gains of about 700 kg/ha for sorghum and 500 kg/ha for cotton from the water retention and early planting in the zai basins. Early innovators, such as Yacouba Sawadogo and Ousseni Zorome, launched their own private extension services to extend application of the new management practices. Later, a coalition of NGO and government support expanded extension efforts more widely (Kaboré and Reij, 2003).

An agronomically analogous planting basin technology called 'conservation farming' was developed independently in southern Africa by private commercial farmers. Driven by recurrent drought, high fertilizer and fuel costs, commercial farmers from the Zambia National Farmers Union (ZNFU, Zambia's commercial and medium-scale farmer organization) sent delegations to the USA, Australia and Zimbabwe in the 1980s to investigate low-tillage technologies for commercial farmers. The ZNFU subsequently became the prime mover in developing an appropriate minimum tillage package, not only for mechanized large-scale commercial farms but also for smallholder hand-hoe agriculture. The hand-hoe analogue of minimum tillage systems was introduced to Zambia in 1995 by a Zimbabwean farm manager brought in as a consultant to the ZNFU (Oldrieve, 1993). Inspired by the notion of 6–8 t maize yields under hand-hoe cultivation, the ZNFU established a Conservation Farming Unit (CFU) in late 1995 to adapt the hand-hoe basin system to Zambian conditions and actively to promote it among smallholders. The conservation farming (CF) system they advocate involves:

- Dry-season land preparation using minimum tillage methods (either ox-drawn rip lines or hand-hoe basins laid out in a precise grid of 15,850 basins/ha).
- No burning but, rather, retention of crop residue from the prior harvest.
- Early planting with the first rains.
- Planting and input application in fixed planting stations.
- Nitrogen-fixing crop rotations.

Recent on-farm investigations – based on measurement of 310 plots in an erratic rainfall year – suggest that CF management practices raise maize output by 1.1 t and cotton by about 400 kg/ha (Haggblade and Tembo, 2003). With support from the Lohrho cotton company (now Dunavant) and a variety of donors, the CFU began extension smallfarmer efforts in 1996, and in 1998 the Ministry of Agriculture and Cooperatives adopted CF as official policy of the Government of Zambia.

Private traders have largely fuelled the rapid expansion of horticulture exports from Kenya over the past three decades. Given increasing market concentration among importing companies, this growth has required growers to meet stringent quantity and quality standards. Contract farming arrangements – in which exporters supply improved seed, fertilizer and fungicides – have afforded the most common formula for enabling smallholders to participate in these growing export markets. In recent years, concern with food safety and certification standards in European markets have required even more careful coordination between exporters and their contract farmers. Many exporters supply spraying services in smallholder fields in order to meet certification requirements. In Kenya, as in Zimbabwe, Zambia and elsewhere, the quality demands of the international marketplace have required the introduction and monitoring of new technology down through the entire supply chain (ECI, 2001; Minot and Ngigi, 2003).

Implications

Sustainable results or roller coaster rides?

In most instances, output gains in African agriculture have more closely resembled

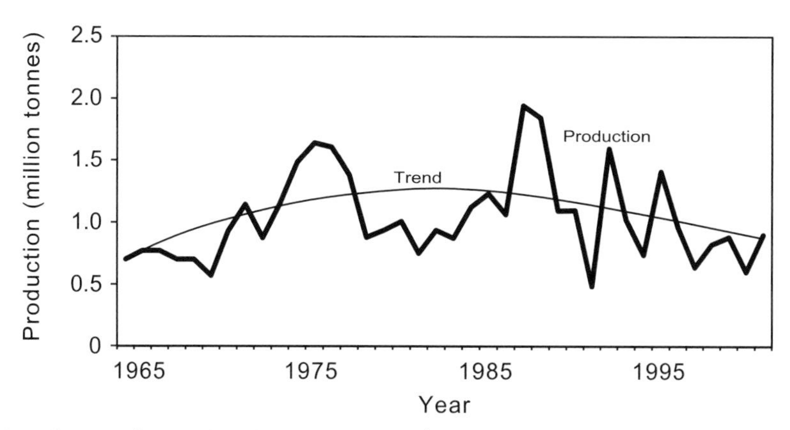

Fig. 8.2. Trends in Zambian maize. (Source: FAOSTAT data, 2004.)

roller coasters than rocket ships (Figs 8.2–8.4, Table 8.4). A combination of several factors accounts for these uneven results.

Pests and diseases

In domesticated agricultural systems, humans assume responsibility for the reproductive success of crops and animals in the face of ever-evolving pests and diseases (Fig. 8.2). In these settings, sustained well-functioning agricultural research systems become not luxuries but, rather, necessities. They govern not only the growth of agricultural systems but their very survival. Uganda's recent loss of 80% of their national cassava crop within 6 years illustrates this danger most clearly (Fig. 8.3).

Yield plateaux

In open pollinated crops such as maize, yields from improved hybrids fall off rapidly unless farmers purchase new seeds each season. Even with well-maintained hybrids and improved close-pollinated species, new seeds lose their edge over time as plant diseases and pests evolve. In spite of a succession of maize breeding successes, the grey leaf spot disease currently poses serious problems for maize farmers in southern Africa. Western wheat researchers indicate that half of all research focuses on yield maintenance to combat rust diseases. Similarly with vegetatively propagated clones like bananas and cassava, which retain identical genetic properties from one generation to the next, mutation by pests and diseases – like the mosaic virus, the cassava mealybug and brown streak disease – causes yield losses over time, sometimes quite rapidly. Together, these biological realities mean that agricultural research systems must turn out a steady stream of innovations simply to maintain yields over time (Evans, 1993; Evanson and Gollin, 2003).

Fluctuating public funding for agricultural research

Yet the public funding which drives this stream of technical innovation has proven uneven over time. Donor funding, which peaked in the 1980s, has fallen by half over the past decade. Similarly, national government funding has fallen from 0.8% of agricultural GDP in 1980 to 0.3% in the 1990s (Pardey *et al.*, 1997). In Zimbabwe, public funding for agricultural research has fallen by 39% in real terms over the past 20 years (Smale and Jayne, 2003). In Kenya, government and donor funding for maize research fell by half between the late 1970s and the late 1980s. As a result, the number of new maize varieties released fell from 13 in the 1963–1974 period to two and six, respectively, in the two successive decades (Hassan and Karanja, 1997).

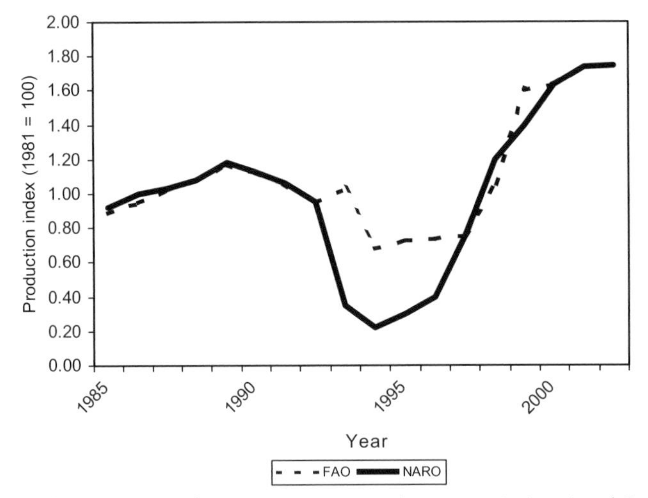

Fig. 8.3. Cassava production in Uganda. (Source: FAOSTAT data, 2004; University of Greenwich, 2000.)

Changing incentives

In some cases, rapidly changing incentives have instigated technological backsliding and output declines. Throughout much of East and southern Africa, eroding input and output price incentives since the early 1990s have contributed to a reduction in hybrid seed and fertilizer use as well as a significant fall-off in maize production (Table 8.4). Similarly, the devaluation of the CFA franc in 1994 triggered a doubling of cotton production in Mali over the next several years, through a combination of increased area and yield stemming from increased input use (Fig. 8.4). Among Kenya's horticultural exporters, stiff competition in export markets has necessitated a rapid sequence of product shifts in order to sustain overall export growth (Minot and Ngigi, 2003).

Declining soil fertility

As a result of rapid population growth, cultivated land per capita has fallen by 40% since 1965, from 0.5 to about 0.3 ha per person (Cleaver and Schreiber, 1994). At the same time, land quality has fallen. Nutrient balances over the past 30 years suggest that Africa has sustained annual net losses of nitrogen, phosphorus and potassium and that the resulting soil mining may account for from one-third to as much as 80% of farm income in some locations (Van der Pol, 1992; Smaling *et al.*, 1997). In this new environment, African farmers will have to pay increasing attention to soil fertility in order to sustain agricultural output growth (Sanchez *et al.*, 1997).

What contributes to sustained technical change?

To draw positive lessons from the past requires looking more closely at the upswings in African agriculture's frequent roller coaster rides. Over periods ranging from one to five decades, farmers in selected locations have sustained output growth with a steady infusion and dissemination of new technology –

Table 8.4. Maize production growth (annual growth rates). (Source: Smale and Jayne, 2003.)

	Period of years	Success growth	Period of years	Uncertainty growth
Kenya	1965–1980	3.3%	1990–2000	–1.5%
Malawi	1983–1993	3.1%	1994–2000	4.4%
Zambia	1970–1989	1.9%	1990–2000	–2.4%
Zimbabwe	1980–1989	1.8%	1990–2000	–0.2%

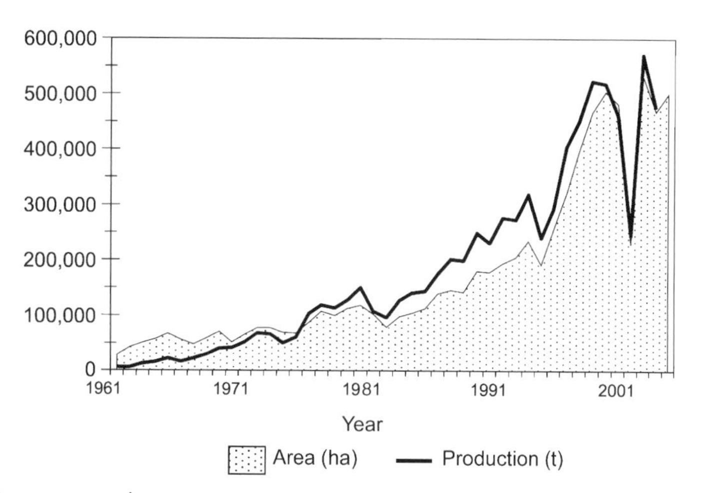

Fig. 8.4. Malian cotton production, 1961 to 2003. (Source: CMDT annual reports in Tefft, 2004.)

in cassava, maize, cotton, dairy and horticulture. In general, these upswings share the following common features.

Governments who care

High-level political commitment has regularly proven necessary for sustaining agricultural growth. It translates directly into favourable policy environments and budget allocations to agricultural support institutions and related infrastructure. Even in the exceptional case of Kenyan horticulture, where government remained largely on the sidelines, its stable investment climate and exchange rate provided the underpinnings necessary for sustained export growth. The experience of maize in East and southern Africa and cotton in West Africa offer the clearest examples of sustained commitment and implicit social contracts between government and the farm community.

Yet in recent decades, farm lobbies have lost much of their political clout. Of the successes that IFPRI's analytical team investigated in detail, only two – maize in East and southern Africa and dairy in Kenya – were driven primarily by domestically funded African government agencies. It seems unlikely a coincidence that in both cases powerful commercial farm lobbies laid claim to government resources and policy protection instrumental to the rise of commercial maize and dairy interests, and which subsequently benefited smallholders as well. In Mali, as well, small farmer lobbies have emerged and played a significant role in restructuring incentives in favour of small farmers (Bingen, 1998). In general, effective farmer organizations and lobbying will probably prove necessary in articulating farm sector needs to government and successfully mobilizing political support for financing agricultural research and support institutions.

Long-term funding for agricultural research

Beginning in 1932, Zimbabwe's famous maize research team enjoyed government funding for 17 years before releasing their first commercial hybrid and for 11 years thereafter before they released SR52, the break-out hybrid that launched Zimbabwe's first Green Revolution, spread rapidly to surrounding countries and still serves as a mainstay in breeding lines in the region today (Eicher, 1995). With cassava, IITA was able to capitalize on over 25 years of colonial research at the Amani Research Station in Tanzania and in 7 years, from 1971 to 1977, turn out a series of transforming cassava varieties in the TMS series (Nweke *et al.*, 2002). In francophone Africa, governments have invested in cotton research consistently since 1949 (Coton et Développement, 1998).

Good pay and working conditions

Small teams of highly motivated, well-paid scientists devoted their career to maize research in Zimbabwe. Four senior maize breeders managed the hybrid maize research in Zimbabwe over a period of 56 years, from 1932 to 1988 (Eicher, 1995). Yet in Zimbabwe, as elsewhere, pay scales and working conditions have declined substantially in the past several decades. Across Africa, spending per scientist fell by 34% between 1961 and 1991 (Pardey *et al.*, 1997). Today, key Nigerian cassava scientists are working in Brazil. In the face of new technical skill requirements, improved training, pay and working conditions will be necessary to retain productive public research systems.

Regional partnerships

In many parts of Africa, small countries cut across common agroecological zones. So research in one country is applicable to its neighbours. In these situations, regional research collaboration holds the potential to considerably reduce costs and at the same time expand the benefits of individual country research. The cases of maize, cassava and cotton all illustrate the considerable gains to be had from regional research collaboration. The Africa-wide cassava collaboration, backstopped by IITA, offers perhaps the best example of the considerable benefits of regional cooperation and exchange. Uganda averted widespread hunger and possibly a famine in the 1990s by importing mosaic-resistant cassava varieties from IITA. Likewise, the mealybug attack across country boundaries demanded regional control efforts. Here again, IITA's close collaboration with a series of national programmes brought about control of the cassava mealybug using biological pest control with natural predators and resulted in an eye-popping benefit cost ratio of 149 (Norgaard, 1988). Maize germplasm has been shared across Africa, through the offices of CIMMYT, IITA and national breeding programmes. The cotton research network in West Africa has provided compelling evidence of the financial benefits of regional research collaboration. A small country like Mali has received five out of six improved cotton varieties from sister research stations outside of the country, clearly illustrating the gains to be made from regional collaboration.

Strategies for sustaining technological advance in African agriculture

Develop regional partnerships

Where small countries cut across common agroecological zones, African researchers can achieve considerable cost savings by sharing research with their neighbours. NEPAD, regional economic commissions and regional research centres, are well positioned to support such moves.

Adopt output rather than input targets

Donors, African governments, NEPAD and many researchers currently use budget expenditure targets in setting goals for government commitments to agriculture and other sectors. Yet history counsels caution in measuring performance using inputs rather than outputs. Over many decades, the Soviet Union spent 25% of its budget on agriculture while achieving notoriously ineffectual results (Brooks, 1990). The experience of state farms in Ghana and grand schemes in Tanzania and elsewhere suggest that spending targets may prove poor proxies for output growth (de Wilde, 1967; Eicher, 2003). A move to output targets, by both donors and national governments, would offer several benefits. It would focus attention on aggregate results in key priority commodities. In addition to signalling funding needs for critical support institutions, a focus on output performance would necessitate paying careful attention to the supporting policy environment and infrastructure necessary to effect significant growth in key commodity sub-sectors.

Use donor resources for strategic investments, not recurrent costs

Donor resources prove welcome in financing one-time investment costs, particularly expensive staff training in new skill areas such as biotechnology. They can also usefully assist in providing the financial glue necessary to link regional research networks.

But donor funding of recurrent operating costs for national research systems is fraught with danger. Where donors finance recurrent research on priority commodities, they condition governments to forget the need to shoulder their own priority investments. When the donor funds inevitably dry up, they leave gaping holes in priority areas along with anaemic farm lobbies unpractised in making necessary claims on their own governments. Among our case studies, withdrawal of donor funding has contributed to roller coaster rides in Kenyan, Zambian and Malawian maize research as well as in cassava research in both Madagascar and Zambia. As Eicher (2001) notes, 'Africa's experience has shown that erratic project aid can undermine the indigenous discipline that is needed to build fiscally self-sustaining research systems.'

A ban on donor financing of recurrent national research costs offers the prospect of a more stable flow of agricultural research funding. However, in an era where donors currently fund roughly 40% of recurrent agricultural research costs in sub-Saharan Africa (Pardey *et al.*, 1997), the forgoing of these external resources would require a substantial collateral boost in own government resources to agriculture. In July 2003, the African Heads of State committed themselves at the AU Summit to just such a resource boost when they endorsed the NEPAD proposal to boost agricultural spending to 10% of recurrent budgets (NEPAD, 2003) over the next 5 years, up from its current 6% level (Fan and Rao, 2003). A switch to domestic financing will require a boost in government financing in the short run but should strengthen capacity and sustainability of the research system over the medium and long term.

Build a political constituency for agriculture

Ultimately, national governments must shoulder the responsibility for increasing agriculture research funding. They must focus scarce public funds on priority commodities and provide remunerative salaries together with operational costs sufficient to maintain research and extension systems. Given acute pressures on already stretched budgets, this is a difficult prescription to fill. It cannot be tackled in isolation but will need to be addressed incrementally within the framework of overall civil service and budget reform. NEPAD's recent success in lobbying for increased government funding commitments for agriculture – from 6% to 10% of government spending – represents an important breakthrough on this score (NEPAD, 2003). To help build on this momentum and sustain it, the formation of articulate farm groups and lobbies will probably prove necessary to sustain government attention on agriculture over the long run in the face of myriad pressing concerns.

Policy makers must recognize that investments in agricultural research are not optional. Pests and diseases evolve continuously. The under-funding of domestic agricultural research constitutes unilateral disarmament. It guarantees sluggish agricultural performance. And in the absence of low, stable world food prices and a ready supply of foreign exchange, this translates into falling real incomes and growing poverty, in both rural and urban areas. Long-term support for agricultural research represents a necessary ingredient for sustained agricultural progress, in Africa as elsewhere.

Conclusions

In the end, the upswings on Africa's agricultural roller coaster suggest that sustaining

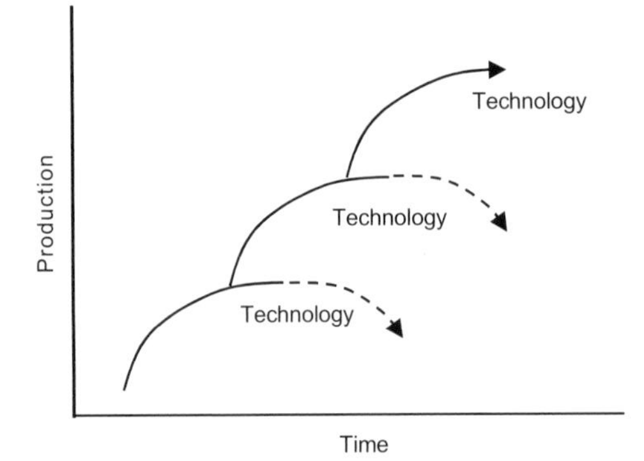

Fig. 8.5. Sustained technology change.

the upward trajectories will require two fundamental ingredients:

- High-level political commitment to agriculture, coupled with
- A sustained stream of technical innovation (Fig. 8.5).

Conversely, examination of the downswings reveals that the dips occur because of sputtering technological pipelines – often exacerbated by rapid mutation of pests and disease, donor departures and insufficient attention to soil fertility – coupled with vacillating policy regimes which translate into erratic institutional support and farmer incentives. These results, though derived from a different set of comparisons from those of the *Afrint* studies, converge in their common emphasis on the importance of political commitment and sustained technological advance.

Notes

[1] Several other organizations – including the International Water Management Institute (IWMI), Free University of Amsterdam, the University of Reading and New Partnership for Africa's Development (NEPAD) – have also commissioned investigations of successful episodes in African agriculture (see Reij and Steeds, 2003 and Wiggins, 2000).

[2] See Nweke (2003), Haggblade and Zulu (2003), Ahmed *et al.*, (2003), Ngigi (2003), Minot and Ngigi (2003), Smale and Jayne (2003), Kaboré and Reij (2003), Haggblade and Tembo (2003), Kwesiga *et al.*, (2003) and Place *et al.* (2003).

References

Ahmed, M.A.M., Ehui, S. and Assefa, Y. (2003) Dairy development in Ethiopia, environment and production. Technology Division Working Paper, forthcoming. Background Paper No.6. International Food Policy Research Institute and Successes in African Agriculture Conference, Washington, DC.

Arulussa Overseas Ltd (1997) *The Implications of Minimum Tillage as Practised by Cotton Producers: Final Report of the Cotton Production and Management Systems Survey*. Lonrho, Lusaka.

Beghin, J.C., Roland-Holst, D. and van der Mensbrugghe, D. (2002) Global agricultural trade and the Doha Round: what are the implications for north and south? Working Paper 02-Wp 308. Center for Agricultural and Rural Development. Iowa State University, Ames, Iowa.

Bingen, R.J. (1998) Cotton, democracy and development in Mali. *Journal of Modern African Studies* 36, 265–285.

Brooks, K. (1990) Agricultural reform in the Soviet Union. In: Eicher, C.K. and Staatz, J.M. (eds) *Agricultural Development in the Third World*. The Johns Hopkins University Press, Baltimore, Maryland.

Byerlee, D. (1994) Maize research in sub-Saharan Africa: an overview of past impacts and future prospects. Economics Working Paper 94–03. Mexico, D.F.: CIMMYT.

Carr, S. (2001) Changes in African smallholder agriculture in the twentieth century and the challenges of the twenty-first. *African Crop Science Journal* 9, 331–338.

Chigunta, F., Herbert, R. and Mkandawire, R. (2003) National environments for agricultural policy. Background Paper No.15. Successes in African Agriculture Conference, Pretoria.

Cleaver, K.M. and Schreiber, G.A. (1994). *Reversing the Spiral: the Population, Agriculture and Environment Nexus in Sub-Saharan Africa*. The World Bank, Washington, DC.

Coton et Développement (1998) Contresens et contre-vérités sur les filières cotonnières Africaines. No. 26

de Langhe, E., Swennen, R. and Vuysteke, D. (1996). Plantain in the early Bantu world. In: Sutton, J.E.G. (ed.) *The Growth of Farming Communities in Africa from the Equator Southwards*. Azania, Nairobi, pp. 147–160.

de Wilde, J.C. (1967) *Experiences with Agricultural Development in Tropical Africa*. The Johns Hopkins University Press, Baltimore, Maryland.

Dembélé, S. (1996) *Recherce cotonnière au Mali: presentation sommaire de quelques resultants saillants*. Institut d'Economie Rurale, Bamako.

Deutsche Gesellschaft für Technische Zusammenarbeit (GTZ) (1999) *Factoring HIV/AIDS into the Agricultural Sector in Kenya*. GTZ, Nairobi.

Diao, X., Diaz-Bonilla, E. and Robinson, S. (2003) *How Much Does It Hurt? The Impact of Agricultural Trade Policies on Developing Countries*. International Food Policy Research Institute, Washington, DC.

Diarra, S.B., Staatz, J.M., Bingen, R.J. and Dembélé, N.N. (2000) The reform of rice milling and marketing in the Office du Niger: catalysts for an agricultural success story in Mali. In: Bingen, R.J., Robinson, D. and Staatz, J.M. (eds) *Democracy and Development in Mali*. Michigan State University, East Lansing, Michigan.

Ebony Consulting International (ECI) (2001) *The Kenyan Dairy Sub-sector: a Study for DfID Kenya*. DfID, Nairobi.

Eicher, C.K. (1995) Zimbabwe's maize-based Green Revolution: preconditions for replication. *World Development* 23, 805–818.

Eicher, C.K. (2001) Africa's unfinished business: building sustainable agricultural research systems. Staff Paper No.2001–10. Department of Agricultural Economics, Michigan State University, East Lansing, Michigan.

Eicher, C.K. (2003) Flashback: fifty years of donor aid to African agriculture. Background Paper No.16. Successes in African Agriculture Conference. Pretoria.

Elwell, H., Chiwele, D. and Freudenthal, S. (1999) *An Evaluation of SIDA and NORAD Support to the Conservation Farming Unit of the Zambia National Farmers Union*. SIDA, Lusaka.

Evans, L.T. (1993) *Evolution, Adaptation and Yield*. Cambridge University Press, Cambridge, UK.

Evanson, R.E. and Gollin, D. (eds) (2003) *Crop Variety Improvement and its Effect on Productivity: the Impact of International Agricultural Research*. CAB International, Wallingford, UK.

Fan, S. and Rao, N. (2003) Public spending in developing countries: trends, determination and impact. EPTD Discussion Paper No.99. International Food Policy Research Institute, Washington, DC.

FAOSTAT data (2004) Food and Agriculture Organisation of United Nations. http://www.fao.org/waicent/portal/statistics_en.asp

Gabre-Madhin, E. and Haggblade, S. (2003) Successes in African agriculture: results of an expert survey. MSSD Discussion Paper No.53. International Food Policy Research Institute, Washington, DC.

Gabre-Madhin, E. and Haggblade, S. (2004) Successes in African agriculture: results of an expert survey. *World Development* (32) 5 745–766.

Gerhart, J. (1975) *The Diffusion of Hybrid Maize in Western Kenya*. International Maize and Wheat Improvement Center (CIMMYT), Mexico City.

Haggblade, S. (2004) *Successes in African Agriculture, Building for the Future: Findings of an International Conference*. InWEnt, Feldafing, Germany.

Haggblade, S. and Tembo, G. (2003) Conservation farming in Zambia. Environment and Production Technology Division. Working Paper, 108. International Food Policy Research Institute, Washington, DC.

Haggblade, S. and Zulu, B. (2003) The cassava surge in Zambia and Malawi. Background Paper No.9. Successes in African Agriculture Conference, Pretoria.

Hassan, R.M. and Karanja, D.D. (1997) Increasing maize production in Kenya: technology, institutions and policy. In: Byerlee, D. and Eicher, C.K. (eds) *Africa's Emerging Maize Revolution*. Lynne Rienner Publishers, London.

Hazell, P., Johnson, M. and Temu, A. (2003) Global environment for African agriculture. Background Paper No.17. Successes in African Agriculture Conference, Pretoria.

Herren, H.R. and Neuenschwander, P. (1991) Biological control of cassava pests in Africa. *Annual Review of Entomology* 36, 257–283.

Howard, J.A. (1994) *The Economic Impact of Improved Maize Varieties in Zambia*. Department of Agricultural Economics. Michigan State University, East Lansing, Michigan.

IFPRI (2003) Analyzing Successes in African agriculture: The DE-A-R Framework. Background Paper No.18. Successes in African Agriculture Conference, Pretoria.

Kaboré, D. and Reij, C. (2003) The emergence and spreading of an improved traditional soil and water conservation practice in Burkina Faso. Environment and Production Technology Division Working Paper, 116. International Food Policy Research Institute, Washington, DC.

Kwesiga, F., Franzel, S., Mafongoya, P., Ajayi, O., Phiri, D., Katanga, R., Kuntashula, E. and Chirwa, T. (2003) Improved fallows in Eastern Zambia: history, farmer practice and impacts. Environment and Production Technology Division Working Paper, 108. International Food Policy Research Institute, Washington, DC.

Legg, J.P. (1999) Emergency, spread and strategies for controlling the pandemic of cassava mosaic virus disease in East and Central Africa. *Crop Protection* 19, 622–637.

Mack, R. (1970) The great African cattle plague epidemic of the 1890s. *Tropical Animal Health and Production* 2, 210–219.

Manyong, V.M., Kling, J.G., Makinde, K.O., Ajala, S.O. and Menkir, A. (2000a) *Impact of IITA-improved Germplasm on Maize Production in West and Central Africa. IITA Impact Series.* International Institute of Tropical Agriculture (IITA), Ibadan, Nigeria.

Manyong, V.M., Dison, A.G.O., Makinde, K.O., Bokanga, M. and Whyte, J. (2000b) *The Contribution of IITA-improved Cassava to Food Security in Sub-Saharan Africa. IITA Impact Study*. International Institute of Tropical Agriculture (IITA), Ibadan, Nigeria.

McMaster, D.N. (1962) Speculation on the coming of the banana to Uganda. *Journal of Tropical Geography* 16, 57–69.

Minot, N. and Ngigi, M. (2003) Are horticultural exports a replicable success story? Evidence from Kenya and Cote d'Ivoire. Environment and Production Technology Division Working Paper, forthcoming. International Food Policy Research Institute, Washington, DC.

Morris, M.L. (2001) *Assessing the Benefits of International Maize Breeding Research: an Overview of the Global Maize Impact Study*. Part II of the CIMMYT 1999–2000 world maize facts and trends. CIMMYT, Mexico City.

NEPAD (2003) *Comprehensive Africa Agriculture Development Programme*. NEPAD, Midrand, South Africa.

Ngigi, M. (2003) The case of smallholder dairying in Eastern Africa. Environment and Production Technology Division Working Paper No. 118. Background Paper No.5. International Food Policy Research Institute, Washington, DC and Successes in African Agriculture Conference, Pretoria.

Norgaard, R.B. (1988) The biological control of cassava mealybug in Africa. *American Journal of Agricultural Economics* (May), 366–371.

Nweke, F. (2003) New Challenges in the cassava transformation in Nigeria and Ghana. Environment and Production Technology Division Working Paper, forthcoming. International Food Policy Research Institute, Washington, DC.

Nweke, F.I., Lynam, J.K. and Spencer, D.S.C. (2002) *The Cassava Transformation: Africa's Best-kept Secret*. Michigan State University Press, East Lansing, Michigan.

Oldrieve, B. (1993) *Conservation Farming for Communal, Small-scale, Resettlement and Cooperative Farmers of Zimbabwe: a Farm Management Handbook*. Rio Tinto Foundation, Harare.

Otim-Nape, G.W., Bua, A., Baguma, Y. and Thresh, J.M. (1997) Epidemic of severe cassava mosaic disease in Uganda and efforts to control it. *African Journal of Root and Tuber Crops* (abstract) 2.

Otim-Nape, G.W., Bua, A., Thresh, J.M., Baguma, Y., Ogwal, S., Ssemakula, G.N., Acola, G., Byabakama, B., Colvin, J.B., Cooter, R.J. and Martin, A. (2000) *The Current Pandemic of Cassava Mosaic Virus Disease in East Africa and its Control*. Natural Resources Institute, University of Greenwich, Chatham, UK.

Pardey, P.G., Roseboom, J. and Beintema, N.M. (1997) Investments in African agricultural research. *World Development* 25, 409–423.

Place, F., Franzel, S., Noordin, Q. and Jama, B. (2003) Improved fallows in Kenya: history, farmer practice and impacts. Environment and Production Technology Division Working Paper, 115. International Food Policy Research Institute, Washington, DC.

Plowright, W. and Ferris, R.D. (1962) Studies with rinderpest virus in tissue culture: the use of attenuated virus as a vaccine for cattle. *Research in Veterinary Science* 3, 172–182.

Provost, A. (1982) Scientific and technical basis for the eradication of rinderpest in intertropical Africa. *Revue Scientifique et Technique de l'Office International des Epizooties* 1, 619–641.

Reader, J. (1997) *Africa: a Biography of the Continent*. Vintage Books, New York.

Regmi, A. and Gehlhar, M. (2003) Consumer preferences and concerns shape global food trade. *Food Review, USDA* 24, 2–8.

Reij, C. and Steeds, D. (2003) Success stories in Africa's drylands: supporting advocates and answering skeptics. A paper commissioned by the Global Mechamism of the Convention to Combat Desertification. Center for International Cooperation, Vrije Universiteit Amsterdam, Amsterdam.

Sanchez, P., Shepherd, K.D., Soule, M.J., Place, F.M., Buresh, R.J., Izac, A.-M.N., Mokwunye, A.U., Kwesiga, F.R., Ndiritu, C.G. and Oomer, P.L. (1997) Soil fertility replenishment in Africa: an investment in natural resource capital. In: Buresh, R., Sanchez, P.A. and Calhoun, F. (eds) *Replenishing Soil Fertility in Africa. Soil Science Society of America (SSSA), Special Publication No.51. SSSA*, Madison, Wisconsin.

Schoenbrun, D.L. (1993) Cattle herds and banana gardens. *African Archaeological Review* II, 39–72.

Scott, G.R. (1985) Rinderpest in the 1980s. Progress. *Veterinary Microbiology and Immunology* 1, 145–174.

Simmonds, N.W. (1959) *Bananas*. Longman, London.

Smale, M. and Jayne, T. (2003) Maize in Eastern and Southern Africa: 'seeds' of success in retrospect. Environment and Production Technology Division Working Paper No. 97. International Food Policy Research Institute, Washington, DC.

Smaling, E.M.A., Nandwa, S.M. and Janssen, B.H. (1997) Soil fertility in Africa is at stake. In: SSSA (ed.) *Replenishing Soil Fertility in Africa: SSSA Special Publication No.51*. Soil Science Society of America, Madison, Wisconsin, pp. 63–80.

Staal, S., Delgado, C. and Nicholson, C. (1997) Smallholder dairying under transactions costs in East Africa. *World Development* 25, 779–794.

Tambi, E.N., Maina, O.W.L., Mukhebi, A.W. and Randolph, T.F. (1999) Economic impact assessment of rinderpest control in Africa. *Revue Scientifique et Technique de l'OIE* 2, 458–477.

Tefft, J. (2004) Mali's white revolution: smallholder cotton from 1960 to 2003. Background Paper No.10 presented to the NEPAD/IGAD Conference on Agricultural Successes in the Greater Horn of Africa, Nairobi, November 22–25, 2004.

Thirtle, C., Lin, L. and Piesse, J. (2003) The impact of research-led agricultural productivity growth on poverty reduction in Africa, Asia and Latin America. *World Development* 31, 1959–1975.

Topouzis, D. (2003) *Addressing the Impact of HIV/AIDS on Ministries of Agriculture: Focus on Eastern and Southern Africa*. FAO and UNAIDS, Rome.

University of Greenwich (2000) Saving a Nation Beseiged by the Cassava Mosaic Virus Epidemic. An application nominating the National Agricultural Research Organization of Uganda (NARO) for the King Baudouin International Development Prize. University of Greenwich, Greenwich, UK.

Van der Pol, F. (1992) *Soil Mining: an Unseen Contributor to Farm Income in Southern Mali*. Royal Tropical Institute, Amsterdam.

Wamwayi, H.M., Kariuki, D.P., Rossiter, P.B., Mbutiia, P.M. and Macharia, S.R. (1992) Observations on rinderpest in Kenya 1986–1989. *Revue Scientifique et Technique de l'Office Internationale des Epizooties* 3, 769–784.

Weatherspoon, D. and Reardon, T. (2003) The rise of supermarkets in Africa: implications for agricultural systems and the rural poor. *Development Policy Review* 21, 333–355.

Wiggins, S. (2000) Interpreting changes from the 1970s to the 1990s in African agriculture through village studies. *World Development* 28, 631–662.

9 The Role of the State in the Nigerian Green Revolution

Tunji Akande
Agriculture and Rural Development Department, Nigerian Institute of Social and Economic Research (NISER), Ibadan, Nigeria

Despite the growth of other sectors of the economy, Nigeria is largely an agrarian economy. While agriculture dominates the economy, food supply runs short of demand and over the years the state has intervened in various ways to initiate and propel a necessary agricultural development process.

This chapter explores the agricultural intensification process in Nigeria within the general framework of Green Revolution. It is a study of state interventionist policies in agriculture before, during and after independence. Emphasis is laid on the forces of change, within the contemporary realities of the Nigerian state. The chapter draws attention to the cacophony of policies implemented in agriculture with apparent lack of endogenous capacity to effect the desired changes in the agrarian system, that is, the lack of adequate resources for the state and the ill-preparedness of the farming communities for the new changes. The contemporary food production efforts of the state are motivated by the desire to locate food supply within the domestic economy, curtailing food imports and saving foreign exchange earnings for other development needs, and building a food-secure nation.

The chapter is in five parts. The introductory part is followed in section two by a consideration of the national economy and agricultural sector within the purview of the state rationale for intervention. Section three traces and analyses the evolution and nature of state intervention in agriculture, particularly the trajectory of state involvement since 1960 to the present time. In section four, we consider the limits of public policy gauged by pitching achievements against efforts and resources committed to agricultural development. The concluding part summarizes the main findings, identifies the challenges which agricultural planning still faces and considers how these can be addressed to generate desired results.

Conceptual Issues

Drawing on Asian experiences, the Green Revolution model explored emphasizes that a Green Revolution is a state-driven, market-mediated and small-farmer-based process of agricultural development. In a broader context the Green Revolution is a process of agricultural intensification aimed at increasing domestic self-sufficiency and security in food production, stimulated by state policies which encompass: (i) protection of the domestic market against dumping and low-priced imported food; (ii) facilitating the extension of technology required for farmers to increase their production (seeds, fertilizers, irrigation, etc.); and (iii) price guarantees for

(eds G. Djurfeldt, H. Holmén, M. Jirström and R. Larsson)

farm-gate prices and/or subsidies of input prices, stimulating producers to adopt new technologies.

Similarly to the Asian type, the Nigerian Green Revolution may be perceived as a series of activities and processes inspired, initiated and executed by the state and directed at making the nation achieve self-sufficiency in staple commodities. The Nigerian staple foods are cereals (sorghum, millet, maize and rice), roots and tubers (yam and cassava) and legumes (cowpeas). All these crops are relevant in the context of agricultural transformation in Nigeria.

The primary agent of Nigeria's Green Revolution is the smallholder farmer, cultivating generally less than 5 ha of farmland. The small farmer is designated the 'centre piece' of the Nigerian agriculture because of the predominance of small scale farming and the realization that the small farmer is an efficient and competitive producer (Olayide *et al.*, 1980).

While the Nigerian Green Revolution is market-mediated, the market has not been allowed to operate unfettered. The state is heavily involved in the input market and has played a major role in the provision, distribution and pricing of inputs, particularly fertilizer. The high-yielding crop varieties being introduced require these inputs to achieve optimal yield performance, but the inputs are seldom available in sufficient quantities and at moderate prices. The Nigerian state also operates in the output market, albeit not in such an elaborate way as it does in the input market. The state serves as the buyer of last resort when the market cannot absorb all quantities of commodities placed in the market.

The Nigerian Green Revolution is a continuing process. Since self-sufficiency has not been achieved in any of the food commodities, the state maintains an obligation to interfere in agricultural development in spite of the effort to make the economy private-sector driven. The attempts being made by the state reflect the concern with which the political class views the problem of agriculture and the urgency to make amendments that could lead to improved performance.

The Nigerian state and society

With the amalgamation of Southern and Northern Nigerian protectorates, the colonial state was established by the British in 1914. The principal organs of government were effectively under the control of officials from the imperial state. However, the native system of chiefs and emirs provided a collaboration to ease the colonial governance system and strategy of penetration. For nearly 50 years, the largely agricultural and mineral-based economy was structured to provide raw materials for the industries in the metropolis while also serving as markets for the manufactured products from Britain until political independence was granted in October 1960. For most of its 43 years of existence as an independent country, Nigeria was under military dictatorship, but since May 1999 she has come under a participatory democracy and a market-orientated economy. The new Nigerian leadership is committed to the promotion of food self-sufficiency and food security as priority items on its agenda.

A multi-ethnic society, Nigeria is the most populous nation in Africa, having a population of well over 130 million people who belong to about 350 ethnic groups (Otite, 1990). Each of these groups exhibits unique dietary characteristics and food preferences, with the result that a diverse array of food commodities and food forms pervades the Nigerian food consumption landscape. The huge population, ethnic diversity and pluralism of culture put considerable pressure on the agricultural system to provide the kinds of food and materials required by different groups and classes of the population.

Agriculture in the economy

Agriculture is the largest single sector of the economy, providing employment for a significant segment of the workforce and constituting the mainstay of Nigeria's large rural community, which accounts for nearly two-thirds of the population. The proportion of the gross domestic product (GDP) attributable to agriculture hovers between 30% and 40%, well ahead of mining and quarrying, as well as wholesale and retail trade, which are the other two major contributors to the country's GDP. While agriculture remains dominant in the economy, the food supply does not provide adequate nutrients at affordable prices for the average citizen. For instance, the daily per capita protein intake from animal sources is under 7 g/day, far below the FAO recommended level of 27.2 g per capita/day (Sahib *et al.*, 1997:148). Also, calorie intake is said to be under 2600 calories per capita/day (Igene, 1991).

From the 1970s the Nigerian food sector has been characterized by excess demand over supply due primarily to high population growth rates of about 3.2% per annum, high rates of urbanization, and rising per capita income, stimulated by an oil export revenue boom. Consequently, the pattern of food consumption has been changing rapidly in terms of quantitative and qualitative adaptations to new food preferences and consumption habits.

The principal component of Nigerian agriculture, the crop sub-sector, contributes annually about 30% of the total GDP and about 80% of the agricultural GDP. It also accounts for about 90% of the farming population. By its share size the crop sub-sector provides the bulk of agricultural income. It is the focus in this study.

Nigeria is greatly endowed with abundant natural resources for sustainable growth of crops, particularly its land and climate. About 57% of the total land area is either under crops or pasture and land use involves three broad systems of production namely, rotational fallow systems, semi-permanent or permanent production system and mixed agriculture.

Rotational fallow agriculture, which is common in the south-western and north central parts of Nigeria, is a 'low' form of intensification involving the cultivation of land on a rotational basis under a shifting cultivation system (Boserup, 1965). Semi-permanent or permanent agriculture are progressively intensive systems, where a small piece of land is ultimately cultivated on a continual basis (Boserup, 1965). This system is usually found in densely populated areas of eastern Nigeria, and supported by deliberate spreading of household refuse, animal droppings and ash on the land. Mixed agriculture is an integrated crop/livestock system combining the rotational grass fallow system of crop production with animal husbandry and predominates in the Guinea and Sudan savannah regions of northern Nigeria.

Nigeria has a tropical climate, which favours the production of a variety of industrial and food crops. The amount, incidence and variations of rainfall largely explain the differences in cropping patterns and farm management practices in various agroecological zones of the country. Based on its climate, the country has three distinct ecological regions, which include the humid, sub-humid (with highlands) and semiarid regions (Agboola, 1979). The humid region is in the southern forestry ecology, covering about 1.9 million ha of the country's land area. The region carries the highest population density – about 235 persons/km^2. Tree crops produced here are cocoa, oil palm, rubber, kolanut, citrus and plantain. The major arable crops are roots and tubers (yam, cassava and cocoyam), cereals (maize and rice) and grain legumes or pulses such as cowpea and pigeon pea. Small farms are more predominant in the humid zone where thick forests and high population densities make opening up of large farms very costly and almost impracticable in several areas.

The sub-humid region lies to the north of the humid region and it is the largest, occupying nearly 40 million ha of Nigeria's land area. More than 50% of this is uncultivated land. It is a sparsely populated region, with about 107 persons/km^2. This is the famous 'Middle Belt', which is more aptly described as the 'food basket' of Nigeria on account of the array of crops and the intensity of cropping activities going on in the area. The crops here include yam, cassava, sweet potatoes, sorghum, maize and rice as well as cowpea, soybean, groundnut and onion.

The semiarid region occupies the northernmost parts of the country and covers about 33 million ha of land, stretching from the extreme west in Sokoto to the extreme east in

Lake Chad. It embraces the Sudan and Sahel savannah vegetation zones. Average annual rainfall is about 750 mm and may be as low as 200 mm in its northern limits. With a crop-growing season of between just 100 and 150 days, irrigation is required to cultivate a variety of crops, especially maize, wheat, millet, sorghum, cowpea and groundnut.

The State, Agriculture and National Economy

Role of agriculture in Nigeria's development

Policy makers have over the years expressed the fundamental role of agriculture in Nigeria's development process as consisting of five main functions. First, agriculture is expected to provide food in sufficient quantities and quality for the rising population. The ability to achieve the food need purely from domestic sources enhances the credibility of the Nigerian state and its capability, particularly in eliminating importation, which may imperil the nation's external accounts position. However, a state of autarky is not envisaged as trade is expected to make up for the most essential products that cannot be sourced from domestic effort. The food supply function of agriculture is thus interpreted in terms of self-sufficiency in basic food staples, where Nigeria has clear comparative advantage and where increased production can be achieved.

Second, agriculture is expected to provide raw materials on which to base industrial processing and manufacturing. This is the linkage function expected of agriculture, whereby the fibre output could constitute the base for industrial expansion and development. Historical experiences of Europe and North America indicate that the development of agriculture preceded and actually facilitated industrial revolution and development. However, most of the African countries, including Nigeria, have attempted to precede their agricultural development with industrial expansion ostensibly to catch up quickly with the rest of the world. It has now been realized by leaders in Africa that agricultural growth is *sine qua non* to industrial growth. Consequently most of the countries in Africa, including Nigeria, have begun to pay particular attention to the agricultural sector, hoping to create the necessary linkage between agriculture and industry.

A third function nursed by the agricultural sector is the generation of employment. This is not simply primary occupation in farm level production, but also occupation in value-added activities associated with raw material processing and utilization. Since agriculture houses the bulk of the labour force, most of which is unskilled, the development of the quality of labour itself is a prerequisite for improved agricultural performance. A largely uneducated agricultural population may not be able to adopt improved practices nor apply managerial know-how suitable for achieving agricultural productivity.

The fourth function of agriculture in Nigeria's development process is the generation of income to farm workers and others in postharvest operations, including manufacturing. With expanding agricultural output and availability of surpluses, Nigeria attempts to make her agriculture contribute significantly to development through exports of agricultural products with substantial value added and earnings in foreign exchange.

Thus, food provision, raw material supply, employment generation, income and foreign exchange earnings constitute the instrumental value of agriculture to Nigeria's development. The agricultural sector is also significant for poverty alleviation and equity considerations.

Rationale for state intervention in agriculture

A yawning gap has existed between two schools of thought on the role of the state in economic development. While one school of thought advocated a more activist role, the other favoured more residual and subdued engagement. The debate has often been heated but muddled by the ideological learnings of the protagonists – capitalism, socialism and mixed economic management – during the time when these political systems competed for space and ascendancy. Karl Marx's radical writings seem to have influenced many countries in the first half of the last century. For instance, a large number of East European and newly independent countries displayed faith in central economic planning with the government being the motivator and engine of economic growth. In some cases, the belief in central planning was corroborated by performance outcome and growth was accompanied by a certain degree of equity as incomes were relatively equally distributed and the welfare state, in spite of its inefficiencies, ensured that ordinary people had access to basic goods and services (World Bank, 1996). This tended to strengthen the belief in the efficacy of central planning as a desired development paradigm. The achievements in centrally planned economies were not lost on the Africans, who were somewhat estranged from most capitalist countries during the struggle for independence.

The role of the Nigerian government in agriculture is specifically predicated on the situation prevailing at independence, which promoted state intervention. For instance, agriculture provided nearly two-thirds of government's revenue and foreign exchange earnings in the 1960s. Also, about 70% of the population lived on agriculture. However, the agricultural sector was characterized by little growth of output per capita, low productivity, pervasive illiteracy, static and poorly developed institutions, restrictive markets and unprogressive policy stance. The government was unable to ignore these challenges and saw agriculture as the fulcrum around which the entire national socio-economic development should revolve. Consequently, agriculture was perceived as having a significant role to play in the development process and government intervention was seen as desirable.

Initial conditions at independence and development path chosen

Nigeria emerged from the British colonial rule as an independent federation in October 1960. The political and administrative structure consisted of three regional governments (Northern, Eastern and Western Regions) and the Federal Government. The federal constitution placed agriculture in the concurrent list, meaning that the regions and Federal Government had roles to play in the development of the sector. Each region developed its agriculture independently, laying emphasis on the most important export crops available in its domain, that is, cocoa in the Western Region, palm produce in the Eastern Region and groundnut and cotton in the Northern Region. All the constituent parts of the federation adopted policies favouring plantations and transformatory production processes, which involved the setting up of farm settlements. The colonial policies inherited were not disposed of.

However, soon after independence serious economic and political problems culminated in a civil war between 1967 and 1970. The war disrupted agricultural production and regional trade in food commodities in two principal ways. First, in the theatre of war, that is, the Eastern Region of Nigeria, agricultural land could not be put under cultivation due to battles. Second, agricultural workers abandoned the farms and enlisted in the war efforts. This created labour shortages, a situation which was further worsened by the migration of young school leavers to towns and cities in search of white-collar jobs. In addition, existing markets and trade channels in food commodities were obstructed, creating a disincentive to regional trade and food distribution from areas of abundant supply to areas of need. The chilling experience of lack of food in the beleaguered Biafra during the war aptly demonstrated the disruptions in the regional flow of food commodities.

On the economic front, emphasis was placed on import-substituting industrial development as the framework for economic development. The problem here is that the agricultural revolution, which could provide the basis for industrial development, was not accorded priority consideration in economic development planning. Agriculture was important only to the extent that it provided development funds through foreign exchange earnings of raw agricultural exports; it was not considered as the likely driver of the much-promoted industrial development.

The agricultural strategies pursued immediately after independence were akin to the strategies of the colonial rulers, who had patterned the economy after the needs of the imperial state. The new leaders placed emphasis on export commodities, such as cocoa, groundnut and palm produce, which were important foreign exchange earners. Food crops hardly featured in planning, ostensibly because food supply did not constitute a problem at the time and food export was neither contemplated nor thought feasible.

The strategic approach can be labelled transformatory and focused on the creation of intensive farming systems and establishment of large-scale farms (Olatunbosun, 1971). This took the form of large-scale farms and settlement schemes with the major proportion of the capital investment coming from the public sector and where farmers submitted to a regime of discipline, in terms of the crops they could grow and the husbandry production practices they could adopt.

The first national development plan (1962–1968) launched by the Nigerian state incorporated the transformatory concept in the agricultural strategy. Consequently, the bulk of agricultural investment in all the three regions went to the establishment of plantations and farm settlements, primarily for export crop production. The small farmer and producer of the nation's staples (yam, cassava, sorghum, etc.) was never considered relevant in the extensive plan to transform the agricultural sector.

To a considerable extent, agriculture in the first decade of the transition to independence witnessed the continuation of the colonial strategy of accumulation and exploitation of the Nigerian peasant farmers to provide development funds, through taxes and payment of commodity prices that were inferior to the ruling prices in the international market. Government withdrawals from producers' income were estimated to vary between 20% and 25% during the 1960s (Hill, 1972). The entire amount was committed to 'development', with little or nothing going to the rural sector to help the peasants. The instrument for generating these surpluses remained the marketing boards.

With the military taking over power in 1966, the era of pan-territorial agricultural policy began. This was facilitated by the command structure in the military. The advent of the military also marked the beginning of the dismantling of native authorities and their replacement with local, formal authorities. The courts, police and the prisons were all brought under Federal Government supervision. The power relations were gyrating to the centre, even though the states and local government authorities had specific roles in the constitution, which usually is the first casualty in a military *coup d'état*.

In general terms the development of agriculture during the first decade of independence was perceived only in terms of export crops, while food crops received scanty attention. The Green Revolution efforts at this period were not dispensed to make knowledge, inputs and marketing opportunities available to staple crop producers but to enhance the productive capacity of export crop producers. The policy makers did not see the apparent discriminating practice as having any long-term repercussions on the ability of the nation to feed itself and achieve self-sufficiency in its most readily consumed foodstuffs.

Ascendancy of petroleum and its challenges

As early as 1973, there was a tremendous turnaround of the Nigerian economy, triggered by the dramatic increases in the price of crude petroleum in the international market. Price increases were occasioned by the Arab–Israeli war and the response of the Organization of Petroleum Exporting Countries (OPEC), of which Nigeria is a member. The OPEC quota for Nigeria during the 1970s stabilized at around 2 million barrels/day. This translated to unprecedented export earnings of about US$5 billion in 1975 and nearly US$14 billion at the end of the decade. The petroleum component of export earnings was above 90%, dwarfing the contributions of agriculture, which was once the main foreign exchange earner. With such a huge revenue accruing from petroleum exports alone, the Nigerian development problem became not the lack of investible funds, but of identifying growth poles and investment opportunities which would have significant impact on the entire economy and the welfare of the people. The petroleum earnings strengthened the role of the state, which assumed a commanding control and influence on the economy, engaging in virtually all areas of the economic system including production, services, manufacturing and security. The major instruments of control were the macroeconomic and sectoral policies enunciated to guide the conduct of economic agents. The investment behaviour of the state also changed in the face of unprecedented resources.

In spite of the huge revenue accruing from petroleum, public spending did not favour agriculture. Less than 5% of total government expenditure went to agriculture and urban areas received greater financial attention than agriculture and rural development (NISER, 2001). Besides, expenditure on agriculture fluctuated in both nominal and real terms with significant implications for the development of this sector of the national economy.

Other problems, traceable to the fortunes brought by petroleum, also emerged. The minimum wage was raised to an unprecedented level following a review of salaries and wages. Urban wages rose by over 400% and agricultural wages also soared, rising by about 450%. Nevertheless, rural–urban migration of young people was enhanced, following job openings in construction works and opportunities in the urban informal sector. High urban population growth rates created a demand for more food. Petroleum earnings were found handy to prosecute food imports to satisfy the potentially restive urban population. Both the quantity and value of food imports increased significantly during the 1970s. The percentage of food in total imports bill rose from about 7.6% in 1970 to 19.8% in 1983. In value terms, food imports cost about US$130 million in 1971 and about US$1.8 billion in 1981 (Watts, 1987).

Meanwhile, government was gradually becoming aware of the impending food crisis as a result of three main factors. As noted by Sano (1983), an FAO 1966 report emphasized the long-term problems of food production and the need for Nigeria to effect production plans for both food and export crops. In a similar vein, in 1969 an American consortium from the Michigan University for the study of Nigerian Rural Development (CSNRD) had emphasized the need for Nigeria to jettison the transformatory strategy of large-scale plantations or farm settlements while recommending that the country should concentrate efforts on smallholder farming. Second, there was an unexpected natural disaster, the Sahelian drought of 1972–1974, which caused untold production setbacks and led to local shortages of staple foods. A third factor was the rapid urbanization fuelled by rural–urban migration and a large number of demobilized soldiers after the war. Farming was identified as a likely haven for unskilled labour that was now available.

Emergence of food crisis

Nigeria's food crisis has both remote and immediate causes. The remote causes took their roots in the colonial management and strategy of accumulation, which was reinforced by the acceptance and continuation of this strategy by the founding fathers of independent Nigeria in 1960. More immediate causes included the civil war, with the attendant dislocation of agricultural production. By the time the country emerged from the war as an indivisible entity, the Sahelian drought overtook the nation, decimated livestock and obstructed food crop production and once again exposed the fragile structure of the Nigerian agricultural system. Although petroleum exports brought wealth to Nigeria in the 1970s, the consequences were very negative for farming.

During the 1960s and 1970s the indices of agricultural production revealed a tendency to stagnate, as Nigeria fell behind from the perspective of the world agricultural output and also in the performance of the African region. Increased food output had all along been achieved through an expansion of area cultivated. However, with the probable exceptions of rice and cowpeas, there had been no significant increases in area cultivated in the 1960s and 1970s. Similarly, yield had remained stagnant and output had also been generally low except for rice. The situation was not too different in later years.

The food output projections for the period 1985–2000 show huge national food deficits for all staple commodities (cereals, roots and tubers and legumes), with self-sufficiency ratios falling below 80% in most cases. Similar findings were obtained by Ajakaiye and Akande (1999) in their projection of food supply and demand for the period 1996–2010. There have also been cases of regional and seasonal food deficits as happened, for instance, in the Sahelian droughts of 1973 and 1980 and the recurring phenomenon of cassava output shortages for several years (Akande, 2002). At the household level, the rising food price inflation coupled with declining per capita income has meant inadequate feeding. The composite consumer price index for food had risen from about 348% in 1976 to 1180% in 1994 (CBN, 1994). The nominal farm-gate prices of various food commodities were simply horrendous, particularly between 1990 and 1995. The average rural market prices of maize, for instance, had increased 75 times during this period.

With respect to household income the average annual per capita income was about $1500 in both the 1960s and 1970s. In the 1980s this fell to below $1000 and fell further to about $300 in the 1990s. Today, Nigeria is ranked among the poorest countries in the world, with per capita annual income hardly reaching $300 in spite of its petroleum-dominated economy. Indeed, the 1990 National Demographic and Household Survey conducted by the Federal Office of Statistics (FOS) showed that malnutrition was a serious national problem, with about 43% of Nigerian children under 5 years of age having stunted growth, 36% being underweight and nearly 10% being wasted (Fos, 1993).

Food imports to make up for the deficits in domestic food supply also illustrate the food crisis Nigeria has faced over the years. Imports became quite significant after the mid-1970s as wheat, rice and maize dominated food commodity imports. Whereas annual rice imports barely reached 3000 t, in the 1970–1975 period, imports of the commodity climbed to nearly 320,000 t in 1976–1980 and higher still to about 390,000 t annually in 1980–1985. Maize imports showed a similar pattern before the embargo imposed on maize and rice imports in 1985 dampened the inflow of the commodities to the country since that year (see below). However, the country has resumed importation of rice for which the sum of over US$600 million is now spent annually (Akande, 2003).

Evolution of State Intervention in Agriculture

Prior to independence, the colonial state engaged agriculture in three principal ways. First, the colonial state was strongly committed to the development of smallholder peasant agriculture and opposed large-scale agrarian capitalism. To ensure this, no producer, native or foreign, could have a plantation or farm unit that exceeded 1200 acres. The alienation of land or land rights, which was freely practised before British occupation, was stopped and it became an offence to sell land. The colonial land policy in general reinforced the smallholder structure of the agrarian system. The second line of colonial engagement in agriculture was in the provision of infrastructure. Roads and railways to the hinterland were constructed so that produce could be evacuated more easily to the ports for subsequent exportation to the metropolitan industries. The third form of engagement was the application of colonial science to peasant production as demonstrated by attention paid to research activities aimed at changing the agrarian landscape.

The emergent African leaders at independence continued with the colonial policy of accumulation and peasant extraction. The three regional governments were preoccupied with generating development revenue through the marketing boards. The state was spending the revenue extracted from agriculture through produce and export tax to run the machinery of government, with only a very small fraction going to agriculture to run state-owned farm settlements, which were established to demonstrate and import new farm technologies to school leavers expected to be the vanguard of a new generation of educated farmers. However, as a result of their own quarrelsome and political disagreements, the pioneer leaders were not able to consolidate effectively the direction of the economy, which led to military seizure of state power in 1966. The first major action of the military was to begin the systematic dismantling of the existing state structures including the Native Authority system and a move towards centralization of state power. Agricultural policies then began to lose the regional specificities and assumed pan-territorial posture. The regions and the states that were created later had to follow the direction dictated by the Federal authorities in agricultural development initiatives.

The vision of agrarian transformation nursed by the military was delayed by the civil war. While the war lasted, between 1967 and 1970, the state was engrossed in maintaining peace and stability and endeavouring to 'Keep Nigeria One'. All sectors of the economy were, therefore, mobilized to support the war efforts, with agricultural exports providing the bulk of the funds needed by the state.

The period 1960–1970 may be regarded as a period of agrarian limbo. Not much could have happened because of the fluidity of the political scene and the civil war that ensued. However, the agrarian system of trade was firmly in the hands of the state, which exploited it to finance its posture of 'developmentalism'. The transformational strategy adopted to change the agrarian system was not effective because the farm settlements could not be replicated to cover most farming communities and because of lack of social amenities in the rural areas, which could have served as a stabilizing factor for young farmers to stay in the rural areas rather than migrate to the cities the way they did.

State intervention and agrarian change

Beginning in 1970, with the war over and a new era of peace and stability in place, the Nigerian state began a remarkable agrarian revolution, which lasted until 1985 (Ekong, 1986). This 15-year period could be regarded as the 'golden age' of a new agrarian adventure in which the state was the moving spirit. The financial capacity of the state was strengthened by the oil-boom. With money for investment not constituting a constraint to development, the state felt challenged to develop the agricultural sector. The official position on the best way to increase productivity in agriculture was to introduce modern technology and skill into the sector.

Thus, in 1972 the state introduced the National Accelerated Food Production Programme (NAFPP), which was a technology-based programme designed to make all inputs available to farmers across the length and breadth of the Nigerian Federation. The programme focused on six staple crops – rice, maize, millet, sorghum, cassava and wheat. The International Institute of Tropical Agriculture (IITA) was the technology pathfinder for the programme in the areas of research, service development and input delivery. In order to complement the technology approach, the state embarked on capitalization of agriculture by simultaneously establishing the Agricultural Credit Guarantee Scheme Fund (ACGSF) under the Central Bank, and the Nigerian Agricultural and Cooperative Bank (NACB) to lend directly to smallholder farmers. As it turned out, the beneficiaries of the technology and capitalization process were largely the urban-based capitalist farmers to be found in the civil service and the military, particularly the elite group who showed interest in poultry production. Even though the focus of the agrarian change was the peasant smallholder, less than 20% of the NACB loans between 1973 and 1992 actually reached smallholders (Mustapha, 1998).

The state also intervened in the marketing system in a number of ways. First, the Guaranteed Minimum Price Scheme (GMPS) was established to assure food crop farmers minimum prices and to purchase their products if the open market faltered. The state also established the National Grain Reserve Scheme (NGRS) and constructed silos in different parts of the country as an aspect of food security. Second, state supply companies were set up to trade in rice and meat to dampen food shortages before the situation developed into a crisis. Rice imports reached over 500,000 t as early as 1982 from a level of less than 2000 t in the late 1960s. Maize imports, in a country which had all the conditions for producing the crop, reached about 350,000 t in 1982. A third action was the scrapping of marketing boards and their replacement with commodities boards, which now included one for the food grains.

The state in the mid-1970s also forayed into direct production. Large-scale, capital intensive, mechanized plantations were set up under state-run companies including the National Grains Production Company, the National Roots Production Company, the Bacita and Savannah Sugar Companies, the National Beverages Company and the Bauchi and Western livestock companies. The state ran into trouble in all these ventures. As a result of inept management and corruption, not one single company was run profitably. They all had to be discontinued. Meanwhile, there was an emergence of large-scale individual farmers ready to replace the peasant production system with a capitalist system. These were generally wealthy, well-connected and city-based 'farmers' with no other interest than land speculation and appropriation of the benefits accompanying such other state policies as land nationalization and input subsidy schemes.

Land nationalization came with the Land Use Decree enacted in 1978. Prior to land reform different land regulations operated in the regions of the Federation. In the Northern Region, land belonged to the state, but in the Southern Region various communal, family and individual ownership practices operated. What the Land Use Decree did was to normalize land ownership throughout the country as a means to ease access to land for large-scale production. It turned out that highly placed and influential individuals in the society and bureaucracy used this policy to help themselves to more than their fair share of state land. In addition, the policy led to an emergence of a bimodal structure of production with a capitalized, highly technical and skilled modern sector existing side by side with the smallholder peasant system. Although the modern sector was able to attract most of the privileges emanating from the state, it could contribute only 5% of the total agricultural production (FOS, 1982).

In order further to strengthen the state's effort to modernize agriculture, a generalized subsidy policy covering seeds, fertilizer, herbicides, pesticides, mechanization, fishery and livestock inputs, credit and a host of services was instituted. Most of the special programmes initiated by the state to boost food production since 1970 had relied on subsidy, which was intended to improve the competitiveness and productivity of farming. The level of subsidy ranged between 75% and 100% during most of the 1970–1990 period (CBN/NISER, 1991), while the share of the annual budget going to subsidy hovered between 10% and 33% in the late 1970s to the early 1980s. The problem with the subsidies was not just the burden imposed on the treasury but the beneficiaries were not those intended. Instead the élite cornered the supplies and resold them to farmers at much higher prices than stipulated by the state.

The implementation of technology-based programmes

Research and extension as an aid to increased agricultural production was fully controlled by the state. Public agricultural research institutes (ARIs) number more than 20 and are spread across geopolitical and agroecological zones of the country. Seven ARIs are concerned with advancing improved technologies such as high-yielding crop varieties, extending new knowledge to the farm scene, and controlling for pest and diseases. The ultimate research goal is to achieve high productivity. However, in spite of a series of technological innovations introduced in the late 1970s and the 1980s, and despite farmers' willingness to adopt them, adoption rates have been quite low due largely to non-availability of needed inputs, particularly fertilizer (Falusi, 1986; Erinle, 1994; Idachaba, 2000).

The two most conspicuous technology-based programmes, which have featured prominently in the process, are the River Basin Development Authorities (RBDAs) and the Agricultural Development Projects (ADPs). The RBDAs focused on developing irrigation agriculture in Nigeria and were fully state-controlled. At the take-off in 1976, 11 RBDAs were established in order to develop the agroeconomic potential of available water bodies. The capital allocation to agriculture in the Third and Fourth National Development Plans of Nigeria, 1974–1985, was expended mostly on RBDAs. This intervention was expected to reduce the risk associated with the vagaries of rainfall. The RBDAs engaged in several functions, including development of water resources for irrigation and domestic water supply, control of floods and erosion and watershed management. Later, additional responsibilities of direct food production and rural development were added and the number of RBDAs shot up to 18 (Ayoola, 2001). In 1985, the number was reduced to the former 11 following the creation of another agency, the Directorate of Food, Roads and Rural Infrastructure, which effectively assumed responsibility for rural development activities. At the end the RBDAs were able to make only a modest contribution to the Green Revolution effort of the state. Out of 536,282 ha envisaged to be irrigated in 1985, only 82,305 ha were executed. Forrest (1993) has roundly criticized the RBDA as being a total failure and very costly and inefficient, a position shared by Andræ and Beckman (1985).

The ADPs undoubtedly constituted the most effective programme of the state in the agricultural sector. The ADP was an integrated approach combining technology, extension services, physical inputs, market and infrastructure targeted at promoting knowledge-based agriculture of smallholder producers. The target was self-sufficiency and remunerative farming. The ADPs started as enclave projects and were later expanded to cover whole states. Alkali (1997) has indicated that up to 1986 the ADPs set up 601 farm service centres and constructed nearly 5500 rural roads, 101 dams, 3632 boreholes and 1661 wells. Evbuorhwan (1997) acknowledged the achievements and performance of the ADPs, which were said to have made significant impact on the host communities, particularly reducing poverty by raising the productivity and income of participating farmers. However, the ADP strategy has contributed significantly to Nigeria's external debt because the projects were financed largely through World Bank loans (Williams, 1999).

Structural adjustment and modifications of food policies

By the early 1980s the Nigerian economy had virtually collapsed, with an adverse balance of payments position, severe unemployment, low capacity utilization in the manufacturing sector, negative growth in the agricultural sector, rising inflation and a general depression in the quality of life of the people. The state responded with a battery of fiscal, monetary, exchange rate, trade, wages and incomes policies to reverse this trend. Adopting the liberalization programme – Structural Adjustment Programme (SAP) – in July 1986, the state decided to allow the market forces to guide the conduct of economic exchanges and management. For agriculture, the stated aim of SAP was the elimination of administrative price distortions within the economy and the reallocation of resources in favour of smallholder agriculture. The local currency, the Naira, was allowed to float, while the commodity marketing boards were discontinued. There was commitment to liberalize agricultural input markets, remove input subsidies and stop the concessionary interest rates. The elimination of input subsidy had an immediate impact as farmers faced input price increases of about 300% (CBN/NISER, 1991). The resource-poor farmers were hard hit as they could no longer access fertilizer and pesticides.

Compensatory measures to dampen the harsh impact of SAP on welfare were found in the establishment of the Directorate of Food, Roads and Rural Infrastructure (DFRRI), the

People's Bank, Community Banks, the National Agricultural Land Development Authority (NALDA) and the Fadama Farming Scheme. DFRRI was charged with the responsibility of improving living conditions and economic opportunities in the rural areas while the People's Bank and Community Banks were to make loans available to farmers. NALDA was formed to assist farmers in land preparation and to provide intermediate technology.

The overall impact of SAP on agriculture took various forms. There was an immediate decline in public expenditure on agriculture. The state investments in agriculture in real terms were N23 million in 1970, N629 million in 1985 and N101 million in 1990. Average annual growth rates in total public agricultural expenditure were about 69% in the period 1970–1974, –7.5% in 1975–1980, 5.9% in 1981–1985 and –27.5% during the SAP period, 1986–1990 (Olomola, 1998). The contraction in state expenditure on agriculture and rural development led to corresponding decline if not total collapse of farm service centres, water schemes and rural electrification. Farm labour was expensive and child labour became increasingly prominent. Dike (1998) has opined that SAP promoted a process of agrarian decapitalization in Nigeria. The growth rate in capital stock in agriculture was said to have declined from 10.8% in 1980–1986 to –18% in the 1987–1992 period as tractors and other agricultural equipment depreciated and were never replaced because of astronomical prices. The state was also unable to fund its research institutes at the level it had done 10 years earlier (Sahib *et al.*, 1997). However, SAP was not all hardship. The programme produced some positive impact, especially on food production, which recorded significant growth rates (Table 9.1).

From the beginning of 1994, with the change in government, new policy measures were initiated to replace or at least repair the damages inflicted by SAP. Fertilizer procurement and distribution were to be handled by local governments and not by federal or state governments, as was the case before. The Family Economic Advancement Programme (FEAP) was initiated to stimulate and finance micro-enterprises including cottage industries and agricultural processing. The Family Support Programme was also initiated and focused on alleviating the hardship being faced by the citizens. The entire policy framework of this era was tagged 'guided deregulation', which implies a shift from wholesale liberalization to measures that addressed specific policy objectives.

Table 9.1. Growth rates of total food production in Nigeria, 1970–1993. (Source: Federal Government of Nigeria, 1997:132.)

Period	% Growth	Period	% Growth
1970–1975	–1.74	1986–1990	14.55
1976–1980	–4.41	1991–1993	5.71
1981–1985	9.30		

Post-adjustment policies

In the early 1990s, it was clear that SAP's impact on the economy was far-reaching and bringing in its wake serious socio-economic problems, which had not been anticipated. The effect on prices was too severe, with runaway inflation pervading the entire economic system. Unemployment rates were also rising while the growth rates recorded in the agricultural sector were beginning to slow down. It became apparent that the government would need to tinker with certain aspects of the SAP measures if the gains from the programme were to be sustained and if the social dislocations already created were to be effectively addressed. The government undertook a series of measures, which largely reversed some of the SAP policies, the most conspicuous and far-reaching of which were the shifts in policy described as 'guided deregulation'. This essentially involved instituting new policies directed at checkmating the perceived advantage of foreign firms over local producers in the supply of certain goods and services. The policies were aimed at stimulating output growth, alleviating poverty and reducing unemployment through enhanced private sector participation in the economy. Phased, rather than wholesale, privatization of public enterprises, including those in agriculture, was instituted.

Meanwhile, subsidies were brought back on certain commodities, including fertilizer and petroleum products. Tariff review for agro-allied industries was carried out to enhance local value added and make local products competitive. Commodity-focused expansion programmes were initiated for cassava, rice and poultry products where it was hoped Nigeria could make significant savings by reducing importation of the commodities (rice and poultry) or their derivatives (cassava). While some of these measures were influenced by the urban-based élite farmers, the benefits could also accrue to smallholders involved in the production of the various commodities arising from increased demand.

The policy thrust of this era was also encapsulated in certain programmes which aimed at improving the socio-economic status of the poor, collectively called 'poverty alleviation programme'. Two similar such programmes were initiated, the Family Support Programme (FSP) and the Family Economic Advancement Programme (FEAP). They attempted to dampen the socio-economic repercussions of SAP measures on household living conditions and general welfare. Women were targeted and assisted with micro-credit and other arrangements to enhance their economic activities and to promote acquisition of new skills.

The new civilian administration installed in May 1999 inherited a host of socio-economic problems. The government's initial response was to enunciate certain policy thrusts which would serve as the pillar of its economic programmes. The economic guiding principles, which include macro-economic and sectoral issues, have implications for agriculture and food production. The specific agricultural policy measures undertaken are in the area of input subsidy. First, the government reduced fertilizer subsidy from 50% to 25% and made a general policy pronouncement to the effect that the agricultural input market would be totally liberalized. This was short-lived as the government restored subsidy on fertilizer to the tune of 50%. This ambivalence has continued and fertilizer subsidy to date remains a contentious issue among policy makers, farmers and development analysts. Meanwhile, efforts have commenced for the outright sale of the National Fertilizer Company of Nigeria (NAFCON), which is the largest fertilizer plant in the country. This may be seen as a move to get the private sector to assume full responsibility for fertilizer production, procurement and distribution.

Limits of a Weak State

The foregoing discussions have clearly shown one quintessential feature of the Nigerian agricultural scene – the circumscribing role of the state exercised through a series of policy initiatives. The pertinent questions which need to be addressed are as follows. Has agriculture, through state intervention, been able to fulfil its instrumental value in the development of the Nigerian state? Has agriculture been able to provide sufficient food, raw materials, farm income, employment and foreign exchange earnings through agricultural export? Are there limits to state intervention in the desired agrarian transformation of Nigeria? Have there been changes in the potential for the state to act out its desired role as 'driver' of the attempted Green Revolution? What factors seem to have constrained the effectiveness of state intervention in food production and self-sufficiency?

Food availability at affordable prices has not been achieved. Nigeria actually experienced agricultural production depression during the periods 1971–1975 and 1976–1980 in the order of 1.1% to 2.9% negative growth rates (ILO/JASPA, 1981). Paradoxically the 1971–1980 period witnessed significant state intervention in agriculture through a whole range of projects and policies. Success was achieved during SAP when agricultural output grew at an annual average of 7.9% (Table 9.2). Although the growth rates exceeded the population growth rate of 3.2%, output could not be sustained, possibly as a result of poor state management of the sector. The performance of the improved crops, in terms of productivity and in spite of heavy state subsidy on productivity-enhancing inputs, has been quite modest. As the data in Table 9.3 show, output growth rates tend to reflect the area growth rates, suggesting that increases in output were the direct result of expansion in area cultivated. The adoption of HYVs by the farmers could not be translated to significant productivity levels because of lack of necessary inputs that must accompany cultivation of HYVs. Also, the supply of nutrients from domestic efforts could not match those countries with which Nigeria shares similar socio-economic features. Table 9.4 indicates that per capita supplies of calories, protein and fat per day were only marginally higher in Nigeria than in Indonesia whereas Nigeria's performance in the supply of these nutrients was quite inferior to achievement levels in, e.g., Brazil, Argentina and Malaysia, particularly in fat and protein supply. However, in an African context, Nigeria appears to perform quite well. Compared with the countries selected for the *Afrint* study, per capita supply of these nutrients is higher, in most cases substantially higher, in Nigeria than in the rest of sub-Saharan Africa (apart from the Republic of South Africa). Nevertheless, food-related imports today constitute at least 10% of Nigeria's merchandise imports annually.

The provision of raw materials was also insufficient to keep agroindustrial enterprises running all the year round (Table 9.5). Nigeria was seriously engaged in importation of raw materials to sustain her flour mills and other factories in the early 1980s before importation of wheat, rice and maize was declared illegal. The expected agriculture–industry linkage

Table 9.2. Crop production index in Nigeria, 1961–2000 (1989–1999 = 100). (Source: FAOSTAT data, 2004.)

Period	Average crop production index	Growth rate in crop index (%)
1961–1965	49.74	3.70
1966–1970	56.30	0.94
1971–1975	57.48	–2.87
1976–1980	55.92	–1.09
1981–1985	62.94	3.83
1986–1990	85.32	7.90
1991–1995	121.76	4.69
1996–2000	148.82	2.88

Table 9.3. Annual growth rates of production of staple food crops in Nigeria: 1990–2000 (percentage). (Source: FAOSTAT data, 2004.)

Crop	Area	Production	Yield
Maize	–3.3	–2.8	0.5
Millet	2.6	2.9	0.3
Sorghum	3.6	4.3	0.7
Rice	4.5	0.9	–3.6
Cassava	4.1	3.2	–0.9
Yam	6.5	4.8	–1.7

Table 9.4. Per capita supply of calories, protein (g) and fat (g), selected countries, 2000. (Source: FAOSTAT data, 2004.)

	Vegetable products			Animal products			Total		
Countries	Calories	Protein	Fat	Calories	Protein	Fat	Calories	Protein	Fat
Ethiopia	1974	48.8	13.2	93	6.1	6.3	1887	54.9	19.5
Ghana	2496	39.4	32.2	117	14.6	5.5	2613	54.0	37.7
Kenya	1780	36.8	33.0	258	16.4	15.9	2038	53.2	48.9
Malawi	2107	49.4	23.1	59	3.9	4.4	2166	53.3	27.5
Nigeria	2649	55.4	57.8	93	8.6	5.9	2742	64.0	63.7
Tanzania	1832	38.0	21.9	126	9.7	8.1	1958	47.7	30.0
Uganda	2230	45.3	20.9	151	10.9	10.5	2381	56.2	31.4
Zambia	1805	38.2	27.0	96	8.9	6.0	1901	47.1	33.0
South Africa	2546	50.8	47.8	361	25.8	25.2	2907	76.6	73.0
Brazil	2384	39.2	46.1	618	40.7	42.7	3002	79.9	88.8
Argentina	2171	37.2	36.8	1010	67.4	71.8	3181	104.6	108.6
Indonesia	2292	47.4	41.0	120	11.7	7.3	2412	59.1	48.3
Malaysia	2400	34.5	53.2	518	42.2	30.2	2918	76.7	83.4

Table 9.5. Food industry demand and local supply of raw materials, (000 tonnes) 1995. (Source: Ajakaiye and Akande, 1999:28.)

Commodity	Demand	Supply	Demand–supply shortfall	Shortfall (%)
Maize	3,000	1,500	1,500	50
Sorghum	2,000	1,000	1,000	50
Rice	15,000	500	14,500	96
Wheat	4,000	40	3,960	99
Cassava	500	200	300	60

has remained very weak and the multiplier effects in employment and income generation for off-farm workers such as haulage firms, factory operators and product distributors and marketers have not been realized. The commodity development programmes of government in cassava, rice and poultry are making significant contributions to the domestic self-sufficiency drive.

Given the performance scenario above, it is obvious that there have been limits to the effectiveness of public policy in Nigeria. State intervention has been less effective for a number of reasons. First, the Nigerian state was over-ambitious and failed to take the first necessary steps before embarking on an agricultural modernization process. Sano (1983:83) has argued that high-tech and highly capitalized agriculture is hardly the correct starting point for a process of agrarian transformation, the more so for a country that has to rely on foreign technology imports and foreign investments. He opined that a more appropriate approach could have been a strategy that aimed at modifying the existing agrarian practices and technologies and trying to improve them; that is, a more smallholder-orientated approach would probably have performed better.

Second, the military dictatorship, which controlled the Nigerian state for all but 11 of the 43 years of the country as an independent nation, was not adroit in steering the economy towards prosperity but intensified petroleum prospecting to the disadvantage of the real sector of the economy. Agriculture was itself a casualty of pan-territorial policy measures such that local specificities were not explored. Invariably, the states and local governments got entangled in unhealthy rivalry to get a slice of the national resources in any agricultural programme floated by the Nigerian state.

Third, being on the one hand a victim of rent-seeking and on the other preoccupied with 'keeping Nigeria one', the military failed to define a proper role for the state in transforming the agrarian structure. Most policies introduced tended to ingratiate government involvement in direct production and distribution of inputs. However, many of the parastatals established in the agricultural sector ended up draining government treasury and leaving no lasting effects on the agrarian system. Resources intended for smallholders were diverted to enrich unintended beneficiaries. Worst of all, the culture of political instability fostered by frequent military regime changes drastically affected policy continuity and implementation.

There was also the fundamental challenge of deciding on how to organize the smallholder farmers to achieve significant productivity improvements. The pioneer Nigeria leaders at independence voted for a transformatory strategy and initiated state farm settlement. But the atomistic nature of smallholder farming and the spatial distribution of millions of farmers across the Nigerian landscape undermined farm settlement, apart from technical problems, which confronted the running of the settlements. The cooperative system erected was not successful because it was too heavily controlled by the state, thereby ignoring the sensibilities of the local people and how they manage and administer their society (King, 1981).

The state also faced other constraints such as the technical issues of pest and diseases, inadequate tools and equipment, and processing and storage constraints, which often led to loss of substantial quantities of output produced. Postharvest losses were assessed at nearly 40% of total production in 2000 (Van Buren, 2001).

Conclusion

Nigeria has attempted to execute a Green Revolution in localized areas across the country through a series of policy interventions since the 1970s, but has failed to make the required impact. Partly, this is because the focus has been on technology while, inadvertedly or not, smallholder peasants have largely been by-passed. Since the programme has not spread equally and evenly across all agroecological zones of the country, and given the dichotomy existing between designated 'progressive farmers' and 'other farmers', technology has raised concern over equity in inter-regional and inter-class terms. While the state designated the small farms as the centre of agricultural development and transformation, the rent-seeking behaviour of its officials and the bureaucracy truncated the benefits of the Green Revolution to the small farmers. Influential and town-dwelling 'farmers', aristocrats, input contractors and transport owners constituted the unintended beneficiaries of the policies introduced. The state appears very weak and inefficient in implementing its farm programmes and activities.

The initial stages of Green Revolution in Asia relied on intensive application of abundant labour and improvement of land. Asian Green Revolutions, moreover, were not only small farmer-based, they were also highly indigenous in that they to a large extent relied on domestically produced seeds and fertilizers, etc. Unlike these Asian countries, the endogenization of the capacity to undertake a Green Revolution was not initiated in Nigeria. In its modernization efforts the state relied almost exclusively on external technology such as high technology sprinkler irrigation facilities, imported fertilizers and pesticides. However, the technologies introduced were often not well adapted to smallholders' agricultural and socio-economic situations and, hence, could not be adequately engaged by a farming clientele that is largely dispersed, poor and illiterate.

For most of the period studied, Nigeria's attempted Green Revolution has been neither smallholder-based nor market-mediated. At the same time, the state's role and the depth of its engagement have shifted, depending largely on external circumstances. In the initial years after independence, focus was on exportable cash crops, as food provision was not seen as a problem at the time. Civil war in the late 1960s and drought in the 1970s accentuated the need for food provision. However, the unprecedented revenues from oil-exports from the mid-1970s led to a neglect of agriculture in general and food production in particular. Food imports soared while the state apparatus grew exponentially and the treasury approached near-bankruptcy.

SAP was introduced in the mid-1980s with the intention to reduce the role of the state and create space for markets and private entrepreneurs. Whereas the overall effects of SAP in Nigeria have been diverse and disputed, agriculture-related reforms have been deemed 'arguably the most successful element of the structural adjustment programme' (Van Buren, 2001:757). As previously shown, food production improved significantly during and after reform implementation. Notwithstanding the persisting difficulties for smallholders to access inputs like high-yielding seeds and fertilizers, attention has now become 'focused on the smallholder farmers, who produce some 90% of the food consumed in Nigeria' (Van Buren, 2001:757). To a significant degree this is because the state has in recent years resumed some of its earlier role as 'driver' of the Green Revolution, notably by pursuing a more cautions policy of 'guided deregulation'. It might be too early to say where these role- and policy-reorientations will lead to, and what their long-term effects will be on Nigerian agriculture in general, and on food production in particular. What can be concluded, however, is that important steps now have been taken which may result in the Nigerian attempt at a Green Revolution becoming much more Asian-like

– and, hence, more likely to become successful – than were its predecessors.

New challenges

The role of the state in directing the course and tempo of agricultural development in Nigeria is much more than symbolic. The state is at the heart of agricultural change and food production. However, it is not the only important player. During the pre-SAP periods, the state was carrying out all functions from policies and direct production to coordination of agricultural programmes. The lacklustre performance of the agricultural sector in fulfilling any of its development functions belies the enthusiasm and resources the state committed to agricultural development. The state seems to have realized its limitations and has started, since the time of implementation of SAP, to shed its load of responsibilities by liberalizing the economic system and encouraging the organized private sector to participate in developing the agricultural sector. Currently, a new approach being considered is public–private partnership in the conduct of certain activities in the agricultural sector. In this arrangement, public and private interests and resources are to be mobilized to run certain programmes and services, which may bring about effectiveness and efficiency. For instance, public–private partnership is being planned to manage irrigation water and in operating agricultural commodity marketing.

The place of the market in efficient resource allocation remains valid in order to eliminate the socio-economic impediments that have slowed down the pace of agricultural development. Price policy, investments, intensification or productivity enhancing measures must make economic sense for smallholders. The market system tends to promote allocative efficiency and indicate which crops or crop combinations can prove profitable to the farmer. The challenge of the market is both in the supply of needed inputs and in the disposal of farm surpluses at remunerative prices. During the pre-SAP period the market was highly regulated and controlled by government institutions. The measures undertaken during SAP totally eliminated these institutions while a free market system dominated. It has since been realized that some elements of state intervention are still required, particularly in ensuring standards, fair pricing and in balancing demand and supply configuration both in time and space. The proposed arrangement is to set up various commodity marketing companies to be owned and run by farmers or producer associations, with the government merely facilitating the process and setting rules and regulations expected to guide the operations of such marketing companies. It is assumed that the companies may be able to eliminate most of the limitations that have been associated with agricultural marketing in the past.

The third leg of the Green Revolution tripod, the smallholder, presents a challenge to the Nigerian state. The Nigerian smallholder farmer is ageing and has not benefited significantly from state policies in the last 40 years. The young generation is unwilling to stay in the rural areas if the same pattern of drudgery, which had been the lot of their parents, persists. What this implies is that there should be a deliberate state policy focused on total restructuring of the rural economy to encourage young people to stay in rural areas where they could profitably engage in economic enterprises that include farming and para-agricultural activities. The challenge actually is how to reduce considerably the production costs while ensuring that farmers receive adequate prices for their output. The need for subsidy may be eliminated if the market processes guarantee remunerative and sustainable returns to the farmers. With significant improvement in farm income, an increasing proportion of farmers will be willing to adopt new technologies and improve on their management practices.

References

Agboola, S.A. (1979) *Agricultural Atlas Map of Nigeria*. Oxford University Press, Oxford, UK.

Ajakaiye, D.O. and Akande, S.O. (1999) National Agricultural Research Programme in Nigeria: the efficiency of policy studies. Publication for

NAMRP. NISER Annual Monitoring Research Programme (NAMRP), Ibadan.

Akande, S.O. (2002) The recurring food crisis in Nigeria: what policy response? NISER, Nigerian Institute for Social and Economic Research, Ibadan.

Akande, S.O. (2003) Trade liberalization and the environment: the case of the rice sector in Nigeria. Technical Research Report. United Nations Environmental Project (UNEP), Geneva.

Alkali, R.A. (1997) *The World Bank and Nigeria*. The World Bank, Kaduna, Nigeria.

Andræ, G. and Beckman, B. (1985) *The Wheat Trap*. ZED, London/The Scandinavian Institute for African Studies, Uppsala, The Bath Press, Avon, UK.

Ayoola, G.B. (2001) *A Book of Readings on Agricultural Development Policy and Administration in Nigeria*. T.M.A. Publishers, Ibadan.

Boserup, E. (1965) *The Conditions of Agricultural Growth*. Aldine, Chicago, Illinois.

CBN (1994) Annual report and statement of accounts. Central Bank of Nigeria, Lagos, Nigeria.

CBN/NISER (1991) The impact of SAP on Nigerian agriculture and rural life. Central Bank of Nigeria, Lagos, Nigeria.

Dike, E. (1998) Investment and labour flows in Nigerian agriculture under structural adjustment. In: Tshibaka, T.B. (ed.) *Structural Adjustment and Agriculture in West Africa*. CODESRIA, Dakar.

Ekong, E.E. (1986) *The Nigerian Government Policies on Agriculture and Food Production in 1960's, '70's and '80's: Implications and Prospects*. CODESRIA, Dakar.

Erinle, I.D. (1994) The need for an effective agricultural research programme. In: Shaib, B. (ed.) *Towards Strengthening the Nigerian Agricultural Research Systems*. Ibadan University Press, Ibadan, pp. 129–152.

Evbuorhwan, G.O. (1997) Poverty alleviation through agricultural projects: a review of the World Bank-assisted agricultural development projects in Nigeria. *Bullion Publication of the Central Bank of Nigeria* 3.

Falusi, A.O. (1986) *Use of Fertilizer for Increasing Crop Productivity ADP Experience*. National Productivity Centre, Lagos, Nigeria.

FAOSTAT data (2004) Food and Agriculture Organisation of United Nations. http://www.fao.org/waicent/portal/statistics_en.asp

Federal Government of Nigeria (1997) *Vision 2010*. Lagos, Nigeria.

Forrest, T. (1993) *Politics and Economic Development in Nigeria*. Westview, Boulder, Colorado.

FOS (1993) *Survey of Modern Holdings National Accounts of Nigeria 1981–1992*. Federal Office of Statistics, Lagos, Nigeria.

Hill, P. (1972) *Rural Hausa: a Village and Setting*. Cambridge University/IFDC, Cambridge, UK.

Idachaba, F.S. (2000) Desirable and workable agricultural policies for Nigeria. *Topical Issues in Nigerian Agriculture* Series. Department of Agricultural Economics, University of Ibadan, Ibadan.

Igene, J.O. (1991) The challenges of the industrial processing of indigenous foods. In: *The Nigerian Food Processing Subsector*. Federal Ministry of Industries sponsored by Policy Analysis Department (PAD) NISER, Lagos, Nigeria.

International Labour Organization/Jobs and Skills Programme for Africa (ILO/JASPA) (1981) *First Things First: Meeting the Basic Needs of the People of Nigeria*. ILO Office, Addis Ababa.

King, R. (1981) Cooperative policy and village development in northern Nigeria. In: Heyer, P.R. and Williams, G. (eds) *Rural Development in Tropical Africa*. St Martin's Press, New York, pp. 259–280.

Mustapha, A.R. (1998) *Harvesting SAP: Cocoa, Farming Households, and Structural Adjustment in Nigeria*. Queen Elizabeth House, Oxford, UK.

NISER (2001) Public sector spending and management for rural development. Study commissioned by the World Bank and the Department of Rural Development of the Federal Ministry of Agriculture and Rural Development. The Nigerian Institute of Social and Economic Research, Ibadan.

Olatunbosun, D. (1971) Western Nigeria farm settlements: an appraisal. *Journal of Developing Areas* 5, 56–89.

Olayide, S.O., Eweka, J.A. and Bello-Osagie, V.E. (eds) (1980) *Nigerian Small Farmers: Problems and Prospects in Integrated Rural Development*. University of Ibadan, Centre for Agricultural and Rural Development. Ibadan.

Olomola, A.S. (1998) Structural adjustment and public expenditure on agriculture in Nigeria. In: Tshibaka, T.B. (ed.) *Structural Adjustment and Agriculture in West Africa*. CODESRIA, Dakar.

Otite, O. (1990) *Ethnic Pluralism and Ethnicity in Nigeria*. Shaneson, Ibadan.

Sahib, B., Adamu, A. and Bakshi, J.S. (1997) *Nigeria: National Agricultural Research Strategy Plan: 1996–2010*. Intec Printers Limited, Ibadan.

Sano, H.-O. (1983) The political economy of food in Nigeria: 1960–1982. Scandinavian Institute of African Studies Research Report No. 65. Scandinavian Institute of African Studies, Uppsala, Sweden.

Van Buren, L. (2001) *Nigeria: Economy. Europa Yearbook. Africa South of the Sahara*. Europa, London.

Watts, M.J. (1987) Agriculture and oil-based accumulation: stagnation or transformation? In: Watts, M. (ed.) *State, Oil and Agriculture in Nigeria, 115*. University of California, Berkeley, California.

Williams, R. (1999) *Reforming Africa: Continuities and Change. Africa South of the Sahara 2000.* Europa, London.

World Bank (1996) *World Development Report 1996.* The World Bank, Washington, DC.

10 Why the Early Promise for Rapid Increases in Maize Productivity in Kenya was not Sustained: Lessons for Sustainable Investment in Agriculture

Willis Oluoch-Kosura and Joseph T. Karugia
Department of Agricultural Economics, University of Nairobi, PO Box 29053 Nairobi, Kenya

Maize is the major staple food crop as well as a source of sustenance for the majority of households in Kenya. Its absence is associated with famine in the country, even if other food grains are available. The maize sub-sector experienced a considerable breakthrough in the 1970s, especially in the spheres of varietal development. The use of improved varieties supplemented by purchased inputs, especially fertilizers, increased maize yields in the late 1960s to the mid-1970s. However, given the early promise and in the face of increasing depth and breadth of poverty in Kenya, achieving sustained increases in maize productivity has been an elusive goal. The average yield of maize has stagnated at around 1.7 t/ha – a level representing less than a third of the optimal yield on farms.

If this yield gap is closed or narrowed, food poverty will effectively be eliminated in many households. While climatic factors, such as incidences of drought, may be part of the reason for the decline, policy and institutional related factors stand out as the major reason for not sustaining the increases witnessed in the 1970s. The aim of this chapter is to explain how these factors contributed to the decline in maize output and yields and the subsequent deepening and broadening of food poverty witnessed today. The chapter provides lessons to be learnt to enable a re-focusing of attention on ways to achieve sustainable investment in agriculture in order to improve the livelihoods of the majority of households in Kenya.

Background

Maize is a major food crop and dominates all national food security considerations in Kenya. It accounts for more than 20% of all agricultural production and 25% of agricultural employment. Smallholders produce about 65% of the maize mainly for domestic consumption while large-scale commercial farms contribute the largest share of the marketed surplus. Large-scale producers in the Rift Valley account for 25% of the total production and over 50% of the marketed surplus. When all crops grown in Kenya are considered, maize occupies the largest area and variety of agroecological zones. The area under maize cultivation has stabilized at around 1.4 million ha with limited potential

 The African Food Crisis (eds G. Djurfeldt, H. Holmén, M. Jirström and R. Larsson)

for further expansion (Republic of Kenya, 1997).

Per capita maize consumption averages 125 kg/year (Republic of Kenya, 1983). The national annual production averaging 2.3 million t is far below the annual domestic consumption estimated at 3 million t. This gap implies that even in normal production years the country must import maize. During drought years, production can be as low as 1.6 million t. Imports in the late 1990s averaged about 400,000 t/year. The gap between production and consumption varies across regions and, based on these trends, the country can be categorized into surplus regions, marginal deficit regions, and chronic deficit regions.

In Kenya, the Green Revolution technologies were introduced by the development of high yielding varieties of maize, wheat and rice accompanied by the application of fertilizers and other chemicals in the early 1970s. Research in maize production continues to be accorded high priority among the food crops (Republic of Kenya, 1994). There have been tremendous achievements in maize technology development since the 1970s, especially varietal development following the establishment of maize research programmes in the country. The maize research programmes started in 1955 when the Kitale programme, which focused on the production of late maturing hybrids, was initiated. The establishment of the Katumani programme followed in 1957 and concentrated on production of open pollinated varieties (OPVs) for the dry mid-altitude zones. Subsequently, the Embu and Mtwapa programmes followed producing maize varieties for the moist mid-altitude and lowland coastal areas, respectively (Hassan *et al.*, 1998a). In 1985 the Kenya Agricultural Research Institute (KARI) undertook a Fertilizer Use and Recommendation Project (FURP) to recommend suitable fertilizer types and application rates for specific maize zones.

The adoption of hybrid maize varieties in Kenya in the 1960s and 1970s has been compared to that of the USA (Gerhart, 1975). The rate of adoption was quite rapid although initial adopters tended to be large-scale farmers in the high potential areas. In the semiarid zones, adoption of improved maize varieties remained decidedly low. The adoption of inorganic fertilizers on the other hand closely followed the adoption of improved maize seed in the large farm sector. However, smallholders' adoption of fertilizers lagged behind their adoption of improved seeds and has remained virtually negligible in the marginal areas. Following the release of improved varieties in the early 1960s, there was a dramatic increase in maize production from less than 0.3 million t to a high of 2.5 million t in 1981 (Fig. 10.1). At the same time, there was an expansion in hectarage from below 0.7 million ha in 1963 to 1.4 million ha by 1980 (Fig. 10.2). These increases in output and hectarage were accompanied by an increase in maize yields from an average of 1.5 t/ha in 1963 to about 2 t/ha in 1979 (Fig. 10.3). However, this early promise was lost in the 1980s and as a result decreases have been witnessed in production and yields.

The trends in maize yields have been on the decline especially in the decades following the 1970s. The average yield is about 1.7 t/ha compared with a potential average yield of 6 t/ha. The yield gap provides evidence of wide divergences in crop yields between experimental research station plots or well managed research on farm trials and average yields that farmers typically realize on their farms. This potential could be achieved through increasing the use of improved seeds, fertilizers and appropriate crop husbandry (Republic of Kenya, 1997). Hassan *et al.* (1998b) demonstrate that the current levels of adoption of improved maize seeds and inorganic fertilizers are sub-optimal. The study further notes that in some cases even farmers who once used fertilizers have abandoned the technology (Hassan *et al.*, 1998b). Other research studies show that farmers may adopt new seed varieties, but consistently ignore extension recommendations on improved soil fertility management practices (Rukadema *et al.*, 1981). Consequently, many farmers achieve only a small proportion of the potential productivity gains possible from adoption of new crop technologies.

Maize Production Trends in Kenya

Maize is an important staple food crop for 96% of Kenya's population. The crop is produced by both small- and large-scale farmers in most parts of the country for home consumption while the surplus is sold to meet cash needs of households. For most producers, maize is sold immediately after harvest, and the same households return to the market later in the year to buy maize. About 60% of the rural households are net maize buyers, indicating a dependency on the market for their food supplies. Production varies from region to region with surplus regions being found in the North Rift districts of Nakuru, Uasin Gishu, Trans Nzoia, Kapenguria and Nandi. In these regions, yields are high and range from 2 to 5 t/ha. Large-scale farmers dominate these areas and maize is produced for commercial purposes, although some small-scale farmers grow it for both home consumption and for sale. The second category of maize producing regions consists of the self-sufficient and marginal deficit maize producing regions, which cover most parts of the Western, Nyanza, Central and Eastern provinces. In these regions, small farmers produce maize mostly for subsistence needs with a small surplus being sold to the market. The third category of maize producers are the deficit producing regions in Eastern, North Eastern and Coast provinces where maize is produced for subsistence needs. In these regions, domestic production rarely meets household needs.

Maize production patterns in Kenya are unique in that production can occur all the year round because of the diversity of production conditions. Surplus producing regions have a unimodal rainfall pattern and have only one maize-harvesting season, in October to December. The marginal deficit maize producing areas have also a unimodal rainfall pattern and the main maize-harvesting season is in September to October. Maize deficit areas have a bimodal rainfall, which, however, is not reliable and two harvesting periods can occur (October to November and March to April). These production patterns can ensure that the country has maize supplies all the year round.

Maize production trends in Kenya have fluctuated since independence in 1963. In general, production has lagged behind consumption and deficits are a frequent phenomenon. In the first two decades after independence, maize production increased tremendously (Fig. 10.1). National production increased from less than 0.3 million t in 1963 to about 2.6 million t in 1981. This early expansion in production can be attributed to area expansion, heavy government and donor involvement through subsidization of services and inputs – mainly improved seeds, fertilizers, research, extension and marketing infrastructure.

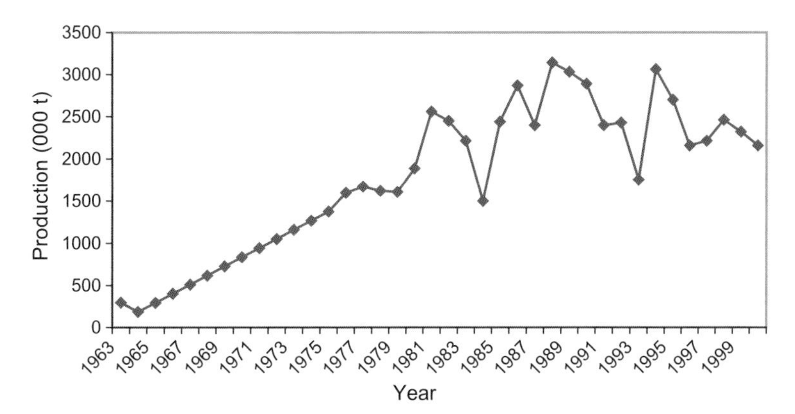

Fig. 10.1. Maize production. (Source: Ministry of Agriculture, 1963–1999.)

During this time, farmers were adopting the newly released high yielding maize seed varieties coupled with the application of fertilizers. Production then declined significantly in 1984 following a drought in that year before it again peaked to a high of 3.1 million t in 1988. In the 1990s maize production has been on a decline hitting a record low of 2.2 million t in the year 2000. The loss of the earlier promise in the latter part of the 1980s and the significant decline in the 1990s can be attributed to a number of factors. These include mainly the scarcity of funds following implementation of Structural Adjustment Programmes (SAPs) in 1986 that implied a reduced involvement of the government in subsidizing production. In addition, unfavourable weather conditions and pest and disease outbreaks have had an adverse impact on production.

In the period 1963–2000, there was a significant expansion in the area planted to maize from a low of 0.7 million ha to a high of 1.6 million ha in 1999 (Fig. 10.2). The area cultivated with maize has, however, stabilized at 1.4 million ha with a minimal potential for expansion. The rapid initial expansion was attributed to the purchase of farms in the former white highlands by indigenous Kenyans, the subdivision of these farms and the eventual cultivation of maize by the formerly landless Kenyans. The purchase of former white highlands was initiated under the first African Land Development plan (ALDEV) and the Swynnerton Plan (1952–1962). However, it was not until after independence that the government, through the Agricultural Finance Corporation (AFC) and the Settlement Fund, offered loans for the purchase of land. Later, efforts were made to expand maize cultivation into the marginal areas resulting in soil mining. Given these trends in area expansion, the decline in maize production cannot be attributed to contraction in area but to changes in yields.

Trends in maize yields in Kenya

In the first two decades of independence, the performance of the maize sub-sector showed great promise for achieving high productivity. The opening up of virgin land for maize cultivation was accompanied by a simultaneous increase in yields partly due to the use of improved seeds and chemical fertilizers. However, as demonstrated in Fig. 10.3, the initial promise in yield growths was not sustained and from the mid-1980s yields declined. From a high of around 2 t/ha in the late 1980s, maize yields have stagnated at around 1.7 t/ha in the last 5 years. This yield level represents an achievement of only 30% of the potentially achievable yield on farms.

Achieving sustained increases in maize productivity in Kenya has been elusive, given the early promise. The maize sub-sector experienced considerable breakthroughs in the 1970s, especially in the spheres of varietal development that led to increases in maize yields. However, this early growth was not sustained in later years and Kenya missed out on the benefits of the Green Revolution

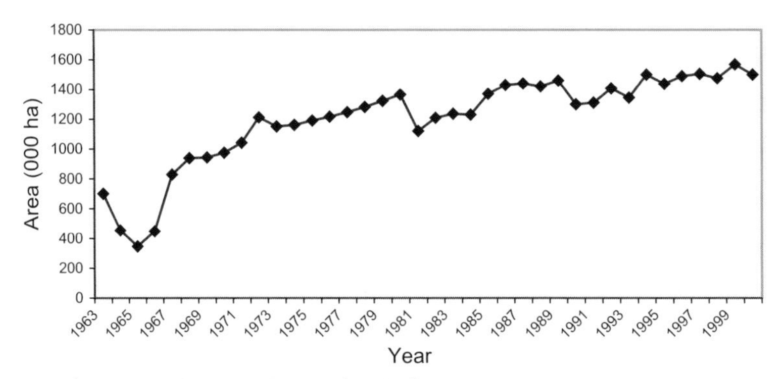

Fig. 10.2. Maize hectarage. (Source: Ministry of Agriculture, 1963–1999.)

technologies in maize production. Both output and yields have generally declined in the decades following the 1970s. Even in years when there was an expansion in output, this cannot be attributed to increases in yield but to increases in hectarage under maize.

While climatic factors, such as incidences of drought, may be part of the reason for the yield decline, there is overwhelming evidence that policy and institutional related factors stand out as the major reasons for not sustaining the increases witnessed in the 1970s.

Weak institutional support for agriculture

Agriculture still offers the best prospect for Kenya's economic growth given the fact that it contributes to about 25% of the GDP and has a multiplier effect of 1.6 compared with the rest of the economy estimated at 1.23 (Block and Timmer, 1994). Similarly, agriculture contributes over 45% of government revenue through agricultural taxation. In view of this, it is necessary to allocate more funds for the agricultural sector within the national budget.

However, the allocation of government expenditure to the sector forms a relatively small share when compared with education and health and has been declining. This greatly contrasts with the three sectors' average contribution to gross fixed capital formation with agriculture taking the lion's share at 6.7%, followed by health at 2.15% and education at 1.9%. Available statistics indicate that on average Kenya used to spend over 10% of its total government budget on agriculture in the first decade after independence (Fig. 10.4).

This declined to an average of 7.5% in the period 1980–1989 and to a record low of 3% in the 1990–2000 period. With the introduction of the SAP reforms, the allocation to

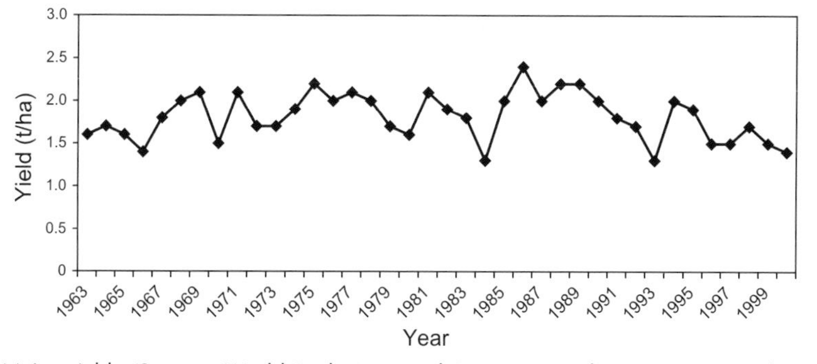

Fig. 10.3. Maize yields. (Sources: World Bank, 1990 and Government of Kenya, Ministry of Agriculture (Annual Reports).)

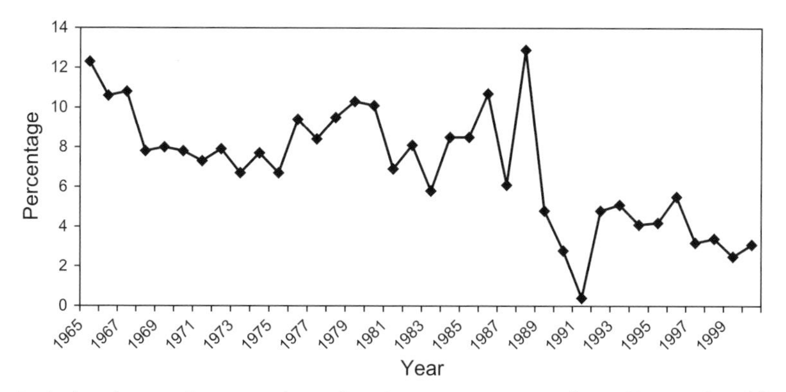

Fig. 10.4. Agricultural expenditure as a share of total government expenditure. (Source: Republic of Kenya, 1965–2001.)

agriculture declined significantly as a result of withdrawal of subsidized services to farmers. In the period 1980–1985 (pre-SAP), the allocation to agriculture was on average 9.3% of total public expenditure. This declined to 7.9% of total during the transitional period 1986–1993 (SAP) and further declined to 3.6% for the post-SAP period 1994–2000. The implication that can be drawn here is that the government gives back to agriculture less than what agriculture contributes to the economy.

However, about 60% of the government's expenditure on the agricultural sector is on recurrent expenditure, which is dominated by salaries (for employees, including the extension officers). Hence, only about 40% is spent on agricultural development, which includes agricultural research and market information, animal health services, crop protection, seed inspection, mechanization services and farm planning services (Fig. 10.5) and has been declining in the recent past. The allocation to development expenditure outweighed that to recurrent expenditure until 1982 when the allocation to recurrent expenditure overtook development expenditure. Thereafter, the amount spent on recurrent expenditure has been consistently higher than that spent on development expenditure except for a few years. This is possibly because of fiscal reforms in which the government emphasized reduction of its public expenditure and found it easier to reduce development expenditure than recurrent expenditure. Most important perhaps, is that most of the development expenditure is funded by donors. The problem with donor funding is that it is usually unstable due to the donors' changing policies and hence is not a sustainable long-term strategy for agricultural development. The instability of donor funding is part of the reason for the observed fluctuations.

Several agricultural institutions supported maize production in the first decade after independence. These included agricultural research institutions, extension service, credit, cooperatives and marketing institutions. Research and extension institutions have since experienced managerial and financial weaknesses that have hindered their performance.

Although Kenya's agricultural research system has been strong, lack of progress in increasing total factor productivity in agriculture suggests that it is operating below its

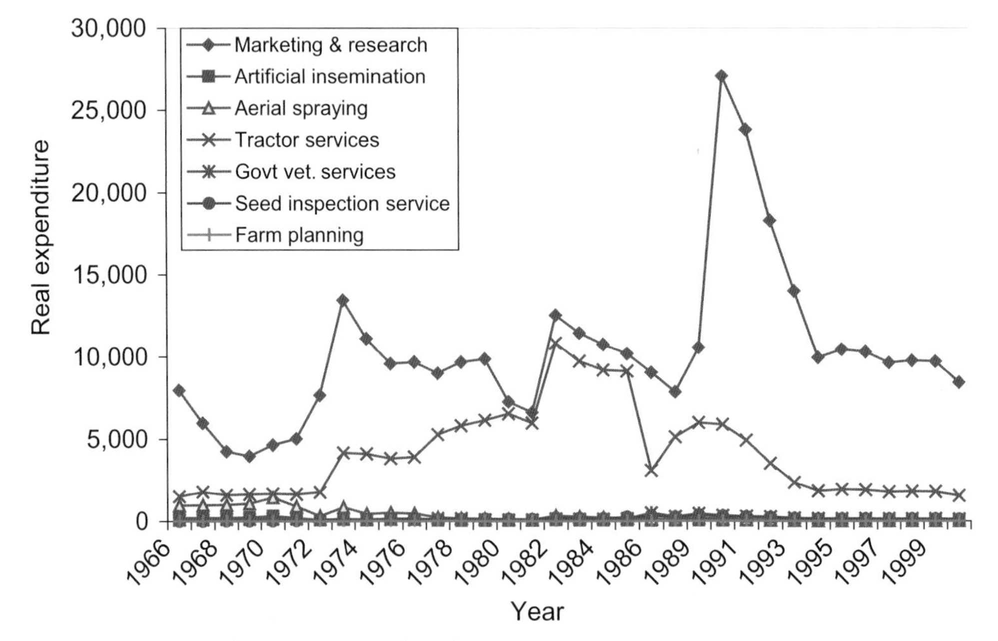

Fig. 10.5. Real expenditure on agricultural productive services (1995 = 100). (Source: CBS, 2000.)

potential. This has been related to weaknesses in priority setting, financing, management and interagency linkages (Nyangito and Karugia, 2002). KARI and the Ministry of Agriculture dominate agricultural research and extension in Kenya. Government expenditure on research has been minimal. Recent analysis (Oluoch-Kosura, 2001) shows a declining trend in efficiency and effectiveness of the Ministry of Agriculture extension services. This may be due to declining budgetary allocations to the sector, lack of clear objectives, failure to identify role of beneficiaries and poorly defined organizational and institutional structures. With low government expenditure on extension after liberalization, the service has almost collapsed.

Karugia (2003), in a study in the Nyeri and Kakamega districts, found that the state was previously responsible for providing free regular extension services during the pre-SAP period. Currently, however, extension services are absent in most villages and where they are available they are often irregularly provided. A number of farmers get the services from private agents at a cost. The cash crop farmers get extension services from their out grower companies (of which none exists for maize) or the buyers of their produce. As most maize farmers do not access this essential service, they are unaware of many of the yield-improving farming techniques.

The weaknesses in research and extension since the introduction of SAP have limited the generation of new technologies and created a weak link between researchers and the farming communities. Although new technologies are available on shelf, the farming communities have not benefited from them since research findings do not flow to the farmers. Weakened public financial support for research and extension has led to a collapse of scientific and institutional cooperation in Kenya – a cooperation which in the 1960s and 1970s created an early success story in maize production.

With regard to credit, several credit schemes were operated by the Agricultural Finance Corporation to support maize farmers. These included seasonal credit in the form of Guaranteed Minimum Return (GMR) and the settlement fund. Before the onset of market liberalization, formal agricultural credit was provided at subsidized rates through the Agricultural Finance Corporation (AFC). However, the parastatal could not recover loans advanced and had to stop lending at subsidized rates.

In comparison with other commercial banks, lending rates of the AFC remained lower and have been more stable. Although there is a legal requirement that banks should lend between 17% and 20% of their loan portfolio to the agricultural sector, the local banking system has remained conservative in lending to agriculture, probably due to risks in agricultural production. The total public credit provided to agriculture is on average estimated at less than 10% of the total credit provided through the domestic financial system.

In agriculture, smallholder farmers and female farmers are at a distinct disadvantage. It is estimated that only one-third of total rural credit is allocated to smallholder producers while the bulk of the credit goes to large-scale farmers. Only a limited number of farmers are able to access credit. A study by Karugia (2003) in Nyeri and Kakamega districts shows that procurement of inputs for use in maize growing (such as fertilizers and hybrid seed) is increasingly being self-financed by smallholders.

Further, the study reveals that liquidity constraints limit demand for key productivity-enhancing inputs. The AFC has a very limited reach in extending credit. It lends to only 1% of rural households. Some 85% of the AFC loans are made in the Rift Valley and 15% in the Central Highlands. Loans were made to farms averaging 19 acres (7.7 ha), compared with a national average farm size of 4.3 acres (1.7 ha), and 73% of borrowers had off-farm sources of income (Argwings-Kodhek, 1999).

In the maize growing villages of Munyuki and Mukuyu in Lugari district, Karugia (2003) found that a number of farmers who used title deeds as collateral for AFC loans lost their land due to loan-repayment failure. Farmers now fear to seek these loans. On the other hand, the AFC prefers advancing loans of large amounts (above Kshs500,000), which are beyond the capacity of small-scale farmers. Opportunities for acquiring credit are, thus, very few if not non-existent for most small-scale maize farmers. Karugia's study also found that in the maize growing villages of Munyuki and Mukuyu in Lugari district, there are no credit organizations, save for women merry-go-round groups which advance small amounts of credit to their members.

A study by Karugia *et al.* (2003)confirms the lack of credit sources for maize farmers and traders. Marketing institutions have also hindered farmers' performance owing to disorganized marketing. Prior to the reforms, maize marketing was undertaken by the state via the National Cereals and Produce Board (NCPB). After the reforms, the private sector was unable to fill the gap left by NCPB and hence minimized market opportunities for maize farmers. For example, farmers in western Kenya are reported to sell maize at very low prices to middlemen and brokers due to poor market access (Karugia, 2003). The NCPB also sets very stringent conditions for accepting maize from farmers. Consequently, most farmers dispose of their maize at very low prices in the private market whose conditions for accepting maize are less stringent than those set by the NCPB. Reforms in maize marketing mean that prices are quite low immediately after harvest and rise significantly toward the middle of the following year. Those who can hold out get a much higher return from their maize production than the typical smallholder, who must sell immediately to meet pressing cash needs.

Despite the increasing dominance of private trade in maize marketing, its expansion and development is limited by lack of proper institutional support. Ikiara *et al.* (1995)found that despite liberalization, maize farmers preferred to market their produce within their home districts. Selling outside would take too long a time. This is as a result of the fragmented nature of private wholesale and retail trade. Inadequate storage facilities during bumper harvests have often meant depressed prices and difficulty in disposing of the surplus maize produced by farmers. This is further compounded by the fact that private traders in addition to having inadequate facilities also lack incentives to undertake storage of maize. The inadequacy of the national holding capacity for maize restricts the procurement, delivery and distribution of the commodity (Republic of Kenya, 1994). The NCPB has been and remains unable to purchase all the (occasional) surplus maize production due to inadequate holding capacity. Such lack of institutional support constitutes a disincentive for farmers to produce more maize.

Confirming the described situation, Karugia (2003) shows that the main channel through which farmers sell the food crops is the private marketing channel. Government procurement is limited – only in one of the ten villages studied (Munyuki) was government involvement reported. In this area the government buys maize through its agent, the NCPB, which has maize silos in the nearby Kipkarren River market. The relative absence of government involvement is not solved by the private market. Karugia *et al.* (2003) show that maize farmers and traders lack the financial and managerial capacities for storage and, consequently, have to sell at a non-favourable price. Improving maize storage requires investments in physical facilities, research and extension for on-farm and off-farm storage, improvement of credit access by farmers and traders, and a reduction in market risk occasioned by policy reversals. Institutional weaknesses and lack of incentives for the farming communities may be the most important factors contributing to the lack of a sustained yield achievement in the maize sub-sector.

Policy failures

Policies on maize production, marketing and pricing have been major concerns for the government of Kenya since colonial times. These policies ranged from government controls on maize production, pricing and marketing up to 1994 when the current policy of liberalized markets was enacted. Unfortunately, market liberalization was not properly sequenced and coordinated and as a result it has had adverse effects on the sub-sector.

The advent of the reforms also introduced new challenges, especially with regard to input quality and market regulation. It also reduced the number of incentives that farmers enjoyed. One example where liberalization of the maize industry was poorly sequenced is in the seed industry where the quality control body for the seed industry, the Kenya Plant Health Inspectorate Service (KEPHIS), appears inadequately equipped to monitor and enforce a quality control for inputs. Problems with quality control in improved seed generation and marketing have discouraged farmers from purchasing hybrid maize seeds (Karugia, 2003). After the reforms, the state was expected to play a reduced role in the maize sub-sector while the private sector was expected to play an increased role and eventually take over the roles that were previously undertaken by the state. However, the private sector was sluggish in performing these duties and has not filled the gap left by the state.

Reforms in the maize sub-sector have also been affected by policy inconsistencies from food self-sufficiency to food security. However, over the entire period, government rules and regulations regarding trade controls have been hindered by a lack of review of the Act of Parliament (Cap 338), which established NCPB as a monopoly in maize trade. Other limitations include limited access to information, limited access to working capital and risks and uncertainties owing to policy changes. The experience of government intervention in the market prior to and after the reforms indicates that the government has been unable adequately to address policy issues in the maize sub-sector.

Prior to liberalization, the government subsidized agricultural inputs such as fertilizers, seeds, pesticides, vaccines, machinery and artificial insemination services. This was intended to ensure availability, adequacy and timeliness of inputs to boost production. This was done by subsidizing inputs, controlling imports and encouraging the formation of cooperatives to achieve economies of scale for member farmers. The government controlled the process through price controls, import licensing and quotas, thereby opening avenues for corruption and direct involvement in distribution. Today, input markets have been opened up but most of the country still lacks an adequate network of markets for agricultural inputs such as fertilizers, seeds, livestock feeds, artificial insemination services and agro-chemicals.

One key area of improvement is found in packaging, where it is now possible for farmers to access small quantities of fertilizers, unlike the pre-reforms period when fertilizer was typically sold in 50-kg bags. Input liberalization has not significantly changed input use and has introduced new challenges in the market. One basic problem relates to the poor quality assurance for all types of inputs. Another relates to the apparently unexplicable reason why fertilizers and other inputs in Kenya are so expensive relative to other countries. This is particularly puzzling considering that fertilizers are zero-rated in terms of import duty. On the other hand output prices have increased albeit marginally, as indicated in Table 10.1. This has created uncertainties for maize farmers since, unlike in the period prior to liberalization, output prices presently face a downward trend and, moreover, now fluctuate greatly over time.

At the same time, Karugia (2003) has reported an upward trend in input costs in the post-SAP period. This is primarily a result of the removal of input subsidies. In many areas private traders mainly provide agricultural inputs. These traders are, however, reported to be offering poor quality inputs. The government, NGOs, and donor agencies are playing a minimal role in input provisioning (Nyangito and Ndirangu, 1997).

Climatic factors

Kenya is characterized by a wide diversity in agroecological zones. The climate varies from a tropical hot and humid coastline to a temperate climate inland and further to a dry climate in the north. Over 70% of the country is arid, receiving less than 510 mm of annual precipitation while rainfall is greatest in the highlands. Similarly, out of Kenya's total land area of 44.6 million ha, only 12% is suitable for arable farming and is classified as high and medium potential. The other 88% is classified as low potential or Arid and

Table 10.1. Real prices per tonne of food crops 1980–2000. (Source: Republic of Kenya, Statistical abstracts (1995–2001) and authors' calculation.)

	Maize		Wheat		Rice	
Year	Kshs	US$	Kshs	US$	Kshs	US$
1980	1263	168	2180	291	2007	268
1981	1189	131	1986	218	1784	196
1982	1070	102	1880	179	1500	143
1983	1364	102	1966	147	1576	118
1984	1463	100	2249	153	1488	101
1985	1391	86	2015	124	2544	157
1986	1458	89	2158	132	1563	95
1987	1441	87	2034	123	2265	136
1988	1386	76	2202	121	2512	138
1989	1295	62	1987	96	2254	108
1990	1537	63.77	2642	109.62	1427	59.21
1991	1463	52.06	2393	85.16	766	27.25
1992	1619	44.72	1911	52.79	399	11.02
1993	2017	29.57	1407	20.63	1307	19.16
1994	2065	46.09	2609	58.23	1976	44.11
1995	1626	29.08	2643	47.28	2086	37.32
1996	1966	35.74	2913	52.96	2988	54.33
1997	2351	37.55	3030	48.40	2735	43.69
1998	2043	33.05	2688	43.49	3354	54.27
1999	2064	28.31	2703	37.07	3292	45.16
2000	2022	25.98	2305	29.62	3251	41.79

Semiarid Lands (ASALs) and is suited for pastoralism, except in areas where irrigation has been developed.

Rainfall is highly unreliable and unpredictable for most of the periods and the country has experienced recurrent, persistent droughts in the years 1974, 1984, 1994, 1996 and 1999, all associated with significant declines in production. Furthermore, soils are poor and highly erodable and the environments are fragile and, as such, expansion of agricultural production into these environments leads to soil mining. The country's agriculture has remained predominantly rain-fed even as it continues experiencing recurrent droughts and chronic food deficits. These diverse climatic factors make it extremely difficult to sustain maize productivity and reap the benefits associated with the Green Revolution technologies across the various environments.

The national average maize yield of 1.7 t/ha is very low by international standards, but yields differ markedly both across the main agro-climatic zones and among different farmers within each zone. In the marginal environments (mainly in the eastern parts and the lake basin) with 30% of the national area under maize, yields range from 0 to 1.7 t/ha depending primarily on weather conditions. The bulk of maize production in these zones involves inter-crops with beans, cowpeas, pigeon peas, sorghum and millets. Most producers use local unimproved seed, and fertilizer application is minimal. In upwards of 3 out of 10 years the maize crop in these zones fails due to drought. Producers persist in marginal production areas because maize is a major food source and is cheaper to produce than to acquire from the market in those years when they do get a crop. Maize also becomes a major cash crop in good rainfall years.

In the mid-altitude zones that cover 55% of the maize area, inter-crops with beans dominate. Yields range from 1.6 t/ha to 2.6 t/ha. Here the range is less dependent on rainfall and reflects more the production practices used by the farmers. Maize yields in the prime production areas of the lower highlands

– the 'grain basket of the north rift' – range from 2 to 6.5 t/ha in normal years. Yields reflect production practices, namely: the quality and timeliness of land preparation and weed control, the use of certified seeds and the quantity of fertilizer used.

Low levels of adoption of improved technologies

Farmers adopted parts of the technology packages introduced in Kenya in the late 1960s and early 1970s but missed out on the synergies to be derived from the use of these technology packages. Input use among farmers, particularly smallholders, has been low and declining. The quantum index for all non-factor inputs has been almost constant and above the price index until 1989, while the price index has been increasing throughout the entire period (Fig. 10.6). Agricultural input prices recorded a dramatic increase reaching 427 in 1994 with a slight decline in 1995 but rose again in 1996. The rapid increase was attributed to inflationary conditions and the weakening of the Kenyan shilling.

The input prices are also sensitive to exchange rate policies because most of the inputs are imported or have large import components. The level of input use has, however, remained more or less constant since the mid-1980s. This is explained by the fact that only a few farmers, mainly those in the large-scale and plantation sectors who can afford to, use purchased inputs at high levels. Hence, despite the increase in input prices, they continue using the same amounts of inputs for as long as it is profitable to do so. Seed maize use on the other hand increased, peaked in 1978 and has been almost constant since. However, farmers decry the high cost of seed which, at times, is of low quality.

The Kenya Seed Company (KSC), a quasi-private company inherited from the colonial government, undertook maize seed multiplication and distribution prior to the reforms. The KSC had the legal monopoly to produce, process and distribute certified maize seed. It was reputed to have developed an extensive and elaborate network of seed marketing which caused the rapid diffusion and adoption of hybrid and composite seed even among the smallholder farmers in remote areas. Fig. 10.7 shows the trends in hybrid maize seed sales by the Kenya Seed Company.

Later, farmers began recycling the seed occasioning a decline in seed maize sales. This was accompanied by land subdivision with the smaller farmers opting not to buy certified seed. Thereafter, hybrid seed sales have fluctuated but have generally been stagnating. By 1967, over 50% of the seed maize sales went

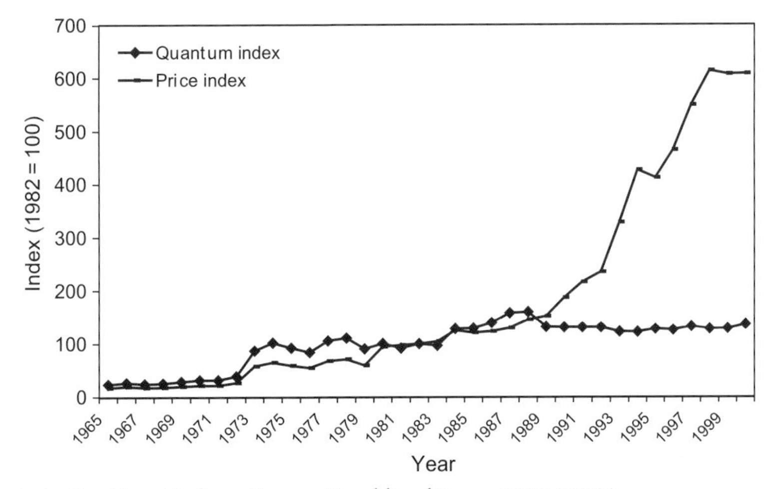

Fig. 10.6. Agricultural input indices. (Source: Republic of Kenya, 1965–2001.)

to small farmers. By 1975, half of all maize farms in high potential areas east of the Rift Valley used hybrids. With the introduction of composites and drought resistant varieties (Katumani Maize), use of improved seed expanded to the medium and marginal areas. Gerhart (1975) also noted that virtually all farms in favourable climatic zones used hybrid maize seed.

The adoption of inorganic fertilizers on the other hand closely followed the adoption of certified maize seeds in the large farm sector. However, chemical fertilizer use in Kenya was mainly confined to large-scale commercial farms and therefore to cash crops such as coffee, tea, sugarcane and maize (Oluoch-Kosura and Chege, 1994). With the removal of restrictions on African farmers to grow cash crops following the attainment of political independence in 1963 and with the need to intensify production on diminishing farm sizes, the pressure to increase fertilizer use increased. Trends for fertilizer use and imports are illustrated in Fig. 10.8.

Fertilizer imports have been increasing over the years and almost equal the total amount used since there is no domestic supply of inorganic fertilizer. Fertilizer application rates have also increased, albeit marginally, and have been constant for most of the years. This can be attributed to the fact that only the large farms that can afford to purchase fertilizers have been applying them consistently. Fertilizer use rose from 38,000 t in 1963 to 200,000 t in 1980. In the 1990s national fertilizer use has fluctuated between 174,000 and 285,000 t but has been below the potential of 600,000 t. Fertilizer use also rose from 7 kg/ha in 1965 to a high of 31 kg/ha in 1986 then fell to 19 kg/ha in 1995 before picking up

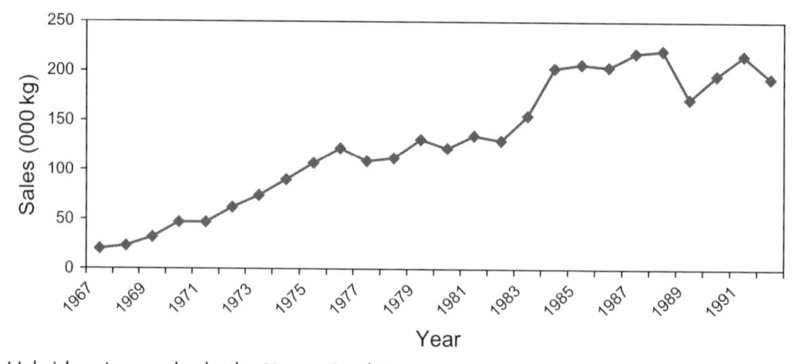

Fig. 10.7. Hybrid maize seed sales by Kenya Seed Company. (Source: Kenya Seed Company, 1967–1992.)

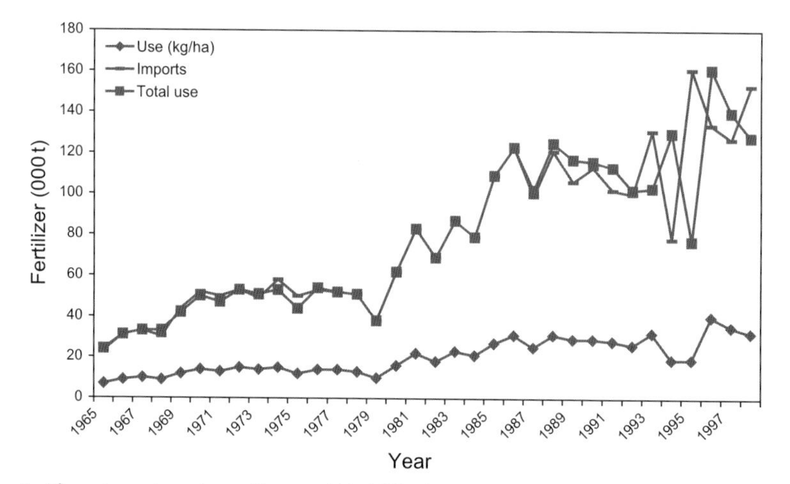

Fig. 10.8. Fertilizer imports and use. (Source: World Bank, 2002.)

again to 30 kg/ha in 1998. However, most of this fertilizer is directed towards cash crops but occasionally spillovers occur to the food crops.

Pesticides are also major non-factor inputs, which are necessary for intensified agricultural production. The pesticide market has evolved over the years as a private market. About 90% of the pesticides imported are already formulated or are for further formulation in the country. The importation policy, however, favours already formulated compounds since they are exempted from import duties while technical grade imports for local formulation face import duties. This does not encourage domestic industries to produce cheap pesticides. Other factors which reduce pesticide use include lack of appropriate technical awareness by farmers regarding return to their use, the subsistence or near subsistence nature of production, weak extension, and poor regulation of imports, manufacturing and distribution, which at times allows for uncertified and sub-standard products in the market.

The use of machinery is not adequately developed in Kenya and most farmers rely on simple hand and animal powered tools. However, mechanization is higher in large-scale farms where tractors and allied equipment are widely used. The use of heavy machinery is encouraged through zero-rated custom duties and value added tax. Hand and animal drawn equipment, however, has customs duties and value added tax charged. Thus, small-scale farmers are disadvantaged in the use of machinery. For example, the high cost of machinery for small-scale maize producers relative to that for large-scale maize producers is partly responsible for the higher profitability of the enterprise for the latter (Nyangito and Ndirangu, 1997). The quality of locally manufactured implements is poor while there is a lack of appropriate machinery for special categories of smallholder production.

Domestic cost of producing maize is high since Kenyan agriculture is heavily taxed. Tariffs and excise duties on diesel, tyres and spare parts raise their prices and make Kenya a high cost producer in comparison with some of the countries it imports maize from. The costs of fertilizer and farm machinery are higher than in the neighbouring countries and are the highest in the region (Argwings-Kodhek, 1999). The cost of transport is high in Kenya due to high taxes on transport equipment and vehicles as well as their spare parts, but also due to the poor state of all categories of roads. Kenyan maize farmers do not use many of the cost reducing technologies available. Although over 70% of Kenya's land is arid, little irrigation is practised and the country continues to rely heavily on rain-fed agriculture. Even in situations where irrigation infrastructure is developed, little success has been achieved. The National Irrigation Board (NIB) is mandated to undertake irrigation development and currently operates six large-scale irrigation schemes out of which only one is currently fully operational with another two being partly operational. These schemes were mainly used for rice production and received heavy government subsidization. With a change of policy in the early 1990s, these schemes collapsed. Even in the small-scale irrigation sub-sector little maize production is done under irrigation since this exercise is not economically viable. Closely tied to low levels of adoption of improved technologies are poor agronomic practices, inadequate land preparation, untimely planting, sub-optimal timing of sales, lack of soil and water conservation and maize cultivation on slopes – all being factors that have been cited as explanations of the declining maize production trends (Hassan *et al.*, 1998b; Argwings-Kodhek, 1999; Karugia, 2003). Maize research must put special emphasis on developing and testing technologies that will help mitigate erosion and conserve water and soils.

Poor infrastructure

A major problem in Africa today, Kenya included, is that markets do not work for the poor. Many poor smallholders cannot access markets due to poor infrastructure among other factors. Access to markets in Kenya by producers is constrained by impassable rural roads and lack of market information. This allows the few traders in local markets to offer producers low prices because of their low bargaining power. The country's road network is about 150,000 km with a fleet of more than 350,000 vehicles (Kariuki, 2001). The state of Kenya's infrastructure has deteriorated to the extent that it has become a hindrance to growth, as it is not feasible for smallholder farmers to deliver their produce to the market. It is currently characterized by a poor state of road network, inadequate railway network, unreliable and costly electricity, inadequate housing and poor quality of water supply, poor telecommunications and a woefully lacking information and communications technology (ICT) infrastructure. Although Kenya's road transport is considered to be impressive in quantitative terms for a developing country, large portions of rural roads are poor and impassable, particularly during the rainy season.

Market infrastructure is dominated by storage facilities for grains owned by NCPB across main towns in the grain (maize and wheat) producing areas. However, at farm level, storage facilities are poor and as a result postharvest losses are high. Kenya continues to rely on an infrastructure base, designed by the colonial masters to serve their interests, that runs from the hinterland to the coast. However, not much expansion of rural feeder roads has been undertaken after independence and most farmers live far away from roads.

Transportation costs in Kenya are high relative to other countries, estimated at US$6/t/km by road and UScts5.8/t/km by rail. Due to the inadequate transportation network, prices often fluctuate substantially from one region to the other, are seasonally volatile and cannot be easily buffered by measures from outside. Even when maize surplus zones have a maize glut, it is not possible to transport the produce to the deficit zones. Similarly, when technical solutions have become available, lack of infrastructure slows down their transmission from research stations to the farmers. Karugia *et al.* (2003) noted that infrastructure constraints (including storage facilities, market centres, financial institutions, and market information and transport infrastructure) have impeded efficient marketing of maize in Kenya.

Lessons for Sustainable Investment in Agriculture

The rapid productivity increases in maize in the 1960s in Kenya were associated with investment in research which yielded improved maize seed varieties. The early adoption of the improved varieties was accompanied with the adoption of inorganic fertilizers and good agronomic practices, including timely planting, weeding, appropriate spacing and crop protection measures and harvesting in a rain-fed environment. These practices resulted in a process of intensification, which contributed to growth in the agriculture sector in the 1960s. The other sources of growth were area expansion and diversification towards higher valued commodities such as horticulture and dairy. The factors which enabled the early adoption of the practices in maize production were not sustained, resulting in a decline in productivity from the mid-1970s. This scenario included erosion of functional institutions and incentives, macro-economic imbalance, poor infrastructure and information networks; and declining investment in research and human capital. Stable and predictable local, national and international institutions are crucial for any form of investment. At the farm level, maize producers in Kenya soon after independence witnessed institutions which supported activities that spurred growth. Subsequently, however, these institutions were eroded, either through mismanagement or lack of transparency and accountability.

These institutions included central and local government, parastatals, cooperatives and NGOs. The farmers, especially the small-scale farmers who had adopted intensification measures, were no longer effectively linked to either input or output markets. The market signals were no longer consistent because of the frequent policy reversals from central government. These led farmers to revert to subsistence maize production. Faced with serious macro-economic imbalance, Kenya was influenced by the IMF and the World Bank to embark upon Structural Adjustment Programmes to restore the balance, beginning mid-1980s. The comprehensive austerity measures adopted by government resulted in cuts in public spending on vital areas supporting growth in agriculture. The sequencing and timing were so poor that the productive sector became disrupted. When the reforms failed to yield the intended results, the government adopted a stop–go approach, which made the situation worse.

Farmers lost confidence in continued adoption of the practices, which earlier had increased yields in maize, and reverted to subsistence production. Reduced public expenditure on research and information reinforced the problem of lack of availability of viable innovations. Where technology was on the shelf, the information system was ineffective either on account of accuracy or timeliness. In cases where the messages were transmitted, it was mainly male farmers who received the message and yet the female farmers were the users of the technology. All these conditions and circumstances help to explain why intensification measures failed to achieve the levels of productivity increases witnessed in Asia.

Similarly reduced public expenditure resulted in low investment in infrastructure and human capital. Road networks are essential in linking farmers to markets. In cases where the infrastructure is weak, access to markets is severely restricted. The result will be farmers reverting to subsistence production and the urge to increase production will be diminished. With an ageing farming population and a low level of investment in human capital, the continued adoption of improved practices cannot be taken for granted. Kenya needs to find its way back to the technology adoption road.

Acknowledgement

The assistance provided by Jonathan M. Nzuma of University of Nairobi and Stephen Wambugu of Kenyatta University in the development of this chapter is gratefully acknowledged.

References

Argwings-Kodhek, G. (1999) Background document for the maize/wheat/fertilizer sector: policy issues facing the maize sub-sector in the North Rift. Tegemeo Institute of Egerton University, Njoro, Kenya.

Block, F.R. and Timmer, P. (1994) Agriculture and economic growth: conceptual issues and the Kenyan experience. *HIID Discussion Paper* No. 27. Harvard Institute for International Development, Cambridge, Massachusetts.

CBS (2000) *Statistical Abstract 1963–2000*. Central Bureau of Statistics Government of Kenya, Nairobi, Kenya.

Gerhart, J. (1975) *The Diffusion of Hybrid Maize in Western Kenya*. Princeton University, New Jersey.

Hassan, R.M., Corbett, J.D. and Njoroge, K. (1998a) Combining geo-referenced survey data with agro-climatic attributes to characterize maize production systems in Kenya. In: Hassan, R.M. (ed.) *Maize Technology Development and Transfer: a GIS Application for Research Planning in Kenya*. CAB International, Wallingford, UK.

Hassan, R.M., Njoroge, K., Njore, M., Otsyula, R. and Laboso, A. (1998b) Adoption patterns and performance of improved maize in Kenya. In: Hassan, R.M. (ed.) *Maize Technology Development and Transfer: a GIS Application for Research Planning In Kenya*. CAB International, Wallingford, UK.

Ikiara, G.K., Jama, M. and Amadi, J.O. (1995) The cereals chain in Kenya: actors, reforms and politics. In: Gibbons, P. (ed.) *Markets, Civil Society and Democracy in Kenya*. Nordiska Afrikainsttitutet, Uppsala, Sweden.

Kariuki, J.G. (2001) *Study on Agricultural and Rural Development Strategy for the East African Community. Country Report Kenya*. Nairobi.

Karugia, J.T. (2003) *A Micro-level Analysis of Agricultural Intensification in Kenya: the Case of Food Staples*. A final report submitted to Lund University, Department of Sociology, Lund, Sweden.

Karugia, J.T., Wambugu, S.K. and Oluoch-Kosura, W. (2003) *The Role of Infrastructure and Government Policies in Determining the Efficiency of Kenya's Maize Marketing System in the Post-liberalization Era*. A research report submitted to the International Food Policy Research Institute (IFPRI) 2020 Vision Network for Eastern Africa.

Kenya Seed Company (1967–1992) Annual reports. Ministry of Agriculture, Government Printer, Nairobi.

Ministry of Agriculture (1963–1999) Annual reports. Nairobi.

Nyangito, H. and Karugia, J. (2002) The impact of globalization and internal policy changes on the agricultural sector in Kenya. In: Bigman, D. (ed.) *Globalization and Developing Countries: Economic Potential and Agricultural Prospects*. International Service for National Agricultural Research ISNAR, The Hague, The Netherlands.

Nyangito, H. and Ndirangu, L. (1997) Farmers' response to reforms in the marketing of maize in Kenya: a case study of Trans Nzoia District. *IPAR Discussion Paper* No 003/97. Institute of Policy Analysis and Research, Nairobi.

Oluoch-Kosura, W. (2001) *Agricultural Input-Output Marketing in Remote Areas. Kenya Case Study*. A Report to the Food and Agriculture Organization, Sub-regional Office for Southern and East Africa.

Oluoch-Kosura, W. and Chege, D. (1994) *Evolution of Fertilizer Policy and Marketing in Kenya.* Agricultural Policies and Food Security in Eastern and Southern Africa: Proceedings of a symposium held at the Kenya Commercial Bank, Nairobi.

Republic of Kenya (1965–2001) Economic survey: various issues. Government Printers, Nairobi.

Republic of Kenya (1983) *National Development Plan 1983–1988*. Government Printers, Nairobi.

Republic of Kenya (1994) *National Development Plan 1994–1996*. Government Printers, Nairobi.

Republic of Kenya (1995–2001) Statistical abstracts. Government Printers, Nairobi.

Republic of Kenya (1997) *National Development Plan 1997–2001*. Government Printers, Nairobi.

Rukadema, M., Mavua, J.K. and Audi, P.O. (1981) The farming systems of lowland Machakos District, Kenya. Report on Farm Survey Results from Mwala location, Technical Report No 1.

World Bank (1990) *Kenya: Agricultural Growth Prospects and Strategy Options*. A World Bank Sector Report. Agricultural Operations Division, Eastern Africa Department.

World Bank (2002) *African Development Indicators*. Washington, DC.

11 From Ujamaa to Structural Adjustment – Agricultural Intensification in Tanzania

Aida C. Isinika,[1] Gasper C. Ashimogo[2] and James E.D. Mlangwa[3]

[1]*Institute of Continuing Education, Sokoine University of Agriculture, Morogoro, Tanzania;* [2]*Department of Agricultural Economics and Agribusiness, Sokoine University of Agriculture, Morogoro, Tanzania;* [3]*Faculty of Veterinary Medicine, Sokoine University of Agriculture, Morogoro, Tanzania*

This chapter examines how government policies and actions have influenced market conditions, other institutions and infrastructure, and how these in turn have impacted on agricultural productivity in general and on food crop intensification in particular. The analysis links up with the overall *Afrint* framework and is based on the premise that preconditions within a country regarding the food situation, agro-ecological conditions, community organization and institutions influence the manner in which various stakeholders react. These include the state at the macro level and farmers and market entrepreneurs at the micro level.

The chapter is mainly based on findings of the Tanzanian leg of the *Afrint* research project, and brings out some macro and micro aspects of agricultural development that have taken place since independence. The micro study was done in two regions, Morogoro and Iringa (Fig. 11.1), focusing on two staple grains, rice and maize. However, additional empirical findings are also drawn from other studies that address the issues at hand, particularly those focusing on the role of infrastructure and institutions in enhancing agricultural productivity. The *Afrint* study compared the situation of food production before and after Structural Adjustment Programmes (SAPs), which in Tanzania were introduced during the mid-1980s.

The chapter is divided into three main sections. Section one gives an overview of the role of the Tanzanian government in the development of markets, institutions and transport infrastructure. The second section assesses the micro-level effects of various government policies and policy instruments. The chapter concludes by summarizing the roles of actors, including private entrepreneurs and farmers, and how these have influenced food crop intensification and self-sufficiency in the country.

A Historical Perspective

Until the early 1970s, Tanzania was generally self-sufficient in food. Thereafter, food imports have been a recurrent phenomenon to meet shortfalls in production from time to time, especially of rice and wheat. Supply and demand analysis between 1993/94 and 1997/98 indicate that on average local production meets demand by 111% for maize, 67% for paddy, 85% for wheat and 97% for sorghum (NEI, 1999). Official records of imports of major grains indicate highly fluctuating trends, as illustrated for maize in Fig. 11.2. Thus Tanzania is still a net importer of rice and wheat. In the case of maize, grain imports were sometimes high (especially between the mid-1970s and mid-1980s) due to poor pricing and institutional policies. It was cheaper for the National Milling Corporation (NMC), a government parastatal which had monopoly in the procurement and distribution of grain, to import maize than to buy locally (Hanak, 1986; Mlay, 1988).

The Arusha Declaration in 1967 marked a major policy shift from a market economy towards a socialist orientation. After 1967, improved food production was associated with two main factors: (i) increased use of chemical fertilizer, which coincided with expanding extension services; and (ii) expansion of area under production. However, the mid-1970s (1973–1976) witnessed a decline in farm production, including that of food crops, mainly due to drought (in 1973–1974) and the massive displacement of rural people into new Ujamaa[1] villages. External factors such as the oil shock of 1973 also changed relative prices such that imported agricultural inputs became more expensive.

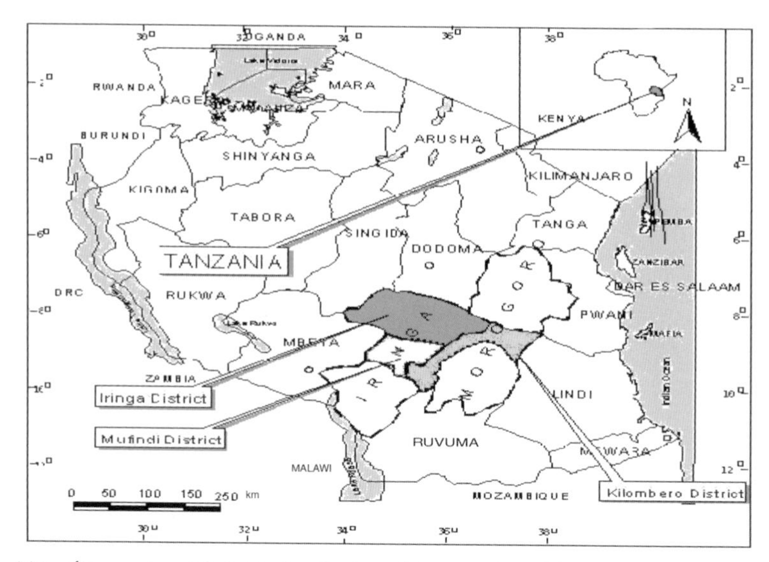

Fig. 11.1. Map of Tanzania – Administrative Regions. (Source: adapted from ESIRIDATA & Maps (1999) Map of Tanzania Administrative Regions.)

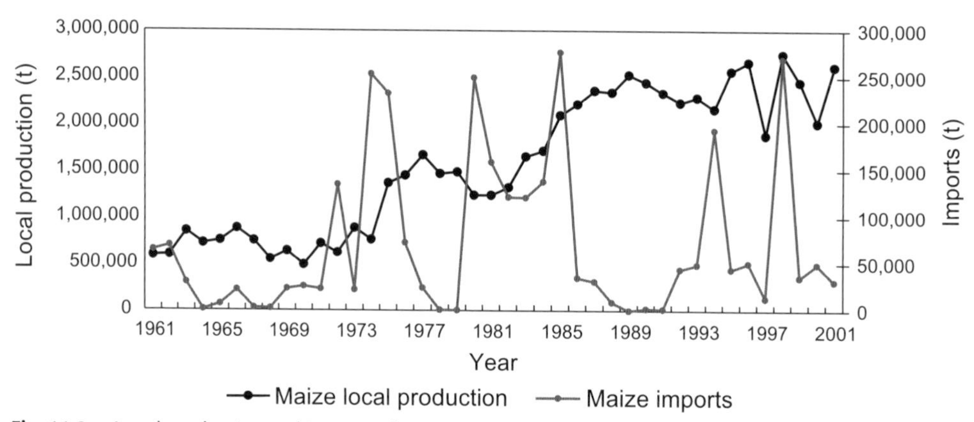

Fig. 11.2. Local production and imports of maize. (Source: Isinika *et al.*, 2003.)

After 1975, a new set of factors came into play. These included subsidized input supply and pricing policies for food and cash crops. In the case of food crops, subsidized inputs were administered under the National Maize Project (1975–1989) covering 13 regions, and in 1982 included the most suitable regions (Bryceson, 1993a). In this respect, we observe some similarity between policies pursued by the government of Tanzania and the Asian model by way of government intervention through subsidized inputs, targeted commodity price and expanded extension services (refer to Chapters 2–4, this volume). Unlike the Asian experience, however, policies in Tanzania were weak on credit supply and administration, as well as on development and maintenance of irrigation and marketing infrastructure. Farmer-owned institutions were suppressed (mostly on account of corruption) in favour of government monopoly institutions, which turned out to be equally corrupt and inefficient. In this respect, government intervention was not favourable to small farmers.

Although the government introduced subsidized agricultural input supply during this period, some scholars (Hanak, 1986; Mlay, 1988) have argued that in the government's quest to promote the industrial sector, mainly through the Basic Industries Strategy of 1976, the agricultural sector became highly marginalized, despite being hailed as the engine of economic growth. For example, maize production in Iringa region, which is one of the leading maize producing areas in the country, shows a declining trend between 1978 and 1980 as well as between 1982 and 1985.

In 1986 Tanzania embarked on an economic transformation towards a market economy. This entailed a number of steps involving liberalization of commodity and financial markets as well as other institutional reforms. The strategy also involved an infusion of donor funds from the World Bank (2000) and other bilateral donors. Analysis of data indicates production gains in the immediate post-liberalization period up to the early 1990s. However, such gains have not been sustained, especially after 1994, when all subsidies for the agricultural sector, both explicit and implicit, were removed.

Following the introduction of market liberalization during the mid-1980s, nominal fertilizer prices started to rise (Fig. 11.3) and, as a consequence, both the number of farmers using fertilizer and application rates fell,

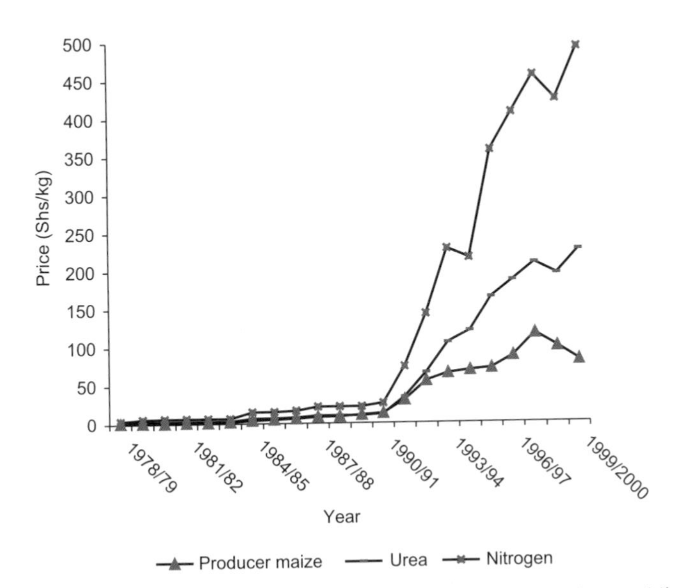

Fig. 11.3. Nominal maize producer and fertilizer prices 1978/79–1999/2000. (Source: Ministry of Agriculture and Food Security (various sources); Tanzania Fertilizer Company (various sources).)

especially after 1994. Although nominal maize prices have increased during the same period, they have done so at a relatively low rate compared with fertilizer prices, and with the phasing out of fertilizer subsidies in the early 1990s the gap between maize and fertilizer price has gradually widened (Fig. 11.3). Respondents of the *Afrint* micro study asserted this deterioration in relative prices.

As a result of all these factors, including fluctuating rains, production trends for both maize and paddy have followed an irregular path (Isinika *et al.*, 2003). In the case of maize, there is a high degree of correlation between annual production on the one hand and annual fertilizer supply (a proxy for access) and area under production on the other. Yield trends have also fluctuated accordingly. The yield of maize rose from 1.0 t/ha in 1971 to about 1.9 t/ha in 1987, an increase of about 90%. In between, significant yield declines were experienced in 1973 due to drought and the oil crisis. Additional yield declines were observed in 1978 and again in 1991/92, both associated with significant declines in fertilizer imports.

Improvement in food production after 1986 has come from both area expansion in response to market incentives and increased yield. In general, yield trends have been fluctuating since independence, probably because of the low and erratic supply of agricultural inputs, especially inorganic fertilizer and improved seed, which have been highly dependent on import grants and donor supported development projects. Nevertheless, a general upward trend of yield can be observed, especially for maize (Fig. 11.4).

In the case of paddy, the most conspicuous yield gain occurred in 1979–1985, that is, until the introduction of SAP. Since then, rice yield has decreased steadily (Fig. 11.4). In Tanzania, 74% of the total paddy area is rain-fed lowland rice, 20% is upland rice and only 6% is irrigated (Kanyeka *et al.*, 1995). Most of the production is small scale with farm units of about 0.5–2.4 ha. Where water management and spacing are optimum, but use of fertilizers and pesticides are low to medium, yields within the range of 1.3–2.4 t/ha have been obtained. There are a few large-scale irrigated farms owned by the National Agricultural Food Corporation (NAFCO). Under the management of Chinese experts, NAFCO rice farms at Mbarali produced up to 8 t/ha. While higher yields from the state farms must have contributed to the higher average national yield level observed at the time, such influence could not have been significant, given the dominance of small-scale rice producers in the country.

Paddy production in Tanzania as a whole has varied between 500,000 and 700,000 t/year during the 1990s (NEI, 1999). According to the World Bank, paddy is the fastest growing food crop, with production having increased fourfold between 1995 and 1998. However, poor weather can reduce production significantly, as was the case during 1991/92 and 1996/97, when production was halved due to drought. Being a tradable crop, local prices are influenced by exchange rates and international prices. Also, rice has a

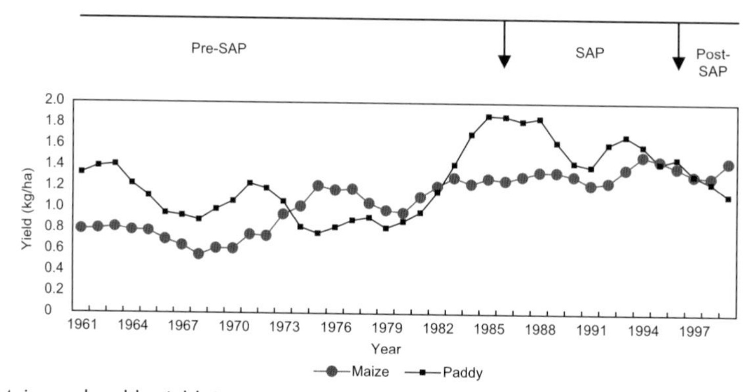

Fig. 11.4. Maize and paddy yield: 3-year moving average (1961–2000). (Source: Isinika *et al.*, 2003.)

high-income elasticity of demand. As incomes rise, demand for rice will increase and create a demand-pull within the local market (World Bank, 2000).

In its effort to address food self-sufficiency, food security and nutritional concerns, the government of Tanzania has formulated and passed various policies and strategies. Among these are The Agricultural Produce Act in 1962 and the Food Security Scheme of 1976, which aimed at reducing the kind of food imports it had experienced in 1971/72. In response the government organized campaigns that exhorted farmers to work harder. These included *Kilimo cha Kufa na Kupona*[2] in 1971 and *Kilimo ni Uhai*[3] in 1972. The government also formed the National Milling Corporation (NMC) in 1973, as a monopoly parastatal for procurement of food crops and storage of national grain reserve. The Strategic Grain Reserve was formed in 1983 under the NMC. Even the villagization campaign (1973–1975) had the ultimate goal of increasing yields of peasant shifting agriculture and raising the total marketed surplus. In 1971/72 the government provided free agricultural inputs and tractor ploughing services to new Ujamaa villages in Iringa region (Bryceson, 1993a).

The first National Agricultural Policy of 1983 had as one of its objectives the advancement of food self-sufficiency and meeting of national nutritional requirements. The National Food Strategy of 1984 aimed at increasing the production of food crops while improving the nutritional status of disadvantaged groups. In 1986 the Food Security Review Mission identified various factors, including drought, flood and low use of industrial inputs, as causes of low productivity leading to food shortage. The Mission proposed remedial measures that included improvement in the delivery of agricultural services.

In 1992, the government, with the assistance of the Food and Agriculture Organization (FAO), formulated a comprehensive Food Security Programme. During the same year the National Food and Nutritional Policy was launched to coordinate food and nutrition activities undertaken by various sectors. The Special Programme for Food Security of 1995 aimed at increasing productivity of major staples through improved agricultural extension services and by involving farmers' groups. Currently (2003) the National Food Security Policy is under review.

Institutional and infrastructure conditions

In Tanzania there have been numerous occasions when the government has interfered with the operations of agricultural markets. As early as 1962 the government introduced the Agricultural Products Act to control prices and marketing. This led to the formation of the National Agricultural Products Board (NAPB), which was transformed into the National Milling Corporation (NMC) in 1973. The NMC and its predecessor (NAPB) had monopoly power in the marketing of grain under a three-tier system involving Cooperative Unions at the intermediate level and Primary Cooperatives at the lower level. In addition the NMC had to maintain a strategic grain reserve. Cooperatives were promoted in order to replace Asian middlemen in crop trading (Bryceson, 1993a,b; Gordon, 1994) and they were given monopoly-buying power at the farm level. Cooperatives were encouraged even where the economic base and adequate local skills were lacking (Hanak, 1986). Rising corruption and inefficiency within cooperatives led to their dissolution by the government in 1976 (Mlay, 1988; Bryceson, 1993a).

Cooperatives were reintroduced in 1984, but without having full autonomy. Therefore farmers and the public in general did not enthusiastically embrace the revived institutions. In 1991, a new Co-operative Act was introduced to provide more autonomy and in 1997 a cooperative policy, which advocates evolution of cooperatives as independent economic entities, became operational. However, cooperatives remain weak due to members' apathy, little working capital and limited or lack of management skills (United Republic of Tanzania, 2001a).

Since independence, the government of Tanzania has striven to pursue policies that were intended to be pro-poor and pro-food production (Bryceson, 1993a). However, the absence of or weak capacity in policy analysis led to many unintended effects, often undermining the primary objectives of government policies. In many cases, political concerns such as party supremacy (Bryceson, 1993a) and a strong egalitarian bias in government policy outweighed economic considerations. For example, pan-territorial pricing was introduced in 1976 in order to achieve inter-regional equity and stimulate production in remote areas through price incentives, but this was done at the expense of production areas that were close to markets. Regional pricing that came into force in 1982 similarly taxed areas that had a competitive spatial advantage while subsidizing remote areas (Ndulu, 1980, in Hanak, 1986; Bryceson, 1993a). Under regional pricing, premium prices were paid for sorghum, millet and cassava in drought-prone areas, which led to accumulation of these less preferred food types in NMC stock (Hanak, 1986; Mlay, 1988). Consequently the NMC sustained heavy losses and became highly indebted (FAO, 1986).

Under the price control regime, crop prices that were set by the government represented a significant implicit taxation of up to 99% and real producer prices eventually became negative for some commodities (Mlay, 1988). Producers responded to policies that were particularly restrictive and unfavourable for cash crop cultivation by switching to production of food crops, which could be sold at higher prices in parallel markets (Isinika, 1995, unpublished dissertation). In the case of maize and rice, this pattern gained prominence from 1979 (Bryceson, 1993a). Pan-territorial prices for food crops were eventually discontinued in 1983 under government-initiated efforts to liberalize the economy under the National Economic Survival Programme (NESP). Further economic and market liberalization followed under the Economic Recovery Programme (ERP) from 1986 onwards.

Excessive taxation of the agricultural sector at local level has also been blamed for having a negative impact on agricultural intensification. A large number of taxes have been collected at various levels. In the *Afrint* study district of Kilombero, for example, farmers lost an opportunity to sell rice and obtain credit from traders due to a high tax rate imposed on the latter by the District Council (Ashimogo *et al.*, 2003). Effective from July 2003, parliament abolished some 40 taxes, which were considered to be a nuisance to the people or not cost effective to collect. However, knowledge regarding which taxes have been abolished or reduced is generally lacking. Pricing policies, which prevailed during the 1970s and 1980s, had other unintended effects. The overvaluation of exchange rates, for example, taxed exports and subsidized imports to the extent that it sometimes became cheaper for the NMC to import maize than to buy locally, with detrimental effects on local production.

The period from 1967 to 1986 witnessed increasing government control in the marketing of agricultural inputs and agro-chemicals as well. Government parastatal organizations, of which there were more than 200, were formed as multipurpose production and marketing agencies. Subsidies for fertilizer and pesticides were introduced in high potential areas between 1976 and 1984. Regional pricing also provided an implicit subsidy until 1989. In addition, the overvalued exchange rate continued to provide an implicit subsidy on fertilizer imports of up to 80% (1989). Due to budgetary pressure, this implicit subsidy was gradually phased out and became nil in 1994/95. Following the removal of subsidies maize production in remote parts of the Southern Highlands decreased by 13%–19% (World Bank, 2000).

Despite such negative effects from an efficiency point of view, explicit and implicit subsidies to the Southern Highlands have had a positive effect on agricultural intensification and on food crop yields in particular. The Southern Highlands, which consumed 35% of all fertilizer between 1973 and 1975, increased their share to 65% between 1989 and 1991 (World Bank, 2000). Consequently, this region became the grain basket of the nation producing about 44% of the total maize grain in 1989/90 and roughly 45% of the grain marketed in Dar-es-Salaam during the first half of the 1990s. However, following the general removal of subsidies, Dodoma has come up as a significant supplier due to its spatial advantage and in 1997/98 provided around 46% of the maize sold in Dar-es-Salaam. Effective from July 2003, the government has restored subsidized fertilizer supply to the Southern Highlands in order to ensure national food security, albeit at a much lower level than previously. The subsidy is administered in the form of a 70–90% compensation of the cost of transporting the fertilizer to target regions; this compensation constitutes about 8% of the retail price at destination.

In the 2003/04 cropping season the government allocated some 54,000 t of different types of subsidized fertilizer worth 2 billion shillings for distribution in the four Southern Highland regions of Iringa, Mbeya, Ruvuma and Rukwa. Iringa region was allocated 31% of this amount. However, it is unlikely that the subsidy will encourage many farmers to increase the intensity of fertilizer use because the resultant price differences between the subsidized fertilizers and other sources of fertilizer in the open market are not significant. In addition, the subsidized fertilizer constitutes only a small fraction of total fertilizer demand in the regions.

Government policy and institutional changes

It has been noted earlier that in the case of the Asian Green Revolution, institutions played a critical role in agricultural transformation towards higher productivity. These include formal institutions as well as informal ones, which take up mobilization and service delivery. In Tanzania, under the first 5-year development plan of 1964–1969 the government introduced the transformation approach, which was to be realized through an expanded extension service (Mattee, 1978), a research system emphasizing food crops (Isinika, 1995, unpublished PhD dissertation) and increased capacity for technical agricultural training (Ngugi *et al.*, 2002). Agricultural transformation was also to be realized through capital-intensive settlement schemes. This, however, failed and had to be abandoned by 1966. The villagization policy, which was implemented during the early 1970s, was another attempt to modernize smallholder agriculture (Bryceson, 1993a). All these efforts led to expansion of the public sector. While expansion of infrastructure was realized mostly through donor funding, expansion of such services put a strain on the recurrent budget. The public sector grew at about 15% per annum, which reduced availability of operational funds and consequently the quality of services (Mlay, 1988).

After independence both agricultural research and extension services changed focus towards food crops. Extension services were expanded in order to have at least one agent in every ward. The expansion of research stations aimed at increasing their spatial distribution across each agroecological zone in order to meet increasing demands for locally adapted technologies. These changes corresponded with expansion of infrastructure and human resources during the 1970s up to the mid-1980s. More than 17 new agricultural research stations have been established since independence often through donor support. Programmes for training farmers were introduced at Farmers' Wings of Agricultural Training Centres under the Ministry of Agriculture. While these were desirable changes, they led to thin distribution of resources, especially for recurrent expenditure, resulting in discontinuation of research programmes as soon as donor funding ceased (Isinika *et al.*, 2003). Moreover, both research and extension services have gone through several changes since independence, some of which have had negative effects on staff morale and efficiency.

Cooperatives had been a dependable source of credit for farmers in many rural areas. After their demise in 1976, the responsibility for providing inputs fell under agricultural parastatals, which had the mandate to coordinate production and marketing of food and export crops. Grains and pulses fell under the jurisdiction of the NMC, which, as stated earlier, became entangled in debt due to government price policies as well as mismanagement. Following the revival of cooperatives, the government is promoting them as potential sources of finance for agriculture. Under the Agricultural Sector Development Strategy, the government intends to facilitate and promote formal linkages between micro-finance institutions (MFI), which include Savings and Credit Associations (SACAs) and Savings and Credit Co-operative Societies (SACCOS) (United Republic of Tanzania, 2001a,b). However, such benefits may be realized in the future. In the interim, however, there are no viable sources of finance for the majority of smallholder farmers since they lack collateral and do not qualify to access resources from formal financial institutions. Many non-governmental organizations (NGOs) currently provide short-term credit, but often at very high interest rates and mostly in urban and peri-urban areas.

Farmers' organizations also provide a forum for negotiation with their business partners as well as government and other institutions. When the cooperatives were abolished in 1976, farmers lost this power. Currently, farmers do not have a unified forum behind which they can rally to lobby or negotiate in their favour. As stated earlier, the cooperative movement is very weak. To fill this void, alternative institutional innovations are emerging in the form of Farmers' Networks. The most prominent is probably *Mtandao wa Vikundi vya Wakulima Tanzania*[4] (MVIWATA), which was founded in 1993, initially comprising 22 innovative farmers from six regions of Tanzania, and with technical support from Sokoine University of Agriculture. Currently the membership stands at about 150 local networks that range from five to 15 affiliated groups. Each group has between five and 100 members. There are about 5000 cardholders but the network is estimated to be reaching about 50,000 smallholder farmers.

MVIWATA has the vision to grow and become a strong farmers' organization that will guarantee farmers' participation and representation in the socio-economic and political decision-making process. MVIWATA has accomplished several achievements, including enhancing farmers' confidence to address their problems, developing sound relationships with extension staff and organizing farmers' visits in Tanzania and abroad. The NGO has also organized a number of national and regional workshops involving more than 1000 farmers. Traders on the other hand do not have any formal institutionalized organization. Bryceson (1993a) discusses at length how prior to market liberalization traders were viewed with distrust by the government and farmers alike. Currently, however, their role is recognized but there have been no concerted efforts to facilitate their operations in terms of credit or improving their institutional efficiency.

Land is another important dimension for crop production. In Tanzania, land ownership and transfer has been by two tenure systems, customary and statutory, which co-exist. These are based on the Land Ordinance of 1923 (United Republic of Tanzania, 1994). In 1999 the government of Tanzania enacted two land laws.[5] Both became operational in May 2001. The government's intention is to give land a monetary value so as to facilitate its use as collateral for credit, even in rural areas. The greatest challenge for the new laws is whether they will be accessible and understood by all citizens, particularly in rural areas, and whether they will be able to provide an efficient, effective, economical and transparent system of land administration.

Generally, most villagers continue to make production and investment decisions based on their own perceptions of tenure security, which are guided by customary law. In this respect the new land laws alone should not be expected to bring about any significant changes in food production intensification. Despite the fact that the legacy of the villagization policy in the 1970s in some places has contributed to the eruption of land disputes, and may increasingly do so in the future, most farmers do not feel that their right to land is currently threatened. About 94% of the respondents in the micro study indicated that they had full control of all the land they cultivated including the right to decide on the types of crop to grow or to change land use.

Based on the foregoing discussion it is evident that until 1986 the government, through its policies and strategies, was highly engaged in stimulating food production and marketing systems for the purpose of achieving egalitarian objectives. Some of the government policies that had an impact on agricultural production include decentralization (1972), villagization (1974–1976), dissolution of cooperatives (1976), the Basic Industries Strategy (1976), Restoration of Co-operatives and Local Governments in 1984 and 1985 respectively, Structural Adjustment strategy (SAP) and the National Economic Survival Programme (NESP) in 1982 and 1983 respectively. They all reflect government's grappling efforts to realize egalitarian goals as espoused in the Arusha Declaration. However, as noted by Mlay (1988), institutional changes, which aimed to promote more intensive agriculture, were often applied in isolation and at sub-optimal levels. Moreover, institutional instability prevailed due to frequent policy and organizational changes (Isinika, 1995, unpublished PhD dissertation). Since the introduction of economic liberalization in the 1980s, the government has embarked on a new set of efforts to enhance the performance of the agricultural sector through streamlined policy making and providing an enabling environment for the private sector. Several policies and implementation strategies including that on land (1995), agriculture (1997) and cooperatives (1997) have been developed. The Agricultural Sector Development Strategy (2001) is expected to transform agriculture from the current subsistence level to a modern market-orientated sector by the year 2025.

Private sector under economic liberalization

Between independence in 1961 and the time of economic liberalization the state exerted control of grain marketing in three main areas: imposition of a single marketing channel, imposition of price controls, and restriction on grain movements. The government's motives for intervention stemmed from a profound distrust of private trade. In response to the enormous fiscal strain imposed by the NMC's losses and mounting donor pressure, the government began a sequential programme of food market liberalization in 1984. In response to liberalization, the role of the National Milling Corporation (NMC), the food marketing parastatal, declined drastically, and the private trade in grain grew rapidly. For example, in 1985 private trade supplied 50% of maize to Dar-es-Salaam. By 1992 this figure had increased to 80–90% (Coulter and Golob, 1992).

There is considerable evidence that grain marketing in Tanzania has become more efficient since liberalization. Private sector margins have declined as traders' transfer costs have fallen. Trucking costs have declined because of greater availability of imported spare parts and improved infrastructure (Bevan *et al.*, 1993). The abolition of restrictions on grain movement means that private traders no longer incur the substantial costs of evading state restrictions. Private sector trade appears to be highly competitive and the number of traders has greatly increased since liberalization. There seem to be few barriers to entry, and there is also evidence that staple grain markets have become more spatially integrated (Gordon, 1994).

Government policy and infrastructure development

Agricultural infrastructure is defined in a broad context to include transport, storage, irrigation, communication and other facilities. This means the entire capital stock necessary to facilitate marketing services in space and time. The government normally assumes leadership in planning and financing infrastructure in order to facilitate marketing services. Moreover, the government is supposed to create a conducive institutional framework to promote an efficient marketing system. This discussion focuses on infrastructure that facilitates marketing, namely transport and storage infrastructure.

Gabagambi (2003) reported that about 60% of the farmers in his sample sold their crops at home, 24% at paved roads and 17% at feeder roads. He reports further,

> because of limited movement and lack of assembly markets where exchange of information about market conditions could take place among farmers and traders, farmers rely on traders for price information as reported by 50.5% of the respondents. Other sources of information included neighbours or friends (34.2%) and travelling (17.3%).

Price determination is by negotiation, but the traders are few and the possibility of collusion to offer low prices cannot be ruled out. In fact both the *Afrint* micro study and Gabagambi found that farmers within a village received almost identical prices, which farmers complained to be too low.

In Tanzania, transport cost has been identified as a major constraint facing the agricultural sector (World Bank, 1994; Gabagambi, 2003). The transport and communication infrastructure expanded during the 1970s and 1980s (Nyerere, 1973; Hanak, 1986). However, little maintenance was done during the same interval, and the condition of the road surface deteriorated significantly. Currently, most of the country's rural transportation and communication infrastructure is in a poor state. Tanzania has 85,000 km of road network about 5.3% of which is paved (MCT, 2003) compared with 14% in Kenya (Gabagambi, 2003). The road network is classified as trunk roads (10,300 km) and regional roads (24,700 km). The rest are district, unclassified and urban roads. All the paved roads fall under trunk roads, which means that none of the regional or district roads are paved. According to Gabagambi, the road density per 100 km^2 is 9 compared with 26 in Kenya and 27 in Uganda, but the vehicle density is comparable to that of Kenya at 4 vehicles/km of road. Since agricultural production takes place in areas which are often inaccessible, the importance of developing transportation and other infrastructure for marketing, communication, processing and storage cannot be over-emphasized.

There are very few comprehensive marketing studies which have estimated the share of transportation costs on consumer prices in Tanzania. However, extrapolating from existing figures, one can make some deductive conclusions. For example, the real average retail price of maize during 1995/96 was 165 Shs/kg, which is equivalent to US$0.28.[6] If maize were to be transported from Iringa or Mbeya to Dar-es-Salaam, which are about 600 and 900 km away, respectively, transport cost per kilogram would be US$0.108 and US$0.162, respectively. Thus transport cost would constitute 49% and 60% of the retail price at the Dar-es-Salaam market for maize from Iringa and Mbeya, respectively. Maro (1999, cited in World Bank, 2000) similarly estimated that during 1994/95 60% of marketing cost between Morogoro and Dar-es-Salaam (220 km away) was accounted for by transport cost. Meanwhile, available estimates from Rukwa region for 2002/2003 indicate transfer cost of up to 90% (Isinika *et al.*, 2003). Other studies have shown that transportation cost in Tanzania is two to five times higher than those in Indonesia and Pakistan (Transport Research Laboratory (TRL), 1995 in NIT, 2003).

Focus group members for the *Afrint* micro study reported improvement in the road infrastructure during the post-SAP period, which has enabled buyers to reach quite remote villages. Some villagers have also been able to take their produce directly to central markets within their vicinity. National data on incremental improvement on rural road infrastructure could not be obtained. However, during and after SAP, Tanzania has witnessed the inception of the Transport Sector Recovery Programme (TSRP), in 1987. This has led to marked improvement and growth of infrastructure in the transport sector (MCT, 2003). For example, respondents in Kilombero district cited improvement of the road from Ifakara to Mlimba under the road maintenance programme. However, respondents also observed that transport costs were much lower during the pre-SAP period when government institutions used to buy directly from farmers, thereby providing an indirect transportation subsidy.

A study by Gabagambi (2003) concluded that farmers' access to transport infrastructure influenced: (i) the number of crops grown, being fewer in easy access areas; (ii) use of fertilizer, being higher in accessible areas; (iii) use of pesticides, being higher in easy and medium access areas; (iv) sources and value of off-farm income, being significantly higher in easy access areas; and (v) market orientation, also significantly higher in easy access areas. The study established that if time and distance to reach an unpaved road were reduced, this would increase aggregate productivity, use of industrial inputs and household incomes. This means there is a direct relationship between improvement in road infrastructure and agricultural intensification.

Currently, the National Transport Policy guides development of transport infrastructure in the country. Effective from 2001/02, Tanzania has decided to allocate local resources for major road construction projects, under which the road from the north-western (Kagera) to the south-eastern (Mtwara) corner of the country is supposed to be surfaced within the next 3–4 years.

The Role of Markets, Institutions and Infrastructure at Farm Level

The *Afrint* micro study was done in Iringa region, a major maize producing area, and in Kilombero district within Morogoro region, which is prominent for rice production. The methodology was designed to capture temporal effects of food crop intensification by asking respondents to compare different parameters of crop production (use of improved inputs, markets, infrastructure) at the time the household heads got married relative to the time when the interview was done in August–October 2002. The sample had 400 respondents from ten villages, five in each region. In this way factors that influenced food crop intensification before and after economic liberalization in 1986 were captured and compared. Before discussing the situation in each of the two regions, a summary of the general situation with respect to food crop production and the role of markets, institutions and infrastructure is given below.

Crop production, marketing and technology use

Results from the *Afrint* micro study show that more than half of all respondents in both regions studied reported that agricultural productivity had declined since the time the households were formed, mainly due to poor production technologies and declining soil fertility. The use of improved technologies is low in both regions. Hand hoeing remains the predominant mode of land preparation. Although hybrid seeds and open pollinated varieties (OPVs) were introduced in the traditional maize growing areas much earlier in the 1970s, 82% of the sampled households said they used traditional maize seed varieties at the time of the survey. For paddy, two varieties, Line 85 and Line 88, were introduced in 2001. Also, the use of chemical fertilizer and pesticides is limited. Due to high prices of inorganic fertilizers some 30–40% of the households apply animal manure and only 26% and 39% used chemical fertilizers and pesticides, respectively.

Lack of access to and affordability of improved seed, fertilizer and insecticides are among the most important constraints to increased crop production, explaining the low yields among farmers. Typical crop yields among survey households are generally low, on average less than 1 t/ha for maize and about 1.6 t/ha for paddy. However, with good cultural practices, including use of improved varieties, fertilizers, pesticides and disease control, some farmers manage to obtain considerably higher yields.

In principle, the government still provides extension services but the quality has deteriorated due to many institutional changes and poor incentives for extension personnel. Some regions critically lack extension officers to such an extent that they have a staff:household ratio of 1:1672 compared with the national average of 1:600. Strong extension services and technical assistance are needed to communicate timely information and new developments in technology as well as suitable resource management to farmers and to relay farmer concerns to researchers. However, due to the prolonged absence and/or poor quality of extension services in some cases, farmers do not exert any effective demand for such services.

Limited market access is another constraint. The main outlets for marketed surplus are private traders to whom over 90% of farmers sold their produce. However, food crops are never grown on the basis of prearranged contracts between farmers and private traders. Farmers' responses from the *Afrint* micro study showed that the most important economic factors constraining maize and paddy production for the market were lack of credit, low and/or fluctuating prices, high input prices, unreliable market outlets, expensive hired labour, lack of knowledge about yield-improving farming techniques and chronic illness in the family.

State initiatives, farmer organizations and credit facilities

According to the Agricultural Sector Development Strategy (ASDS)[7] (United Republic of Tanzania, 2001a), several factors that are influenced by the state have contributed to the modest performance of the agricultural sector in the country. The incentive structure over the past decade (post-SAP) has not encouraged growth or investment in the sector. Agriculture's barter terms of trade that measure the relative change in agricultural producer prices compared with the prices of industrial goods have not changed significantly over time.

Nonetheless, price incentives may have motivated households to sell more produce because over a half of the households surveyed acknowledged that the price for maize and paddy are better now than when the households were formed. This is also supported by the fact that 45% of respondents consider that access to market outlets for agricultural produce is better now compared with previous years. However, any benefits accruing from price incentives were probably outweighed by increases in the price for modern inputs (fertilizers and seeds), and it is not surprising that over two-thirds of the households consider that profitability for both paddy and maize has deteriorated.

The public sector is no longer involved in buying any farm produce in the survey villages and formal agriculture related organizations are rare. One main drawback seems to be the poor management and inefficiency of cooperatives. Therefore, the contribution of farmers' organizations to agricultural intensification appears to have been small. In the pre-SAP period some villages had access to credit through cooperative unions but since the mid-1980s such services have ceased to be readily available and use of credit for development activities is not a regular feature within the surveyed villages. Credit schemes that are on and off are unreliable and therefore unlikely to sustain agricultural development in the long run.

Market infrastructure

With regard to infrastructure, the micro study revealed that the most common means of transport in rural areas was carrying luggage as head-loads on foot (51%) and by bicycle (40%). Use of motorized transport is marginal. Farm-level storage is mostly by means of traditional basket granaries, while traders generally store grain in warehouses that are often poorly constructed, leading to potentially high storage losses from pests and moisture.

Despite evident signs of improved economic performance at the macro level following market liberalization, there are several obvious weaknesses in private sector trade. Private traders still operate at a very small scale and have not been able to integrate their operations vertically. Most new traders come from farming backgrounds and they rely on agricultural earnings to raise their start-up capital (Santorum and Tibaijuka, 1992). In a study that was done in 1988, Bryceson (1993a) classifies traders as mobile intermediaries, stationary wholesalers and retailers. The study established that the majority of traders (98%) did not have their own transport. Traders rely on rented trucks, for which markets are incomplete, in the sense that there is no structured institution for service delivery, which increases transaction costs. Moreover, only 17% of the traders owned storage facilities, and these often amounted to a room within a residential house. Most traders are therefore unable or unwilling (due to high cost) to store stocks beyond the minimum turnover period despite seasonal price increases of 30–40%. This means that farmers carry out most within-year storage, and the Strategic Grain Reserve (SGR) provides the only (although non-significant) storage across crop years.

Access to market information is extremely limited in the Tanzanian grain markets. At the producer level, farmers have very little information on prices prevailing even in nearby markets. Farmers have indicated that their primary source of market information is the marketplace itself, as well as conversations with neighbours and traders. Similarly, grain traders rely on contacts with fellow traders and transporters to obtain market information, which leads to considerable uncertainty in grain marketing. Similarly, quality grading and standardization of grain is limited to visual inspection at the time of transaction. Lack of proper standardization of grain quality results in lack of quality premiums and prices that do not reflect quality differences (Beynon *et al.*, 1992).

All these constraints have slowed the growth of the private sector grain trade after the initial, substantial, growth in the number of traders after liberalization. Generally, there is little evidence that market liberalization has benefited rural people in remote areas of the country. This is largely because the necessary backward and forward market linkages are not fully in place since farmers lack both reliable and cost-effective inputs such as extension advice, mechanization services, seeds, fertilizer and credit on the one hand, and guaranteed, profitable markets for their output on the other (Wangwe and van Arkadie, 2000).

Intensification of maize production in Iringa region

This case study investigated how interactive effects of infrastructure improvement, institutional development and government policy could set a precondition for intensification of agriculture using maize in Iringa and Mufindi districts in Iringa region as examples. Maize is the preferred staple food grain in Tanzania and it dominates the country's grain economy. For a long time Iringa region has depended on an agriculture dominated by maize production. The region's good climatic conditions, its strategic geographical location near major links of transport to major maize markets, and its history of maize intensification are some of the factors that made it suitable as a case in the *Afrint* study.

Given the importance of maize, the region has received various kinds of state support over the years. This history of intensification in Iringa and in the Southern Highlands in general has involved the introduction of improved inputs, particularly HYVs and inorganic fertilizers, complemented with improved agronomic practices such as use of tractors and oxen for ploughing as well as timely weeding and pest management. Most of these changes were state driven and took place in the 1970s, especially under the National Maize Project (1975–1982). Maize yields of up to 4 t/ha were a clear sign that a technological change had taken place.

It should be noted that the introduction of subsidized inputs for maize production in Iringa region also coincided with the construction of the Zambia–Tanzania highway (ZAMTAM), the Tanzania–Zambia Railway (TAZARA) (Bryceson, 1993a) and the construction of Uyole Agricultural Training and Research Station in Mbeya region to cater for the entire Southern Highlands regions. There was also expansion of extension services. Marketing services were provided by state driven cooperatives and the NMC provided guaranteed markets as well as an implicit transport subsidy (Isinika *et al.*, 2003).

Yet these advances could not be sustained. The *Afrint* study revealed that although all sampled households in the region continue to grow maize, yields have dropped significantly to a mere 1.2 t/ha during 1998/99–2001/02, compared with an estimated potential of 2.5 t/ha. These trends are explained by a number of factors. Technology adoption, often considered as one of the key determinants of grain production, remains low. Only a third of sampled farmers in the region currently use oxen to plough land and only 27% use improved seeds. Chemical fertilizer use is also very limited. While about 37% of the farmers in Iringa apply some fertilizer, the average application rate for all farmers is low and stands at 15 kg/ha.

Adoption of modern technology has been greatly impaired by liberalization policies and the removal of subsidies for inorganic fertilizers during the early 1990s, when prices of inputs shot up considerably. Since the introduction of economic liberalization there has been an apparent mismatch between input and output prices (see Fig. 11.3). In the *Afrint* survey more than nine out of ten farmers reported input prices as measured in maize equivalents to have increased since they formed their households. The focus group interviews further revealed that the ratio of price for fertilizer to that of maize decreased from 3:1 during the pre-SAP period to 1:2 in 2002, implying that 1 kg of maize could fetch only 0.5 kg of fertilizer in 2002 compared with 3 kg previously.

According to the survey, about half of the households (50%) currently sell some maize. It is predominately the younger and middle-aged households established in the SAP and post-SAP period who have responded to market signals. Table 11.1 shows that households of the post-SAP period, especially, were less inclined to sell maize when their households were formed than were farmers of the pre-SAP period.[8] This may be an effect of the negative price situation as subsidies on fertilizer were phased out between 1990 and 1994 and young farmers had to cope with production conditions quite different from those of the earlier period. This could also reflect the tendency of younger couples at the time to be preoccupied with subsistence production (and off-farm incomes) rather than with producing food grains for the market. In this sense, there is not much evidence of a food orientated market integration taking place.

As can be seen in Table 11.1 there are only minor differences between the proportions of farmers established in different periods when it comes to *current* maize sales. The table also suggests, however, an increase in marketing by young and middle-aged farmers (Table 11.1).[9]

Of those farmers who sold maize at the time when their households were formed, slightly over a half (54%) consider maize prices to have improved, a finding that can be attributed to increased access to markets, particularly in the post-SAP period. During this period, a growing proportion of young households appears to have explored the market niche for green maize, which fetches higher prices than the conventional maize grain. Despite these positive signs, a number of constraints, among them a high price for farm inputs as well as other institutional and infrastructural factors, continue to impede any further market orientation among rural households. The majority of farmers sell at the farm gate to traders. Respondents indicated that the farmers sold their crops within the immediate vicinity of their household.

Out of five villages surveyed in Iringa region, only two (Ihemi and Kasanga) had regular transport while the others (Kipaduka, Kitelewasi and Isele) relied on irregular transport by traders coming with lorries to collect crops for distant markets. The mean distance to an all-weather road was 5.2 km. The farthest village was located 14 km away from an all-weather road and 55 km away from a permanent town-based market outlet for crops. Only one village had electricity or access to mobile telephone services. As stated earlier, since the collapse of cooperatives, there is no structured marketing institution within villages. Traders travel to rural areas in search of crops without any prearranged

Table 11.1. Proportion of households selling maize. (Source: *Afrint* micro study data (Ashimogo *et al.*, 2003).)

	Time when household was formed							
	Pre-SAP		SAP		Post-SAP		All respondents	
Selling maize	–1985	Present	1986–1994	Present	1995–	Present	Outset	Present
No	51	52	58	46	70	55	58	50
Yes	49	48	42	54	30	45	42	50
Total (%)	100	100	100	100	100	100	100	100
Total (*N*)	73	75	72	74	37	42	182	191

agreements. Thus the whole process is risky for producers and traders alike.

Rice production in Kilombero district

The importance of rice production in Tanzania has already been alluded to. In Kilombero district, rice is the dominant staple food and cash crop, as reflected by the *Afrint* micro study. The overwhelming majority (97%) of the respondents in the district grew paddy.[10] Respondents cultivated about 1 ha of paddy on average and obtained 1.4–1.6 t/ha during the last three seasons prior to the study. Out of this production, 49% was sold, 36% was reserved for domestic consumption, while the rest was used for paying hired labour and other purposes (Ashimogo *et al.*, 2003).

Kilombero district lies within the Kilombero valley, where the Kilombero and Ruaha rivers are among many tributaries that feed into the Rufiji river basin. The history of rice production in Kilombero district dates back to the 19th century. Rice has since spread to other parts of the district and has become a dominant crop. About 89% of the respondents had grown rice also when their households were formed. Lowland rice production seems to be a more recent phenomenon. In the villages surveyed, lowland rain-fed rice was introduced during the 1970s, and coincided with the villagization period (1973–1976), similar to the case of improved maize varieties and fertilizer in Iringa region and the Southern Highlands in general.

Ujamaa and the villagization process had a definite influence on the spread of lowland rice production in Kilombero district. Focus group interviews at Idete village gave an account of how rice production was introduced at the village in 1969 to a group of 26 families that had agreed to live according to Ujamaa principles. The village was given two tractors, fertilizer and two seed varieties (*Super India* and *Rangimbili*). Other varieties (*Super Mwanza* and *Kisegese*) were introduced in 1970. In 1972/73 the village cultivated 60 acres and harvested more than 3000 bags of unmilled rice. This translates to more than 5 t/ha. Unfortunately, the communal institutional framework could not be sustained due to various reasons, including uncritical recruitment of uncommitted village members. However, rice production remains important in the village even though yield levels have declined. The *Afrint* survey showed that farmers who use chemical fertilizers, albeit in small quantities, currently obtain 1.9 t/ha on average. The average yield without fertilizer is 1.5 t/ha.

Efforts to improve rice production in Tanzania go way back to 1935 when the first research station on rice was established at Mwabagole near Lake Victoria. The centre was first transferred to the Agricultural Research Institute Ilonga in 1966 and later to Kilombero Agricultural Training and Research Institute (KATRIN) at Ifakara in 1975. KATRIN is now the main research centre for rice in the country. The focus of rice research has been on breeding, crop husbandry and management as well as on disease and pest control. Breeding work has adapted many varieties to local conditions and over the years a number of improved varieties of rice have been produced and disseminated by these national centres. However, allocation of resources for rice research has not always been on a level with the importance of rice as a crucial food and cash crop.

Although the use of yield improving technologies was introduced at Idete as early as in 1969, *Afrint* data show that adoption levels to date remain very low throughout the district. Fertilizer use for rice production similarly remains low since only one in ten of the farmers surveyed use chemical fertilizer, and when they do, quantities applied are very small. Similarly, with the exception of *Super India*, the adoption rate of improved varieties has been very low. Besides high yield, farmers prefer varieties that are palatable and aromatic, characteristics that are preferred by consumers as well and for which traders therefore offer premium prices.

KATRIN has recently (2001/02) released new varieties (Line 85 and Line 88), which yield up to 7.5 t/ha under good management compared with only 3–4 t/ha for Super varieties (personal communication with researchers at KATRIN (Msomba and Kanyeka)), but their adoption rate is very low. According to the *Afrint* study only about 13% of the households used improved varieties during the most recent season, this being higher than when the households were formed, when only 6.5% of the households used improved varieties. Researchers at KATRIN argue that the low adoption rate is due to the recent introduction of the improved varieties (personal communication with Professor Nchimbi-Msolla). However, discussion with farmers showed that consumers and traders show less preference for these varieties due to lack of aroma and taste. On the other hand, a study in 2002 revealed that villages such as Mlimba, which were closer to the highway, had no difficulty in selling these varieties, reflecting the observation that a wider range of traders venture into villages that are more accessible (Isinika and Mansor, 2002).

Developing better technologies remains very important. However, although KATRIN has a number of technologies available, the institute is starved of operational funds and of research facilities in general. Funding for research was frozen during the mid-1980s. Since 1995 limited funds have become available under the World Bank funded Tanzania Agricultural Research Project Phase II (TARP II), but their release is often untimely and they are inadequate. It was reported that KATRIN now receives about 25% of its required recurrent funds and 50–60% of required operational funds. This modest improvement in funding has nevertheless facilitated the recent release of new improved varieties and more are due for release.

Farmers in the *Afrint* micro study attributed decline in paddy yield levels to the following factors: increasing weeds and pests, declining soil fertility, untimely planning, inadequate or untimely land preparation, inadequate water and poor quality of seed. Those farmers who felt that yield levels had improved attributed the increase to improved tillage, mechanized farming, access to new seed varieties and use of chemical fertilizer. Thus, while the technological preconditions seem to be present for intensifying production in Kilombero district, adoption of existing technologies has been constrained by other institutional factors including poor extension services and lack of credit (both pre- and post-SAP).

For example, about three-quarters (73%) of the respondents did not receive any extension services in 2001 while only three respondents reported having received any form of credit. Low demand for extension services may reflect the low use of modern inputs, this in turn resulting in a low perceived need for extension services. Credit used to be provided by rice traders from Pemba in the mid-1990s, but high taxation by the District Council has made the traders move to other districts (cf: the discussion on taxation above).

Although a majority of farmers (58%) report sales to have increased over time, the increase could not be traced back to a particular period. Indirect evidence, however, points at a possibly positive effect on yields from increased frequency of market interactions and improvement of the transport infrastructure in the post-SAP period. Respondents in both the focus group and the survey interviews reported that market access had improved in the post-SAP period, mainly due to the upgrading of the main road into the area and improved services of the TAZARA railway. It is households formed in the pre-SAP period, predominately, who report improved access to markets, suggesting that some positive changes in this respect have occurred both in the SAP and post-SAP periods (Table 11.2). The recent improvement in infrastructure in the area also supports this view.

At the same time, however, a majority of the respondents (68%) report that prices for modern inputs have gone up. It is predominately farms established in the pre-SAP and SAP periods that have experienced price increases, indicating that most of the price increments have occurred after 1985 (Table 11.3).

Similarly, two-thirds of the households (67%) consider that the profitability in rice production has deteriorated. A majority of the households of all periods share this opinion, although those of the pre-SAP period are slightly less negative than more recently established households (Table 11.4). In this group, 39% of the farmers actually report

Table 11.2. Farmers' access to markets by time of household formation. (Source: *Afrint* micro study data (Ashimogo *et al.*, 2003).)

	Time when household was formed (% of respondents)			
Change in market access	Pre-SAP (–1985)	SAP (1986–1995)	Post-SAP (1996–)	Total
Worse now	32	33	48	38
No change	3	23	26	17
Better now	65	44	26	45
Total	100	100	100	100
	(34)	(39)	(31)	(104)

Table 11.3. Price of inputs in rice equivalents by time of household formation. (Source: *Afrint* micro study data (Ashimogo *et al.* 2003).)

	Time when household was formed (% respondents)			
Change in input price	Pre-SAP (–1985)	SAP (1986–1995)	Post-SAP (1996–)	Total
Prices have gone down	17	4	9	10
No significant change	4	17	44	22
Prices have gone up	78	78	48	68
Total	100	100	100	100
	(23)	(23)	(23)	(69)

Table 11.4. Profitability of rice production by time of household formation. (Source: *Afrint* micro study data (Ashimogo *et al.*, 2003).)

	Time when household was formed (% respondents)			
Change in profitability	Pre-SAP (–1985)	SAP (1986–1995)	Post-SAP (1996–)	Total
Worse now	61	69	71	67
Same	0	5	16	7
Better now	39	26	13	26
Total	100	100	100	100
	(33)	(39)	(31)	(103)

Note: Tables 11.2, 11.3 and 11.4 include only farmers who sold rice at the time when the household was formed. Differences in all tables are statistically significant (Chi^2).

profitability to have improved, a finding that indicates that yields, marketing and profitability in farming are unequally distributed across households.

Conclusions

The government of Tanzania and some international donor agencies have made a sincere and ambitious commitment to promote development through agricultural growth. The preceding narration underlines the fact that intensification relates to issues of food security, agroecological conditions, the nature and structure of institutions, political orientation and the role of the state, government–donor relationships, the functioning of markets, and farmers' responses to different forces and stimuli. In order to delineate the effect of policy shifts, analysis in this study was based on a periodization for the pre-SAP (prior to mid-1980s reforms) to post-SAP periods.

Over the long term, the growth in food production has been one outcome of the government's efforts and perhaps the most successful aspect of Tanzanian agriculture. The trend of growth in total food production has in general matched population growth. In addition, there has been growth in production of food for the market and the traditional distinction between cash crops and food crops has been blurred, as many farmers now derive their main cash income from the sale of food crops. However, malnutrition remains a serious problem, mainly due to inadequate food supplies and low income among some households, rather than market scarcity (Wangwe and van Arkadie, 2000).

Based on the analysis as presented in this chapter, it is evident that the government in Tanzania has striven to drive the development of the production and marketing of food crops even though the outcomes have not always been in the intended direction. The policy shifts, which Tanzania has experienced, from socialism in 1967 to market liberalization from the mid-1980s, have had a direct bearing on the development of institutions and infrastructure with consequent effects on the production of food and self-sufficiency both at the household and national levels. In the recent past, improvement of road infrastructure has been one of the obvious examples of government efforts to spur development.

In terms of institutions, the analysis shows that the pre-SAP period is characterized by expansion of public institutions, including those for agricultural research, extension and training, which were responsible for supporting production. This period was also characterized by frequent policy and institutional changes, which sometimes did not take account of institutional complementarity and government's financial capability. Thus under-funding and under-utilization of installed capacity became characteristic of many service public institutions. Service delivery continued to deteriorate throughout most of the pre-SAP period. Nonetheless, some intensification of maize production occurred in the Southern Highlands during the mid-1970s and the 1980s, but such gains could not be sustained due to reasons that have been previously alluded to. In the case of rice, production intensification has mainly been associated with demographic factors, especially in the Lake Zone.

The cooperative and parastatal marketing institutions also collapsed due to similar reasons (frequent government intervention and disregard for operational efficiency). Despite such shortcomings, however, respondents of the *Afrint* micro study, while appreciating current government efforts to improve transport infrastructure, and therefore market access, are nostalgic over the input subsidies, guaranteed markets for their crops and the implicit transport subsidy which was provided by Co-operatives and Crop Authorities buying from each registered Ujamaa village.

At the micro level farmers lack any form of organization that can allow them to enjoy economies of scale from group action in joint marketing and bargaining. Farmers also confessed that they lack knowledge on how to initiate and organize cooperative activities and the main advantages of group action. A similar situation applies for traders, who provide a critical marketing service, collecting crops from remote villages and transporting them to various markets. Empowerment of farmers and other actors in the agricultural sector is one area where the government could render assistance through its new Ministry of Co-operatives and Marketing in collaboration with civil societies and organizations.

The case study for maize production in Iringa illustrates how interaction between results from research institutions was complemented by good transport infrastructure and government mediation of input and output markets through subsidies and price control. As noted earlier, however, these interventions could not be sustained in the past due to other distortions which were introduced by various government polices and strategies. From this analysis it seems that there are still a number of important institutional factors negatively affecting livelihood and agricultural development. Future efforts to intensify production must therefore include means of overcoming these constraints.

During and post-SAP, the government has continued to assert its role as a lead player to drive national development through policy making and improvement of marketing infrastructure while the goal of food self-sufficiency remains high on its agenda. When the economy was liberalized, farmers in the Southern Highlands, where maize production had become dependent on subsidized inputs, have experienced a decline in production post-SAP. Meanwhile farmers from other parts of the country have used their spatial comparative advantage to raise their marketed surplus. However, the declining trend in yields (Fig. 11.3) due to lower use of purchased inputs has led the government to restore subsidized fertilizer to the Southern Highlands, though at a much lower rate than before subsidies were phased out. It is too early to tell what the effect of this will be on production and its potential distortional effect in terms of spatial resource allocation.

Has policy change in Tanzania been pro-small farmer? There is no direct answer to this question. One could say that since agriculture in Tanzania is predominantly smallholder (over 80%), then implicitly government policies have been designed to address their concerns. This chapter cites examples that could be considered to be favourable (subsidized inputs, expanded extension and research services, improvement in transport infrastructure) and unfavourable (dissolution of cooperatives, removal of subsidies, under-funding of agricultural support services, marketing control pre-SAP). Some households have been able to make use of improved conditions (e.g. market opportunities, infrastructure) in the SAP and post-SAP periods. For the majority of farmers, however, such improvements have been offset by seriously felt constraints in terms of high input prices, lack of credit and low and uncertain monetary returns from a marketed surplus.

With poor sequencing, coordination and under-funding of many government programmes, the net effect on productivity has been limited or unsustainable at most. The most recent government policy document on agriculture, the Agricultural Sector Development Strategy (ASDS), addresses the issue of strengthening the institutional framework, but mainly in terms of strengthening the participation of the private sector in agricultural development (which presumably includes smallholder farmers). Although there are some projects scattered throughout the country addressing various farming concerns, there is as yet no concerted strategy on the ground which could qualify as being specifically focused on small farmers to alleviate their concerns regarding access to credit for production and access to markets. However, the government has taken deliberate steps to improve funding for transport and communication infrastructures, which should reduce transport and marketing costs in the medium and long term.

The preceding discussion indicates that there was government over-mediation prior to SAP. During and after SAP the government's stance has generally been to leave the markets to operate unimpeded with government facilitation by way of infrastructure improvement. However, due to imperfections and inefficiencies in the markets many problems have emerged. In relation to food production and marketing in particular the main problems have been related to reduced use of purchased inputs due to higher prices and farmers in remote areas not having access to inputs or markets for their products. The relative impacts of these distortions to different market participants have, however, not been uniform, with small farmers being more disadvantaged relative to traders, who have more access to information. Partly for these reasons, economic reforms have not translated into improved livelihoods in many rural areas.

While some steps have been taken at the macro level in terms of improved policy-making and infrastructure, the situation at lower levels is lagging behind. Results for the micro study indicate that in general farmers currently have better access to markets, but are using fewer inputs than before due to higher prices. As a result profitability in food crop cultivation has deteriorated. Local institutions for marketing are either very weak or non-existent. Meanwhile local governments, which have the mandate for much of the rural infrastructure, are burdened with financial constraints. The challenge for the government is therefore to strengthen and improve the capacity of formal facilitating institutions, including local governments, which should go hand in hand with empowering local institutions to exert effective demand on services such as research, extension and market information that enhance agricultural productivity. All these are planned for implementation under the Agricultural Sector Development Programme (ASDP). How effective this will be in directing agricultural transformation in Tanzania towards more small farmer orientation, market mediation and food self-sufficiency at all levels remains to be seen.

Notes

[1] Ujamaa means familyhood. It was one of the pillars of President Nyerere´s African Socialism, as outlined in the Arusha Declaration. For more information see Nyerere, J.K. (1968) *Freedom and Socialism. A Selection from Writings and Speeches, 1965–1967*, Oxford University Press, Dar-es-Salaam.

[2] Translated to mean 'Agriculture as a matter of life and death'.

[3] Translated to mean 'Agriculture is life'.

[4] Translated in English to mean 'Network of Farmers' Groups in Tanzania'.

[5] The Land Law No. 4 of 1999 and the Village Land Law No. 5 of 1999.

[6] The mean exchange rates for 1995 and 1996 were 591.65 and 586.59 Shs/US$ respectively.

[7] ASDS is a ten-year national strategy (2005–2015) to transform agriculture to a modern commercial sector by the year 2025.

[8] Significant at $\alpha = 0.05$ using Chi^2.

[9] This increase is not statistically significant and calls for a cautious interpretation. However, the simultaneous response by 65% of the households formed during and after SAP that they had increased their sale of maize since the farm was established suggests that the market situation may have improved somewhat in recent years.

[10] Paddy means unmilled rice. In this report paddy and rice are sometimes used interchangeably.

References

Ashimogo, G.A., Isinika, A.C. and Mlangwa, J.E.D. (2003) Africa in transition: Micro study, Tanzania final research report for the *Afrint* Research Project. Sokoine University of Agriculture, Morogoro, Tanzania.

Bevan, D., Collier, P. and Gunning, J.W. (1993) *Agriculture and the Policy Environment: Tanzania and Kenya*. Organization for Economic Cooperation and Development. Development Centre Studies.

Beynon, J., Jones, S. and Yao, S. (1992) Market reform and private trade in Eastern and Southern Africa. *Food Policy* 17, 399–408.

Bryceson, D.F. (1993a) *Liberalizing Tanzania Food Trade. Public and Private Faces of Urban Marketing Policy 1939–88*. UNRISD in association with James Currey, Mkuki na Nyota and Heinemann, London, Dar-es-Salaam and Portsmouth, UK.

Bryceson, D.F. (1993b) Urban bias revisited: staple food pricing in Tanzania. In: de Alcantara, C.H. (ed.) *Real Markets: Social and Political Issues of Food Policy Reform*. Frank Cass, London.

Coulter, J. and Golob, P. (1992) Cereal marketing liberalization in Tanzania. *Food Policy* 17, 420–430.

Environmental Systems Research Institute, USA (ESIRIDATA) & Maps (1999) Map of Tanzania Administrative Regions.

FAO (1986) *Food Security Assistance Scheme: Report of the Food Security Review Mission to Tanzania*. Food and Agriculture Organization, Rome.

Gabagambi, D.M. (2003) Road infrastructure investment and its impact on agricultural productivity and equity in Tanzania. PhD dissertation. Verlag Grauer, Beuren Stuttgart, Germany.

Gordon, H.F. (1994) *Grain Marketing Performance and Public Policy in Tanzania*. Fletcher School of Law and Diplomacy. Tufts University, Medford.

Hanak, E. (1986) *The Politics of Agricultural Policy in Tanzania*. A draft report for the World Bank study on Managing Agricultural Development. World Bank, Washington, DC.

Isinika, A.C. and Mansor, H. (2002) *Technical Backstopping for the Eastern Zone Client Oriented Research Programme (EZCORE)*. Morogoro, Tanzania.

Isinika, A.C., Ashimogo, G. and Mlangwa, J.E.D. (2003) Africa in transition: Macro study Tanzania. Final research report for the *Afrint* Research Project. Sokoine University of Agriculture, Morogoro, Tanzania.

Kanyeka, Z.L., Msomba, S.W., Kihupi, A.N. and Penza, M.S.F. (1995) Rice ecosystems in Tanzania: characterization and classification. *KATC Newsletter Rice and People of Tanzania* 1,1.

Mattee, A.Z. (1978) Education transformation in Tanzania. Implementing the policy of education for self-reliance in the secondary schools. A Master thesis submitted in partial fulfilment of the requirement for the degree of Master of Science in Continuing and Vocational Education (Agricultural Education). College of Agriculture and Life Sciences, University of Wisconsin, Madison, Wisconsin.

MCT (2003) *Transport Bulletin*. Ministry of Communication and Transport, Dar es Salaam, October 2003.

Mlay, G.I. (1988) *Analysis of Policies Affecting Maize Production and Consumption in Tanzania*. Food and Agriculture Organization, Dar-es-Salaam, Tanzania.

NEI (1999) *Final Reports Crop and Food Studies*: vol. 2: *Food Crops Grains* (draft). Agricultural Sector Management Programme (ASMP). Netherlands Economic Institute on behalf of Ministry of Agriculture and Cooperatives, Dar es Salaam.

Ngugi, D., Isinika, A.C., Temu, A. and Kitalyi, A. (2002) *Agricultural Education in Kenya and Tanzania (1968–1998)* Technical Report No. 25. Regional Land Management Unit (RELMA), Nairobi.

NIT (2003) *Study on Tanzania Road Network*. A report for the World Food Programme. National Institute of Transportation, Dar-es-Salaam, Tanzania.

Nyerere, J.K. (1973) *Freedom and Development*. Oxford University Press, Dar es Salaam.

Santorum, A. and Tibaijuka, A. (1992) Traders[1] responses to food market liberalization in Tanzania. *Food Policy* 17, 431–442.

United Republic of Tanzania (1994) *Report of the Presidential Commission of Inquiry into Land Matters (Vol.) Land Policy and Land Tenure Structure*. Ministry of Lands Housing and Urban Development, Dar es Salaam in cooperation with the Scandinavian Institute of African Studies, Uppsala, Sweden.

United Republic of Tanzania (2001a) Agricultural Sector Development Strategy. National Printing Company – KIUTA, Dar es Salaam.

United Republic of Tanzania (2001b) The economic survey for the year 2000. Planning Commission, Dar es Salaam.

Wangwe, S.M. and van Arkadie, B. (eds) (2000) *Overcoming Restraints on Tanzanian Growth: Policy Challenges Facing the Third Phase Government*. Tanzania Political Economy Series. Mkuki na Nyota Publishers. Dar es Salaam.

World Bank (1994) *Tanzania Agriculture: Country Report*. World Bank, Washington, DC.

World Bank (2000) *Agriculture in Tanzania since 1986: Leader or Follower in Growth?* IBRD and World Bank, Washington, DC.

12 Smallholders and Structural Adjustment in Ghana

A. Wayo Seini and V. Kwame Nyanteng
Institute of Statistical, Social and Economic Research, ISSER, University of Ghana, Legon, Ghana

In this chapter, the role of Ghana's Structural Adjustment Programme (SAP) is discussed with particular reference to intensification of the major staple food crops, namely, cassava, maize, rice and sorghum. For this purpose, it is pertinent to note that the relevant period of the aggregate data analysis (1970–2000) has been classified into pre-SAP (1970–1982), SAP (1983–1992) and post-SAP (1993–2000) periods.

In the post-SAP period, especially, national data indicate that yields (kg/ha) of these food crops have substantially increased. This has occurred in a situation where prices for farm inputs have rocketed and farmers increasingly have had to face competition from cheap food imports, particularly rice. In looking for explanations for this apparent puzzle, the chapter draws on both macro- and micro-level data. The SAP is reviewed from the standpoint of how its policies and programmes have affected the supply, distribution and prices of inputs and outputs in the Ghanaian context.

In relation to the basic Asian model presented in Chapter 1, this chapter highlights the state-drivenness of agricultural intensification in the *pre*-SAP period in contrast to the market driven changes that have prevailed in the *post*-SAP period.

Data Sources

This chapter is based on data and information that was collected in the course of the *Afrint* project. In Ghana, the *Afrint* household survey covered 416 farmers in two regions. It was designed to capture changes over time, including developments in the pre-SAP to post-SAP period through retrospective questions about changes in farmers' production marketing pattern of maize, cassava, rice and sorghum.

The food crops mentioned are cultivated largely in the savannah and semi-deciduous forest agroecological zones. The two regions that were selected for the *Afrint* household survey correspond to these zones. The East Region was selected for maize and cassava and the Upper East Region for sorghum and rice.

SAP Policies in Agriculture

As a result of economic downturn in the 1970s and early 1980s, Ghana embarked on a series of Economic Recovery and Structural Adjustment Programmes from 1983 to 1992. Policy reforms were implemented in three main identifiable phases (see, for example,

Stryker, 1990; Alderman and Shively, 1996; Seini, 2002b for details). The need for agriculture to lead any sustained overall economic growth in all three phases of reforms was recognized and emphasis was therefore placed on the sector. In the initial *phase of stabilization* (1983–1985), price incentives for the production of food, industrial raw materials and export commodities were restored as short-term measures (Commander *et al.*, 1989). In the second phase (1986–1989), usually referred to as the *growth phase*, increased productivity and internal price stability were emphasized. In the final phase, termed *liberalization* (1989–1992), reforms involved abolishing the guaranteed minimum price for maize and rice and the removal of all subsidies for agricultural inputs (Seini, 2002a). The period after 1993 to date is referred to in this chapter as the post-SAP period.

The reform process included the deregulation of commodity and service markets to reduce domestic price distortions, as well as the liberalization of export and import markets (Stryker, 1990). The agricultural development strategy was set out in the Medium Term Agricultural Development Programme (MTADP) (World Bank, 1991). With deregulation and trade liberalization, there are now no formalized public market institutions for food crops in Ghana. There are, however, trader associations in respective markets, covering specific commodities. It is generally believed that through the associations, the traders are able to collude and fix prices. They also are able to control the flow of the commodities into their respective markets and erect barriers against others entering the market. It is pertinent to note also that the establishment of a largely market driven economy in Ghana implies that the use of modern inputs for the purpose of food crop intensification is now demand driven, in contrast to the era of subsidies, which were aimed principally at encouraging their use.

Overall Impact of SAP Policy Reforms

The SAP policies and strategies had impacts on food crop production and on yields, use of improved planting materials and fertilizer availability and use. These impacts can be compared between the pre- and post-SAP policy regimes. The trends in the production of the target crops of cassava, maize, sorghum and rice are shown in Figs 12.1 and 12.2. In terms of volume, the major cereal crop produced in the country has been maize followed by sorghum. Until 1989, millet (not in figure) ranked persistently in the third position and since 1990 rice has occupied the third position.

Generally, total production of maize, cassava and sorghum declined steadily from 1970 to very low levels in 1983 (pre-SAP

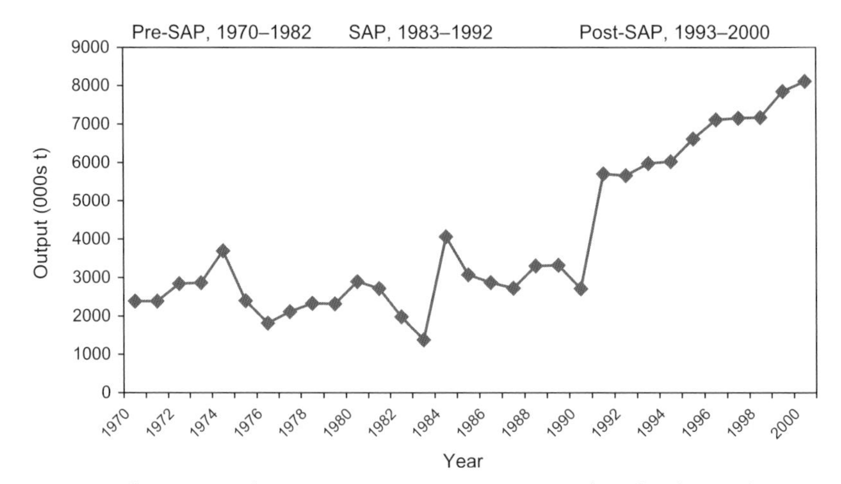

Fig. 12.1. Output of cassava in Ghana, 1970–2000. (Source: Ministry of Food and Agriculture, 2001.)

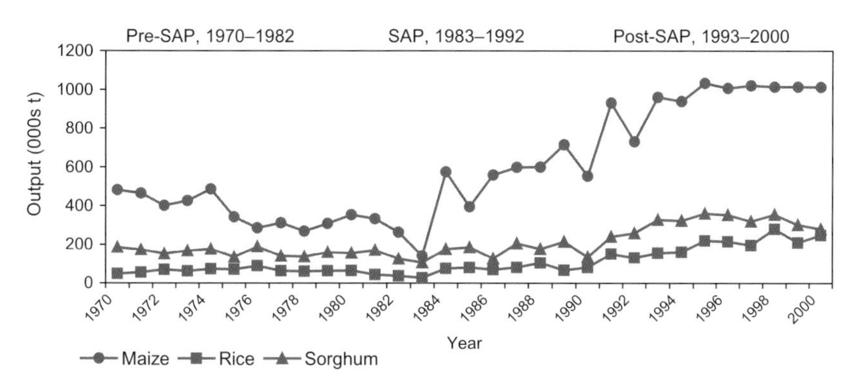

Fig. 12.2. Output of cereals in Ghana, 1970–2000. (Source: Ministry of Food and Agriculture, 2001.)

period). In Fig. 12.1, it is clear that the production of cassava declined in the 1970s, increased somewhat in the 1980s and showed a steadily upward trend in the 1990s. In 10-year periods, the production of maize, as shown in Fig. 12.2, declined in a fluctuating fashion from 482,000 t in 1970 to 309,000 t in 1979, showing a decline of 36%. In the 1980s, that is, in the era of Structural Adjustment policies, the production increased by 102%. In the post-SAP period (from 1993), the production of maize continued the upward trend and increased by 83%. However, in the second half of the 1990s, the annual production of maize stabilized around 1,000,000 t.

Sorghum production decreased by about 14% in the 1970s but showed a remarkable recovery under SAP, when production increased by about 98%. Like other cereals, sorghum production has continued to fluctuate but at a higher level in the post-SAP period than before (Table 12.1). In contrast to the other crops, production of rice increased by 29% in the 1970s but in a fluctuating fashion. In the 1980s (the SAP period), production maintained a marginal increase of only 5%. In the 1990s (the post-SAP period), however, the upward trend for rice gained momentum and an increase of 159% was recorded.

It is significant to note the similarity between the trends in output in Figs 12.1 and 12.2 with the trends in area cultivated in Fig. 12.3. The similarity seems to confirm the general view that increases in the production of staple food crops are due largely to area expansion rather than to increases in yield.

Table 12.1. Crop yields per hectare, 1970–1999. (Source: Ministry of Food and Agriculture, 2001.)

	Average yield (t/ha)		
Crop	1970s	1980s	1990s
Cereals			
Maize	1.0	1.1	1.4
Rice	1.0	0.9	1.8
Sorghum	0.8	0.7	1.0
Millet	0.6	0.6	0.8
Roots and tubers			
Cassava	7.9	7.2	11.2
Yam	5.3	5.2	11.7
Cocoyam	4.7	4.5	7.6

Nevertheless, increases in area cultivated in the post-SAP period suggest positive responses to the liberal market orientated policies of this period.

Land productivity as measured by crop yield per hectare is generally very low in the country. Yields per hectare were particularly low in the 1970s and 1980s when they averaged about 1 t for maize and rice, and 0.7 t for sorghum (Figs 12.4 and 12.5 and Table 12.1). In the 1990s, the yields per hectare increased with maize averaging about 1.4 t/ha, rice 1.8 t/ha and sorghum 1.0 t/ha (Table 12.1). However, yields in the *Afrint* survey were considerably lower (maize 1 t/ha; rice 1 t/ha; sorghum 0.5 t/ha; and cassava 3.8 t/ha), suggesting that there could be considerable outliers of the higher magnitude in computing aggregate average national yields. Nevertheless, in isolated cases in farmers' fields, higher yields per hectare have been

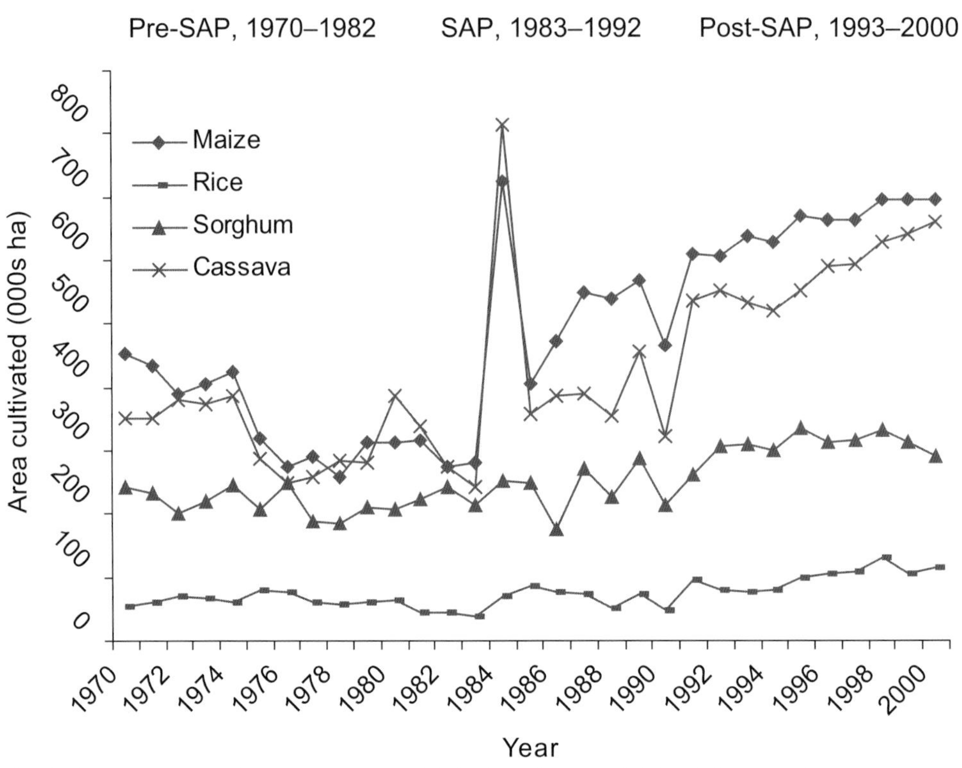

Fig. 12.3. Area cultivated under staple crops 1970–2000. (Source: Ministry of Food and Agriculture, 2001.) Note: The jump in area cultivated of food crops in 1984 can be explained by the devastating famine that hit the country in the year that preceded it. Almost every household found it safer to produce its own staple food, even in backyard gardens.

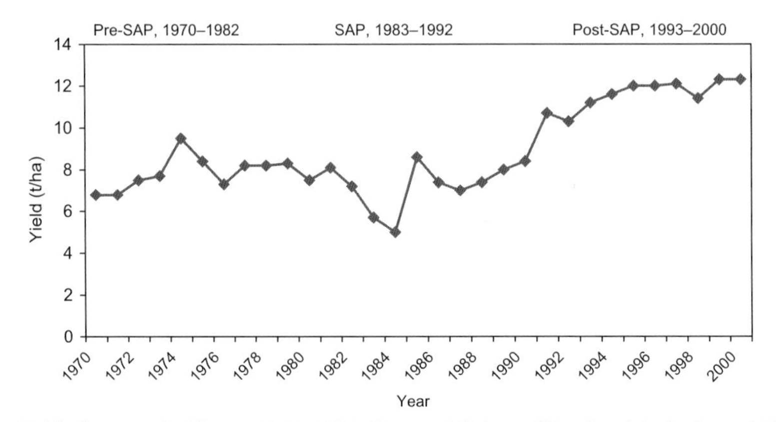

Fig. 12.4. Yield of cassava in Ghana, 1970–2000. (Source: Ministry of Food and Agriculture, 2001.)

achieved, that is, 5 t for maize, 3 t for rice, and 2 t for millet and sorghum (Ministry of Food and Agriculture, 2001).

The yield of cassava per hectare also showed marked increases in the 1990s as compared with the 1970s and the 1980s (Table 12.1). The yield which averaged 7.9 t in the 1970s, increased to about 11.2 t in the 1990s. Like the cereals, higher yields per hectare have been achieved on isolated farmers' fields for cassava, that is, 28 t/ha (Ministry of Food and Agriculture, 2001).

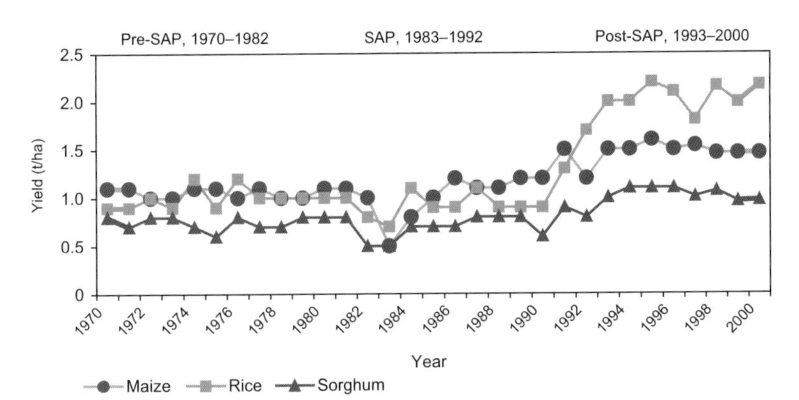

Fig. 12.5. Yield of cereals in Ghana, 1970–2000. (Source: Ministry of Food and Agriculture, 2001.)

Figures 12.4 and 12.5 and Table 12.1 make it clear that increases of output of food crops in the post-SAP period are due not only to increases in area cultivated (extensification) but also to increases in yield (intensification). As we will discuss further below, cassava has been a major target of research in the West African sub-region, particularly at the International Institute of Tropical Agriculture (IITA), and has benefited in terms of yield increases. It is also to be noted that average yields of maize, rice and sorghum have considerably improved in the post-SAP period with increases in the range of 25% to 80% compared with the pre-SAP period.

Availability of modern inputs

Reforms in the agricultural sector had some immediate impacts. On average, the prices of agricultural pesticides increased in excess of 40% per annum between 1986 and 1992 (Asuming-Brempong, 1989, 1994). Fertilizer importation and sales, for example, which hitherto had been under a government programme, were privatized, and domestic prices of fertilizer, depending on type, had doubled or tripled between 1990 and 1992. The private sector response to the privatization policy was sluggish with negative distributional consequences (Jebuni and Seini, 1992).

Fertilizer is the most important of the modern inputs required for food crop intensification. However, Ghana lacks the means to produce fertilizers and therefore imports all her requirements annually. The annual volume of fertilizers imported has fluctuated widely in the 1970–2000 period (Fig. 12.6). Average fertilizer imports in the pre-SAP period were about 34,500 t per annum. Lack of capacity of the private sector to import probably accounts for the fluctuations in fertilizer imports in both the SAP and post-SAP periods. The average volumes of fertilizer imports seem to have decreased in the latter two periods. The average volume of fertilizers imported in the SAP period was 29,062 t per annum, about 16% lower than in the pre-SAP period. In the post-SAP period, the corresponding import was 23,594 t, about 32% lower than the pre-SAP annual average. Throughout, fertilizer imports have remained unstable with wide fluctuations, as shown in Fig. 12.6.

The volumes of fertilizers imported annually have been highly inadequate given recommended application rates for the staple cereals. Per hectare, average availability was only 43 kg in the pre-SAP period, and decreased further to 35 kg and 25 kg in the SAP and post-SAP periods, respectively. The low rate of fertilizer availability is corroborated by the low application rates for cereals found in the *Afrint* micro survey (see below).

Technology and extension for intensification

Useful research outcomes have been achieved over the years by agricultural research

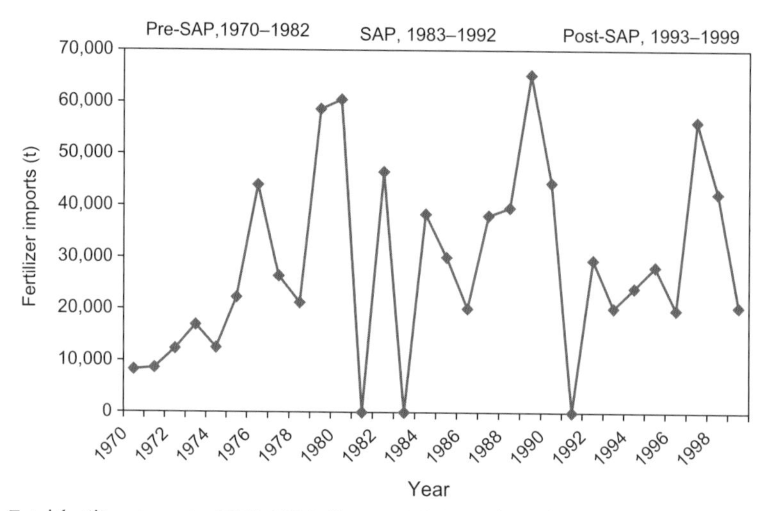

Fig. 12.6. Total fertilizer imports, 1970–1999. (Source: Ministry of Food and Agriculture, 2001.)

institutions. The Crops Research Institute (CRI) in particular has made a considerable contribution to increases in food production in Ghana. An area of major contribution is with maize where total production is estimated to have increased from 296,000 t in 1977–1978 to 1 million t in 1997–1998 (Aryeetey, 2000). This increase was due, in part, to the development of three high yielding hybrids named Dada-ba, Mama-ba and CIDA-ba. These hybrids potentially yield between 5.0 and 7.3 t/ha and have been adopted widely by farmers. The yield of these hybrids compares favourably with Obatampa, an earlier streak-resistant variety with a potential yield of 4.5 t/ha.

The CRI has also developed improved rice varieties over the years. The latest variety that has been released to farmers and is widely cultivated is Sikamo, a high-yielding, disease-resistant rice variety with a potential grain yield of 5.5 t/ha. On cassava, the CRI has developed three improved varieties that are highly resistant to pest and disease and yield three times more than the old cultivars, which had an average yield of about 10 t/ha (Aryeetey, 2000).

Donor support in research and technology development in Ghana has been substantial and increasing, particularly since the SAP era. The largest donor-sponsored initiative has been the National Agricultural Research Project (NARP), a joint Government of Ghana (GOG)/World Bank project. The project, which is ongoing, aims at restructuring agricultural research services in order to make them more efficient, effective and self-supporting. Mainly through the initiative of the Consultative Group on International Agricultural Research (CGIAR) there have been considerable regional and sub-regional linkages in agricultural research. The CGIAR, through IITA in Ibadan, Nigeria, instituted the Collaborative Study on Cassava in Africa (COSCA). The COSCA was able to bring together the various units that make up agricultural research within and outside the participating countries. IITA also undertakes collaborative work in the development of improved disease- and pest-resistant crop varieties, particularly roots and tubers.

The public sector agricultural extension in Ghana is responsible for the dissemination of improved technologies to farmers. For a very long time, the extension system was characterized by a scattered system with various departments of the Ministry of Food and Agriculture (MOFA) providing their own services. The system was found to be largely ineffective and has been replaced with a unified extension system that started in 1988 (Al-Hassan *et al.*, 1998). The unified extension system has brought together all extension activities of the MOFA under the Department of Agricultural Extension Services (DAES). One front line staff (FLS) delivers agricultural extension messages to all crop and livestock farmers in a given area. The extension services delivery method is a modified Training and Visit (T&V) with a mechanism for effective research–extension–farmer linkage (Fiadjoe *et al.*, 1997).

The major extension messages since the unification have been planting of crops in lines with appropriate population densities, regular weeding and application of inorganic fertilizers and other agrochemicals, early harvesting of maize to reduce insect infestation on the field, and treatment of maize and cowpeas with chemicals to reduce storage losses (Al-Hassan, 1997).

Agricultural credit

Government credit policy in the 1970s to the mid-1980s (largely in the pre-SAP period) was at discriminatory sectoral interest rates. This was a control mechanism ensuring that priority sectors gained access to credit (Aryeetey, 2000). The policy was based on the presumption that the market rate, if universally applied, would ration out some priority sectors. The three priority sectors of agriculture, export trade and manufacturing benefited from this policy.

The political determination of interest rates was abolished in 1985 as part of the SAP. Since then, loans to small-scale farmers have virtually evaporated; this is partly attributed to the high rate of default by such farmers. The default rates in loan repayments ranged from 24% in some districts to 74% in others. Nevertheless, in order to encourage agricultural intensification, various efforts are being made to promote access to credit. Many such efforts are spearheaded by the International Fund for Agricultural Development (IFAD) smallholder schemes and by a number of non-governmental organizations (NGOs). For example, in the survey districts of Kassena-Nankane and Bolgatanga, the Adventist Development and Relief Agencies (ADRA) assists farmers with inputs, i.e. seeds (soybean), fertilizer, and seedlings (free) for its agroforestry programme. ADRA picks produce, deducts input costs, and pays to the farmer what is left. The credit in kind is also repaid in kind. In addition, there is a Technoserve (an American NGO) scheme that includes credit to construct storage facilities, and crop sales in the lean season for increased profit.

As part of attempts to finance agriculture, rural banks were introduced in Ghana in the early 1970s. Their presence was observed in the districts surveyed. Besides mobilizing savings in the catchment areas of their location, they also handle project funds. The rural banks handle mainly funds from the Smallholder Credit and Input Project (SCIMP), an IFAD project that advances input credit to food crop farmers in the transitional zone of some regions in Ghana. The Kassena-Nankane and Bolgatanga districts are part of wider regional schemes that include credit components. The Land Consolidation and Smallholder Rehabilitation Project (LACOSREP) encourages farmer groups to open group accounts with the Agricultural Development Bank (ADB) or the Ghana Commercial Bank (GCB). A credit committee in each community, selected by the community, endorses groups for credit and is directly responsible for loan recovery. The Nara Rural Bank at Paga in the Kassena-Nakane district is an additional source of agricultural credit in that district.

Despite these initiatives, self-finance and loans from relatives and friends remain the principal sources of finance for small-scale farmers (Seini, 2002b). These sources have long dominated the rural micro-finance. Often proceeds from one commodity or livestock are used to finance another commodity. Off-farm income and remittances from family members working in the urban areas are other sources to finance farm enterprises.

Smallholder Intensification – Micro-level Data

At macro level, it is reasonable to hypothesize that rising yields can be attributed to an overall increase in farmers' use of improved technologies. However, the aggregate evidence that could back up this hypothesis is scanty. First of all, national statistical data of farmers' use of modern inputs is hard to come by. Second, existing information gives the impression that a large proportion of small-scale farmers in Ghana continue to use traditional planting material despite the fact that some improved cereal and cassava varieties have been released. Similarly, few farmers use chemicals to control weeds. Also, at the farm level, available official data do not provide a clear picture of whether or not policy reforms under SAP have significantly impacted positively on smallholder food crop intensification.

In our search for explanations to the documented surge in national food crop production and yields that have taken place in Ghana in the SAP and post-SAP period, we will use *Afrint* survey data, drawing on farmers' perceptions of changes that have occurred since their households and farm units were established. Table 12.2 summarizes the distribution of farm respondents across the two regions covered in the survey and with respect to the period when their farms and households were established. The mean age of the respondents (farm managers) corresponding to these periods is also given in the table and will provide a crude counter-check to the structural changes that we attempt to distinguish. The number of respondents are almost equally divided between the East and Upper East regions and their distribution across the time periods covered is similar in the two regions.

Nature of land preparation and agricultural practices

Agriculture in Ghana is predominantly on a smallholder basis, although there are some large farms and plantations, particularly for cocoa, rubber, oil palm and coconut and, to a lesser extent, for rice, maize and pineapples. About 60% of the farm holdings do not exceed 1.2 ha and, cumulatively, 85% of the holdings do not exceed 2 ha. The main system of farming is traditional with hoes and cutlasses being the major farm implements and which constitute a major constraint to the expansion in farm size. While generally there is little mechanized farming, the use of bullocks and tractors is becoming more common, especially in the northern savannah zone (Ministry of Food and Agriculture, 2001). The *Afrint* field survey data are largely consistent with this picture.

The small farm size reflects the nature of land preparation, which is largely dependent on the tools used. In practice, all farmers growing maize and cassava (East Region) use traditional hoes and cutlasses for land preparation. In the case of sorghum (Upper East Region), however, the majority of farmers (57%) currently use draft animals for land preparation. Similarly, in rice cultivation, the frequent use of tractors (36%) and animal traction (28%) for land preparation is noteworthy (Tables 12.3 and 12.4). In general, land preparation with tractors offers a higher level of efficiency and timeliness in production. The use of draft animals is an intermediate technology that is also more efficient than the traditional hoe and cutlass. Timeliness of operations is essential for food crop intensification.

Generally, animal traction seems to have increased over the pre- to post-SAP period, as can be seen when the respective SAP columns are compared. A comparison across the columns labelled 'present' offers a crude control of the age or life cycle effect and allows us to

Table 12.2. Mean age of respondents and their percentage distribution by region and period. Total no. of households: 416. (Source: *Afrint* survey data, Seini and Nyanteng, 2003.)

Period when household was formed	Mean age (years)	Region		
		Eastern Region	Upper East Region	Total
Pre-SAP (–1982)	54	44	45	44
SAP (1983–1992)	44	26	25	26
Post-SAP (1993–)	35	30	30	30
Total	45	100	100	100

distinguish some of the structural changes that we are looking for. In the case of sorghum it is predominantly the older farmers (with farms established in the pre-SAP period) who shifted from hoe cultivation to animal traction, while it is the younger aspiring farmers (with farms established in the SAP and post-SAP period) who have increased their use of tractors (Table 12.3). The latter increase is probably related both to the life cycle phase of these farmers and to the improved structural conditions for tractor cultivation in the late post-SAP period. They are also probably just employing the technology that is akin to their era as the younger generations tend to be more abreast with current practices than their older counterparts.

The trend of increased animal and motorized traction is even more pronounced in the case of rice (Table 12.4). The use of handheld tools has generally declined. The older respondents have largely replaced hoe cultivation with animal traction, while farmers of the SAP and post-SAP periods seem to have replaced the hoe, and to some extent also the bullock, with tractors. By and large, tractorization involving small-scale farmers in the Upper East Region is a phenomenon that has accelerated in the SAP and post-SAP period, possibly fuelled by changes in the marketing conditions for these crops, especially rice. By comparing the columns for 'present' one can see that it is predominantly the young and middle-aged farmers who are adopting the more labour efficient farming practice involved in tractor ploughing.

Related to the nature of land preparation are the non-industrial or 'traditional' agricultural practices that accompany crop cultivation and which tend to be associated with the agroecological zones. In the forest zone where maize and cassava are mainly grown (East Region), the major cultural practices include, for both crops, fallowing (80%), intercropping (23%) and crop rotation (28%). These practices have, by and large, remained

Table 12.3. Land preparation method for sorghum, Upper East Region. Per cent users now (2002) and at the time when the farm/household was established. (Source: *Afrint* survey data, Seini and Nyanteng, 2003.)

	Time when farm/household was established							
	Pre-SAP		SAP		Post-SAP		All respondents	
Method for land preparation	–1982	Present	1983–1992	Present	1993–	Present	Outset	Present
Hoe/cutlasses	59	42	46	35	41	31	50	37
Oxen ploughing	36	56	50	53	57	61	46	57
Tractor ploughing	2	1	2	10	0	5	2	5
Other	2	1	2	2	2	3	2	2
Total	100	100	100	100	100	100	100	100

Table 12.4. Land preparation method for rice, Upper East Region. Per cent users at present (2002) and at the time when the farm/household was established. (Source: *Afrint* survey data, Seini and Nyanteng, 2003.)

	Time when farm/household was established							
	Pre-SAP		SAP		Post-SAP		All respondents	
Method for land preparation	–1982	Present	1983–1992	Present	1993–	Present	Outset	Present
Hoe/cutlasses	59	44	40	27	44	29	50	35
Oxen ploughing	24	35	29	22	25	24	25	28
Tractor ploughing	17	20	31	51	31	47	25	36
Other	0	1	0	0	0	0	0	1
Total	100	100	100	100	100	100	100	100

stable throughout the pre-SAP to post-SAP period. Also in rice and sorghum cultivation (Upper East Region), farming practices have remained fairly stable over the entire period but are different from those of the forest zone. For example, fallowing is practised by only a few farmers in the savannah zone (12% by sorghum farmers, 2% by rice farmers). In the case of sorghum, a majority of farmers use intercropping (76%) and apply animal (80%) and green (37%) manure for restoring soil fertility. In rice cultivation, however, these practices are of much less significance, relatively speaking. Only a quarter of the rice farmers apply animal manure (24%), and even fewer (14%) use green manure.

Use of modern inputs – seeds and planting material

The types and nature of planting materials used can be assumed to be crucial to smallholder intensification, i.e. increasing yields. In relative terms, the number of farmers currently using improved planting material has increased for maize, cassava and rice, while no change can be detected for sorghum (Tables 12.5–12.8). In the case of maize, 38% of the respondents use some kind of improved seeds. Throughout the pre- to post-SAP period there has been a steady upward trend in adoption rates of improved maize varieties, as can be seen from Table 12.5.

While the trend of adopting improved maize seeds is present in all respondent groups, it is the now predominantly middle-aged and older farmers of the pre-SAP period who have experienced the shift from traditional to high yielding varieties. To a great extent, the younger groups of farmers (with households established in the post-SAP period) are already familiar with improved varieties when they take up their farms. It is noteworthy that, contrary to the situation in eastern and southern Africa, there seems to have been no loss of momentum in the adoption rate of improved maize seeds in Ghana in the SAP and post-SAP periods (see Oluoch-Kosura and Karugia, Chapter 10, and Isinika *et al.*, Chapter 11, this volume).

The adoption of high-yielding and pest- and disease-resistant cassava varieties (TTM),

Table 12.5. Type of seed used in maize cultivation. Per cent users at present (2002) and at the time when the farm/household was established. (Source: *Afrint* survey data.)

Maize – type of seed	Time when farm/household was established							
	Pre-SAP		SAP		Post-SAP		All respondents	
	–1982	Present	1983–1992	Present	1993–	Present	Outset	Present
Traditional	98	63	84	65	68	58	85	62
Improved/OPV	2	31	10	25	18	25	9	27
Hybrid	0	6	6	10	14	18	6	11
Total	100	100	100	100	100	100	100	100

Table 12.6. Type of cassava variety planted. Per cent users now (2002) and at the time when the farm/household was established. (Source: *Afrint* survey data.)

Cassava – variety planted	Time when farm/household was established							
	Pre-SAP		SAP		Post-SAP		All respondents	
	–1982	Present	1983–1992	Present	1993–	Present	Outset	Present
Traditional	98	70	92	89	91	84	94	79
Improved	2	30	8	11	9	16	6	21
Total	100	100	100	100	100	100	100	100

developed by IITA in Ibadan, Nigeria, shows a somewhat different pattern (Table 12.6). First, adoption rates are much lower than for maize. Secondly, the main adopters seem to be the older households that were established in the pre-SAP period. Third, the timing is different, with the main increase in adoption occurring in the 1980s and thereafter levelling off. This is puzzling in view of the fact that the TTM varieties were not released until the early or mid-1990s. It is possible that the figures in the table prior to the post-SAP period reflect the increased use of local varieties recommended by extension staff rather than the adoption of TTM. If this is the case, it constitutes a major source of data error in the cassava survey material, which must be treated with caution.

For sorghum, adoption rates of improved varieties are marginal (Table 12.7). In this case, the *Afrint* survey did not trace any changes in the adoption rates over the period studied, possibly because relatively less progress has been made on this crop by NARS and the international research community compared with maize, cassava and rice.

Rice, finally, presents an interesting pattern (Table 12.8). More than a third of all farmers interviewed in the Upper East Region currently use improved varieties of the kinds listed in 'Technology and extension for intensification' above. However, the main difference in adoption rates is between farmers of the pre-SAP period and farmers of the SAP period, whose adoption of HYVs jumped from 17% at the outset to 37% with the onset of SAP. In the post-SAP period, adoption rates have levelled off or possibly decreased somewhat.

Use of modern inputs – inorganic fertilizer and pesticides

Good planting material often needs to be complemented by inorganic fertilizer and pesticides if smallholder food crop intensification is to bear positive results. The information on the application of inorganic fertilizers is presented in Tables 12.9–12.11. The information shows that a minority of farmers currently use inorganic fertilizers and in very small quantities. The share of farmers currently using fertilizer on maize is 30%, on sorghum 15%, and on rice 40%.

Table 12.7. Type of sorghum seed used. Per cent users at present (2002) and at the time when the farm/household was established. (Source: *Afrint* survey data, Seini and Nyanteng, 2003.)

	Time when farm/household was established							
	Pre-SAP		SAP		Post-SAP		All respondents	
Sorghum – type of seed used	–1982	Present	1983–1992	Present	1993–	Present	Outset	Present
Traditional	98	96	94	98	98	98	97	97
Improved/HYV	2	4	6	2	2	2	3	3
Total	100	100	100	100	100	100	100	100

Table 12.8. Type of rice seed used. Per cent users at present (2002) and at the time when the farm/household was established. (Source: *Afrint* survey data, Seini and Nyanteng, 2003.)

	Time when farm/household was established							
	Pre-SAP		SAP		Post-SAP		All respondents	
Rice – type of seed used	–1982	Present	1983–1992	Present	1993–	Present	Outset	Present
Traditional	83	66	63	65	67	65	73	65
Improved/HYV	17	34	37	35	33	35	27	36
Total	100	100	100	100	100	100	100	100

The amounts applied are very modest indeed, the average application on maize is 5.4 kg/ha, on sorghum 2.3 kg/ha, and on rice 20.5 kg/ha. Application rates on cassava are negligible.

Interestingly, however, the share of farmers using inorganic fertilizer on maize has doubled in the SAP and post-SAP periods compared with the pre-SAP period. The increase in the proportion of farmers using inorganic fertilizer coincides with the onset of SAP and is consistent for all maize farmers regardless of age.[1]

Sorghum shows a pattern that is opposite to that of maize. The share of farmers using fertilizer on sorghum has steadily declined from 25% pre-SAP to 7%–10% in the post-SAP period (Table 12.10).[2] Since the difference between the groups as regards their present application is small, this decline reflects a structural change over the time period studied.

In general, fertilizer application is higher in rice cultivation than for the other two crops, both in terms of the proportion of farmers using fertilizers (40%) and the average amount applied (20.5 kg/ha). However, the share of farmers who applied fertilizer when they established their farms has been more or less constant over the pre- to post-SAP period (Table 12.11). It is the younger farmers, predominantly, who in the post-SAP period have increased their use of fertilizer on rice (from 35% to 47%), probably responding to a

Table 12.9. Share (per cent) of farmers using inorganic fertilizer on maize currently and at the time of the reference year. Total sample: 168 farmers. (Source: *Afrint* survey data.)

Inorganic fertilizer on maize – application	Time when farm/household was established							
	Pre-SAP		SAP		Post-SAP		All respondents	
	–1982	Present	1983–1992	Present	1993–	Present	Outset	Present
Not applied	86	72	69	71	70	68	77	70
Applied	14	28	31	29	30	32	23	30
Total	100	100	100	100	100	100	100	100

Table 12.10. Share (per cent) of farmers using inorganic fertilizer on sorghum currently and at the reference year. Total sample: 198 farmers. (Source: *Afrint* survey data.)

Inorganic fertilizer on sorghum – application	Time when farm/household was established							
	Pre-SAP		SAP		Post-SAP		All respondents	
	–1982	Present	1983–1992	Present	1993–	Present	Outset	Present
Not applied	75	85	84	80	93	90	83	85
Applied	25	15	16	20	7	10	17	15
Total	100	100	100	100	100	100	100	100

Table 12.11. Share (per cent) of farmers using inorganic fertilizer on rice currently and at the reference year. Total sample: 158 farmers. (Source: *Afrint* survey data.)

Inorganic fertilizer on rice – application	Time when farm/household was established							
	Pre-SAP		SAP		Post-SAP		All respondents	
	–1982	Present	1983–1992	Present	1993–	Present	Outset	Present
Not applied	64	66	60	58	65	53	63	60
Applied	36	34	40	42	35	47	37	40
Total	100	100	100	100	100	100	100	100

growing market demand for this crop (and in the process increasingly using tractors in cultivation).

The share of farmers using pesticides in the production of the selected food crops is much lower than the application of inorganic fertilizers, although the tendency is that application rates have increased over the period covered, particularly for maize and rice. Generally, the use of pesticides is insignificant in the production of food crops in the country.

Irrigation

Irrigation has been practised for a very long time in the country. Public attempts to provide irrigation in Ghana started in the early 1960s. Currently 22 major irrigation projects are implemented throughout the country with a total irrigable area of 11,000 ha. This is about 0.1% of the total agricultural land available or 0.2% of the area currently cultivated (Ministry of Food and Agriculture, 2001). Each project is less than 1000 ha in size except the Tono and Kpong irrigation projects. The public irrigation projects have been used to irrigate rice and vegetables, particularly tomatoes. The long-term investments by government in agricultural lands continue to be mainly in the form of irrigation facilities. Emphasis has been on small-scale irrigation dams that can irrigate between 20 and 30 ha as well as provide water for livestock in the long dry season, particularly in the northern savannah.

Although the use of irrigation at present is not a decisive feature for increasing crop production in the districts surveyed, there is reason to believe that irrigation in reality is more common than official figures indicate and that it is growing in significance. Much of the irrigation that takes place is small scale and managed on a household rather than community basis. The *Afrint* survey found that more than 3% of the total land cultivated by the respondents is irrigated, most of it for rice and vegetables. Of the land devoted to vegetables, 14% is irrigated. About 14% and 15% of the farmers growing rice and vegetables reported using irrigation on some or all of their land. Of the rice farmers who irrigated their farms, 35% used the facility to grow more than one crop in a year. In contrast, only 4% of the maize farmers and 2% of the sorghum farmers used irrigation (Table 12.12).

Crop yields

Although the yield levels estimated through the *Afrint* survey (Table 12.13) are lower than those reported by the Ministry of Agriculture (see Table 12.1), general trends are fairly consistent with those presented from official sources (Table 12.1 and Figs 12.4 and 12.5). Potential yield is the mean yield obtained by the 5% best performing farmers. The yield gap is the difference in per cent between the potential yield and overall mean yield.

In the case of maize there has been a steady increase in the proportion of farmers reporting increasing yields over the pre-SAP to post-SAP period with the main change occurring between the pre-SAP and SAP

Table 12.12. Proportion of farmers (per cent) using irrigation for different types of crops. (Source: *Afrint* survey data, Seini and Nyanteng, 2003.)

Crop	Farmers using irrigation
Maize	4
Cassava	0
Sorghum	2
Rice	14
Other food crops and vegetables	15

Table 12.13. Current mean and median yield of the major staple crops (t/ha), potential yield (t/ha) and yield gap (%). (Source: *Afrint* survey data, Seini and Nyanteng, 2003.)

	Maize	Cassava	Sorghum	Rice
Mean yield	1.2	4.3	0.5	1.0
Median yield	0.7	3.2	0.4	0.8
Potential yield	5.2	18.5	1.3	2.9
Yield gap (%)	−77	−77	−62	−66

Table 12.14. Proportion (per cent) of farmers per period who report increasing yields (as opposed to stagnant or decreasing yields). Total no. of respondents: 124–178. (Source: *Afrint* survey data, Seini and Nyanteng, 2003.)

	Time when farm/household was established			
Crop	Pre-SAP (–1982)	SAP (1983–1992)	Post-SAP (1993–)	All respondents
Maize	35	56	66	51
Cassava	35	55	59	48
Sorghum	37	25	32	33
Rice	28	18	27	25

period, and possibly gaining momentum because of the lower age of farmers in these groups. A similar pattern can be seen for cassava, where SAP is associated with a marked upward jump in the proportion of farmers reporting yields to have increased.[3] In all, about half the number of respondents report increasing yields of maize and cassava (Table 12.14). The change in yields for sorghum and rice is more difficult to interpret and shows no distinct or statistically significant trend. Yields appear to have remained rather steady throughout the period studied. At least in the case of rice, the figures in the table are probably an under-reporting. Based on the evidence we have put forward on the ongoing mechanization, the relatively higher adoption rate of inorganic fertilizer and high yielding seed varieties, the more frequent irrigation and, as we shall see in the next section, the spurt in marketing of rice under SAP, it would have been reasonable to expect that more than a quarter of the farmers would have reported a yield increase.

The farmers interviewed considered the use of inorganic fertilizers to be an important factor explaining yield increases. For maize, cassava and rice, the use of improved varieties was also stated as important. For sorghum and rice, mechanization was reported to have some impact on yields.

Of the factors accounting for decreased yields, farmers viewed declining soil fertility as by far the most important factor, and one that negatively affected yields for all crops. In the case of sorghum, bad weather and poor seeds were other, but less pertinent, reasons. For rice, poor seeds and inadequate water made a negative difference.

Marketing of food crops

One factor that can be assumed to drive smallholder intensification is access to market outlets. Access and proximity to market outlets is likely to influence market transactions. The *Afrint* survey showed that 79% of the maize growers and 81% of the cassava growers were within 5 km of a market outlet. On the other hand only 33% and 23% of the sorghum and rice growers had this advantage. Whereas most households in the southern districts appear to have market outlets close to their communities, most households in the northern districts do not have the same proximate outlets for their food crops. It is, perhaps, also a reflection of the general development of the two areas in terms of infrastructure as the northern parts are generally less developed than the southern parts.

For more than 80% of the maize and cassava growers, private traders constitute the main market outlet. For producers of sorghum and rice, own piecemeal disposal (retailing) of food crops is about as common as selling to private traders, particularly in the northern savannah zone where marketable surpluses for sorghum and rice are small.

Neither farmer cooperatives nor state marketing are significant actors in crop marketing. With the exception of rice, for which 4% of households indicated a state company as the main market outlet, no state company currently serves any other food crop. Perhaps this is an indication of the successful disengagement of the state in the participation of marketing activities under the Structural Adjustment policy reforms. The marginal participation of the state in rice marketing is as a result of the continued existence of the Irrigation Company of the Upper Regions (ICOUR), which is still involved in the production and marketing activities of farmers in the irrigation areas.

A few households reported involvement in contract farming. In most cases this reflects informal arrangements made with traders and could better be described as sponsored farming rather than contract farming proper. The trader normally contributes towards the financial requirements during the farming season and gets the exclusive right to buy all or part of the produce, often at a great disadvantage to the farmer in terms of price (Seini, 2003).

Just as access to market outlets in principle may motivate the smallholder to intensify, the conditions of exchange and the functioning of markets also matters. In this regard, policy processes in relation to market institutions and changes in regulated markets influence agricultural intensification. In the period prior to SAP, various food-marketing institutions existed but broadly had the same objectives of promoting food production through pricing and marketing policies, and of ensuring effective distribution of food. The last surviving of these institutions was the Ghana Food Distribution Corporation (GFDC), which concentrated its efforts on marketing of maize and rice alongside with private traders (P.W. Armah, 1989, unpublished PhD thesis). The private traders offered market prices. The GFDC was the government's major food agency, purchasing maize and rice to support its minimum guaranteed producer prices.[4] As part of Ghana's trade liberalization policy, the guaranteed minimum producer prices were abolished in 1990 and the GFDC is no longer operational.

In the *Afrint* household survey, market dynamics for food crops were studied with respect to how farmers perceived the changing conditions of sales, output prices, market outlets and modern input prices. Table 12.15 reports the percentage of farmers of each period who have experienced an increase in the amount of the four staple crops marketed between the present (2002) and when the household was formed (the pre-SAP, SAP or post-SAP period).

For maize, there has been a steady increase in crop sales over the entire period, probably driven by both changing market conditions and the aspirations of the younger and middle-aged households (Table 12.15). For cassava, the main increase in marketing is located in the SAP period, after which there is a levelling off. Also in this case, the age factor is likely to combine with structural conditions to produce this change.[5] For sorghum, it is difficult to interpret the changes due to a small sample size. The apparent spurt in marketing for farmers of the SAP period, for example, is not statistically significant. Similarly, for rice, there is no distinct pattern with respect to the periods examined. About two-thirds of the farmers report yields to have increased in the course of the farm life span, a pattern that is fairly constant over the entire period.

It should be noted, however, that Table 12.15 is limited to those farmers who at present or at the reference period were producing for the market. When we look at all farmers, we obtain a more dynamic picture of rice cultivation (Table 12.16). In this case the number of farmers producing a market surplus of rice rose from 28% at the outset in

Table 12.15. Proportion (per cent) of farmers who sell on the market and report increased sales since the household was established. (Source: *Afrint* survey data, Seini and Nyanteng, 2003.)

	Time when farm/household was established			
Crop	Pre-SAP (–1982)	SAP (1983–1992)	Post-SAP (1993–)	All sellers (no. of respondents)
Maize	48	62	76	62 (182)
Cassava	57	76	76	67 (203)
Sorghum	77	81	63	75 (63)
Rice	68	68	67	68 (93)

the pre-SAP period to 41% and 45% in the SAP and post-SAP periods, a change that is statistically significant. It is interesting to note that the main thrust in marketing involves farmers who established their farms in the SAP period.

As for the other crops, the picture in Table 12.16 is consistent with that of Table 12.15. As can be seen in Table 12.16, virtually all maize and cassava farmers produce for the market and have done so for a long time. What is new is that quantities marketed for a large share of the farmers have increased in the SAP and post-SAP periods (Table 12.15). While for maize, cassava and rice there is evidence of market dynamism in the form of increased sales and/or more farmers entering the market, there is no such evidence in the case of sorghum.

In Table 12.17, we look at how those farmers now selling for the market have experienced changes in market outlet since their household was formed. Improvements in market outlets are necessary for increased production and the lack of positive dynamics is likely to serve as a disincentive.

It is noteworthy that nearly three-quarters of the maize farmers of the pre-SAP period have experienced an improvement in market outlets since they took up farming, thus giving evidence to the increased market integration and market development that have occurred in the SAP and post-SAP periods. For cassava, the improvement in marketing outlets is equally striking, but in contrast with maize, it seems to have gained momentum in the SAP period. For both sorghum and rice, the majority of farmers perceive marketing outlets to have improved, however, without reference to any of the periods examined. The improved market outlets for all the food crops can be part of a generally positive change in favour of food crop intensification.

Interestingly, most maize and cassava farmers consider producer prices to be lower at present compared with when they established their farm. In the case of maize, the abolition of the minimum guaranteed producer price in 1990 could be a factor. The lower prices now may be due to increased supply *vis-à-vis* demand. It appears, however, that lower prices may have been compensated

Table 12.16. Proportion (per cent) of farmers selling crops when the household was formed and at present. Total no. of respondents per crop: 160–195. (Source: *Afrint* survey data, Seini and Nyanteng, 2003.)

	Time when farm/household was established							
	Pre-SAP		SAP		Post-SAP		All respondents	
Crop	–1982	Present	1983–1992	Present	1993–	Present	Outset	Present
Maize	90	98	96	98	86	91	90	96
Cassava	91	94	94	98	83	88	90	93
Sorghum	21	25	21	33	19	28	21	28
Rice	28	52	41	70	45	57	37	58

Table 12.17. Proportion (per cent) of farmers reporting improved market outlet since their household was formed. (Source: *Afrint* survey data, Seini and Nyanteng, 2003.)

	Time when farm/household was established			
Crop	Pre-SAP (–1982)	SAP (1983–1992)	Post-SAP (1993–)	All sellers (no. of respondents)
Maize	71	55	50	61 (172)
Cassava	77	67	40	64 (165)
Sorghum	81	89	75	82 (33)
Rice	67	63	67	69 (61)

for by productivity gains since average yield for all the food crops under consideration is on the upward trend, particularly in the post-SAP period.

For sorghum and rice, the story is somewhat different. Here, about 60% and 50% of the households considered that farm-gate prices had improved. Possibly, there exists a price incentive to which rice and sorghum farmers could be encouraged to respond provided complementary and affordable inputs are made available. The price dynamics for rice seem to be emitting mixed feelings on the part of the farmers despite the fact that more than 50% of them are of the opinion that prices are generally higher now than in the formative years of the households.

Constraints to intensification

Perhaps the most important constraint to smallholder intensification relates to food crop marketing and where substantial changes have occurred since the introduction of Structural Adjustment. The constraints are of two types, namely, those factors that constrain food crop marketing and those factors that constrain households from producing surpluses for the market. Factors that constrain food crop marketing include low and fluctuating prices, high transport costs, untimely payments, unreliable market outlets, high input prices, unavailability of inputs and lack of credit.

Consistently, a vast majority of the farmers (85–90%), regardless of the type of crop, considers input prices, and particularly that for fertilizer, to have increased since the onset of SAP. Associated with the high input prices is the limited cash resources owned by the farmers and which could be used for investing in agriculture. When asking farmers to rank a number of factors constraining production for the market, lack of credit surfaced as the single most important factor, as reported by more than 40% of the farmers, regardless of crop type. Whereas low and fluctuating prices constrain marketing of maize and cassava, as reported by 35% and 38% of the farmers, high input price is an important constraining factor in the case of sorghum and rice, as reported by 24% and 26% of the households.

Other factors constraining increased production for the market include high cost of hired labour (maize), and the already mentioned small resources farmers have for buying costly inputs (all crops). Lack of farm capital is a dominating constraint also in the case of sorghum and rice production. This is hardly surprising as the Upper East Region is one of the poorest regions in Ghana. Thus, 69% and 74% of the households that cultivate sorghum and rice, respectively, reported that lack of capital to buy inputs and to pay for land preparation services were major constraints.

It is noteworthy that despite these constraints, including the increase in input prices as reported by a majority of the farmers, adoption rates of improved seed and inorganic fertilizers have continued to increase in the SAP and post-SAP periods. Most farmers seem aware of the importance of these inputs for improving crop productivity. The very low quantities of fertilizers applied, despite the fact that an increasing number of farmers adopt fertilizers, is probably what explains the yield gap between the majority and the élite of farmers (Table 12.13). We have not dealt with the socio-economic aspects of this yield gap in the case of Ghana but there is reason to believe that it cannot be disassociated from current policies. As demonstrated in Chapter 7 of this volume, the yield gap not only illustrates the potential for increased productivity generally but it is also telling evidence of the kind of constraints we have described above and which seriously circumscribe smallholder farming in Ghana and elsewhere (see Larsson, Chapter 7, this volume).

The micro study supports the hypothesis of increasing yields and of a market dynamism invoked in maize, cassava and rice production since SAP, but less so with sorghum. These seem to stem from structural changes relating to factors such as mechanized land preparation, momentum in the adoption rate of improved seeds, application of inorganic fertilizers and access to market outlets. On the whole, however, it is largely younger aspiring farmers (with households established in the post-SAP period) who have increased their use of tractors and who also are familiar with improved varieties of these crops as they take up their farms. Sorghum on the other hand has been less dynamic, mainly because it is still largely a subsistence crop with few improved varieties in place. Thus, control for the effect of age shows that in the case of sorghum it is predominantly the older farmers (with farms established in the pre-SAP period) who shifted from hoe cultivation to animal traction, a lesser scale of technology than tractor mechanization.

Conclusions

Deductions made from the micro analyses lead to a number of conclusions on the conditions of smallholder food crop intensification in Ghana. The recorded increases in crop output and marketing over the SAP and post-SAP period underline the fact that policies may have had positive effects. Eventually, Structural Adjustment policies established a largely market driven economy in Ghana. The use of tractor and animal drawn implements for land preparation, particularly for rice and sorghum cultivation, has increased appreciably since the onset of SAP and the current period (2002). This enhances the efficiency and timeliness in production and promotes food crop intensification. In addition, the proportion of farmers who currently use improved and hybrid planting materials has increased substantially for maize and appreciably for rice and cassava. Also, in spite of macro evidence that the use of modern inputs such as inorganic fertilizer may have decreased following the removal of subsidies, the micro level evidence indicates that an appreciable and growing proportion of farmers uses these inputs, albeit in very small quantities following their high prices. The evidence also indicates increases in yields which farmers mainly attribute to the use of inorganic fertilizers.

The motivation for intensification stems from the pressure on the primary resources of land and labour as well as from improved marketing outlets for food crops. In particular, the changing conditions of sales, prices, market outlets and modern input prices are crucial to production for the market and for food crop intensification. At the same time, the most important constraint to smallholder food crop intensification relates to food crop marketing in the form of input prices and the lack of credit and other resources needed for farm investment.

Thus, on the whole, there is some evidence of smallholder food crop intensification at both the macro and micro level. Whereas the macro level evidence relates to increases in the aggregate yields of food crops, the evidence at the micro level seems to suggest that the level of intensification is limited, as indicated by the small proportion of farmers involved in the main attributes for intensification. Nevertheless, what has been demonstrated here is the presence of a dynamic situation in which increased marketing seems to drive agricultural intensification through increasing adoption of modern inputs and, for some crops, an increasing rate of mechanization.

Notes

[1] The differences between the groups in fertilizer application rates is significant at 0.10 level but not at 0.05 level when tested with Chi^2, one reason being the rather small sample size of 168 maize farmers. When farmers of the SAP and post-SAP periods are merged, the difference from the pre-SAP farmers is significant at 0.01 level.

[2] The change recorded is statistically significant at 0.02 level (Chi^2).

[3] Differences for both maize and cassava are significant at 0.01 level (Chi^2).

[4] Farmers were free to sell above the quoted guaranteed minimum prices and could resort to selling to the GFDC only in times of difficulty.

[5] Significant at $P = 0.02$ and 0.003, respectively (Chi^2).

References

Alderman, H. and Shively, G. (1996) Economic reform and food prices: evidence from markets in Ghana. *World Development* 24, 521–534.

Al-Hassan, R. (1997) *Facilitating Factors for Increased Agricultural Production.* A discussion paper prepared for a workshop on Accelerated Growth Strategy for Agriculture. MOFA, Accra.

Al-Hassan, R., Canacoo, V., Srofenyo, F. and Sakyi-Dawson, O. (1998) *Equity Implication of Reforms in the Financing and Delivery of Agricultural Extension Services*. ODA Policy Research Programme. R6470 CA. Department of Agricultural Economics, University of Ghana, Legon, Ghana.

Aryeetey, E. (2000) A diagnostic study of research and technology development in Ghana. *ISSER Technical Publication* No. 60. University of Ghana, Institute of Statistics, Social and Economic Research, Legon, Ghana.

Asuming-Brempong, S. (1989) Ghana rice policy. *Africa Development* 14, 67–82.

Asuming-Brempong, S. (1994) Exchange rate liberalisation and input-subsidy removal. In: Breth, S.A. (ed.) *Issues in African Rural Development 2*. Winrock International Institute for Agricultural Development, Morrilton, Arkansas.

Commander, S., Howell, J. and Seini, W. (1989) Ghana 1983–87. In: Commander, S. (ed.) *Structural Adjustment and Agriculture*. James Currey & Heinemann, Portsmouth.

Fiadjoe, F.Y.M., Al-Hassan, R. and Canacoo, V. (1997) National Agricultural Extension Project: Beneficiary Assessment of the Agricultural Extension Service. University of Ghana, Faculty of Agriculture, Legon, Ghana.

Jebuni, C.D. and Seini, W. (1992) Agricultural input policies under Structural Adjustment: their distributional implications. Cornell University, Cornell Food and Nutrition Policy Programme. Working Paper 31, Ithaca, New York.

Ministry of Food and Agriculture (2001) *Agriculture in Ghana: Facts and Figures, Policy Planning Monitoring and Evaluation*. Ministry of Food and Agriculture, Accra.

Seini, A.W. (2002a) Political instability and agricultural policy dynamics in Ghana. *Journal of Law Politics in Africa, Asia and Latin America* 35.

Seini, A. W. (2002b) Agricultural growth and competitiveness under policy reforms in Ghana: *ISSER Technical Publication* No 61. ISSER, University of Ghana, Legon, Ghana.

Seini, A.W. (2003) Agricultural diversification and information strategies in vegetable production and marketing in the Brong Ahafo Region of Ghana. *Legon Journal of Sociology* 1.

Seini, W. and Nyanteng, W.K. (2003) *African Food Crisis: the Relevance of Asian Models. Ghana Report on Village and Household Survey. Afrint* Research Project, University of Ghana, Institute of Statistical, Economic and Social Research, Legon, Ghana.

Stryker, J.D. (1990) Trade, exchange rate and agricultural pricing policies in Ghana. In: *World Bank Comparative Studies: the Political Economy of Agricultural Pricing Policy*. World Bank, Washington, DC.

World Bank (1991) *Ghana: Medium Term Agricultural Development Strategy* (MTADS). An Agenda for Sustained Growth and Development (1991–2000). Report No. 8914-GH. World Bank, Washington, DC.

13 Green Revolution and Regional Inequality: Implications of Asian Experience for Africa

Keijiro Otsuka and Takashi Yamano
Foundation for Advanced Studies on International Development (FASID), GRIPS/FASID Joint Graduate Program, Tokyo, Japan

Introduction

In view of the explosive population growth, stagnant grain yields, and near-exhaustion of cultivable land frontiers in Asia in the 1950s and 1960s, there was desperate fear that serious food shortages and consequently widespread famine would inevitably occur in the near future in Asia (Barker and Herdt, 1985). As is well known, however, such pessimistic projection has turned out to be far off the mark, owing to the success of the Green Revolution which has taken place in tropical areas of Asia since the late 1960s. Roughly speaking, rice yield doubled and, coupled with increased rice cropping intensity, rice production tripled in the tropics of Asia over the last three and a half decades. As a result, rice production increased more rapidly than population and the real rice price currently constitutes only one-third of the level of around 1970, the dawn of the Green Revolution (Pingali *et al.*, 1997).

The current situation in sub-Saharan Africa is not too different from that in Asia several decades ago: population grows at an annual rate of nearly 3%, whereas grain production increases much more slowly, as discussed in Djurfeldt, Chapter 1 and Holmén, Chapter 5 (this volume). Furthermore, grain yields have been largely stagnant or even declining in some areas for the last few decades. Considering that cultivable land is becoming scarce in many parts of the continent (Otsuka and Place, 2001), there is no question that if the current trend continues, widespread famine cannot be avoided in sub-Saharan Africa. In order to achieve food security in this region, grain yields must be increased in a sustainable fashion. Yet, as shown in Holmén, Chapter 5 (this volume), there is no clear symptom of increasing grain yields, judging from country-level statistics. Furthermore, there is no clear development strategy to improve grain yields in this region.

This chapter contributes to the notion of this volume by arguing that there are a number of useful lessons that sub-Saharan Africa can learn from the experience of the Green Revolution in Asia. First of all, we would like to emphasize that the Asian Green Revolution is a consequence of conscious and massive effort to develop fertilizer-responsive, high-yielding varieties of rice and wheat for favourable production environments, particularly in the early phase of the Green Revolution. This is because it is scientifically much easier to develop new varieties for more favourable production environments. Second, while it is true that large gaps in productivity emerged between favourable and unfavourable production environments because of the differential impacts of the Green Revolution, the regional income inequality thereby generated was much less pronounced than generally believed because of the interregional factor market adjustments, such as migration from unfavourable to favourable areas (David and Otsuka, 1994). Third, the development of improved varieties for unfavourable areas took place later, as it is based upon varieties developed earlier for favourable areas (Evenson and Gollin, 2003). Fourth, it is important to recognize that the Asian Green Revolution was initiated by the international agricultural research institutions, such as the International Rice Research Institute (IRRI) and International Maize and Wheat Improvement Center (CIMMYT). This is because new high-yielding varieties are characterized by global or regional public goods, which are useful for adoption in wide areas across country borders and as breeding materials in national agricultural research programmes. As such, only international research organizations have appropriate incentives to carry out the core research.[1] In sub-Saharan Africa, too, activities of international breeding institutions are indispensable to realizing major yield gains; yet such research is deplorably weak.[2] Fifth, it cannot be over-emphasized that the Asian Green Revolution has been supported by the increased application of chemical fertilizer. None the less, the strategy to facilitate the increased application of fertilizer is unclear in the context of sub-Saharan Africa.

The first purpose of this chapter is to substantiate the arguments briefly developed above based on the empirical evidence in Asia, so as to draw useful lessons for sub-Saharan Africa. The second purpose, based partly on the lessons from Asia and partly on the observed changes in farming systems in East Africa, is to propose a promising strategy to realize a new 'Green Revolution' in East African highlands which utilizes manure produced by cross-bred dairy cows.

The organization of this chapter is as follows. The following section, section two, compares the performance of food production between Asia and Africa and advances a hypothesis that can explain the contrasting difference in food production performance between the two regions. Section three characterizes the Asian rice Green Revolution in India and the Philippines. In section four we present an analysis of income difference between the favourable areas where the Green Revolution took place and unfavourable areas where it did not take place, or took place only to a limited extent. Section five is devoted to an exploration of the possibility of a Green Revolution in sub-Saharan Africa based on the data recently collected by the FASID and its collaborators.[3] The last section discusses the policy implications of this study.

A Comparison between Asia and Africa

In order to grasp the extent of the difference in the performance of food sectors between Asia and Africa for the last several decades, this section compares rice and maize yields per hectare between the two major regions.

Rice and maize yields

Most likely the largest beneficiary from the Green Revolution among Asian countries is Indonesia. Indeed, the average rice yield per hectare increased from less than 2 t in the early 1960s to nearly 4.5 t at present (see Fig. 13.1). A closer examination reveals that the yield growth slowed down considerably in the late 1980s, because of the exhaustion of the yield potential of the Green Revolution technology (see Jirström, Chapter 3, this volume; Hayami and Otsuka, 1994). As will be demonstrated later, the rice yield stagnation also began around the mid-1980s in the Philippines, where IRRI is located.[4]

The yield growth began later in India, after taking a considerable period to transfer the technology developed in South-east Asia to South Asia. Thus, as discussed in Djurfeldt and Jirström, Chapter 4, (this volume) a major yield growth began almost a decade later in India and it continued up to 2000. It is important to emphasize that this technology transfer was made possible by adaptive national research programmes in India, which

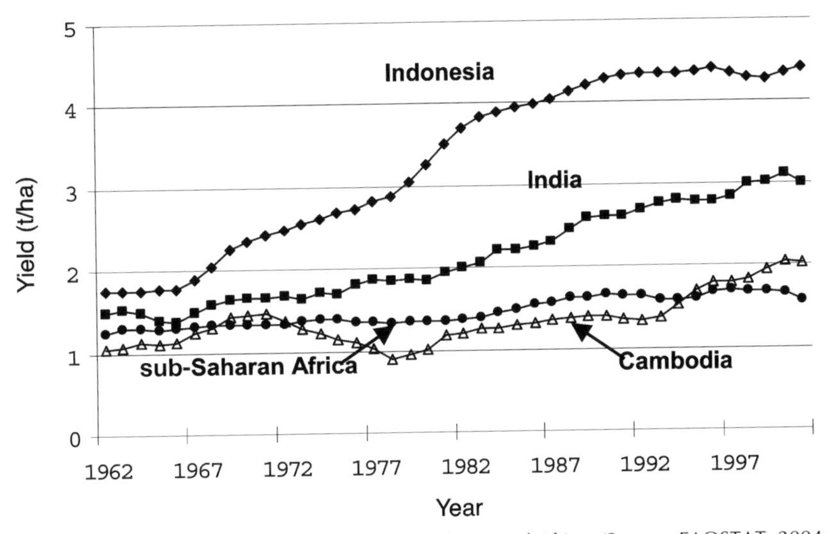

Fig. 13.1. Changes in rice yields in selected countries of Asia and Africa. (Source: FAOSTAT, 2004.)

attempted to assimilate the technology developed by IRRI.

Rice yields have been low and stagnant in both Cambodia and sub-Saharan Africa. After the peace was restored in Cambodia in the early 1990s, however, the rice yield began increasing owing to the activated rice research programme, in which IRRI was involved. Rice research still has to keep improving the Green Revolution technologies. The success of yield growth is limited in Cambodia, partly because many areas in this country are prone to drought, in which the existing modern varieties (MVs) are not highly productive. Such experience clearly illustrates the critical importance of agricultural research for the improvement of agricultural productivity.

It seems to us that the case of sub-Saharan Africa represents another example in which weak research leads to feeble yield growth. In the early 1960s, before the Green Revolution, yields in sub-Saharan Africa were not significantly different from those in Asia. Near constancy of yields in this region thereafter can be attributed largely to the lack of improved rice technology over the last several decades.[5]

Roughly the same story can be found in a comparison of maize yields between Asia and sub-Saharan Africa (Fig. 13.2). While maize yield in Thailand was higher from the beginning, the yields gap between Pakistan and sub-Saharan Africa was small in the early 1960s. The gap widened gradually thereafter, which can be explained by the introduction of high-yielding maize MVs in Pakistan and their absence or ineffectiveness in sub-Saharan Africa.

The Sequence of Green Revolution

The success of the Asian Green Revolution and its absence in sub-Saharan Africa is clearly illustrated by Fig. 13.3, which shows how yield changes with an increase in fertilizer application. Traditional varieties (TVs) are low-yielding because they are tall and have a weak stem so that plants easily lodge as the weight of the grain increases due to increased application of fertilizer. In contrast, MVs have short stature and strong stems, so that plants do not lodge easily when a large amount of fertilizer is applied. This explains why the yield curve of early MVs (called MV1) is located far above that of TVs in Fig. 13.3.

MV1 are potentially high-yielding but susceptible to pests and diseases (David and Otsuka, 1994). According to Otsuka *et al.* (1994a) and Jatileksono and Otsuka (1993), the yield impact of MV2, which are resistant to multiple pests and diseases and were

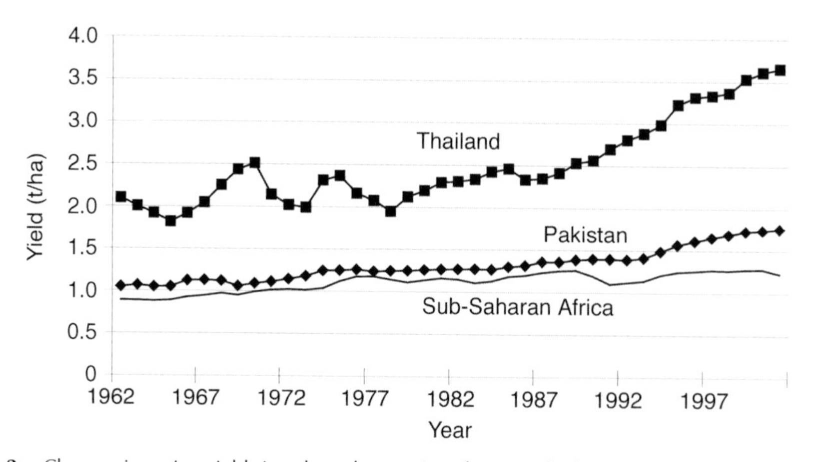

Fig. 13.2. Changes in maize yields in selected countries of Asia and Africa. (Source: FAOSTAT, 2004.)

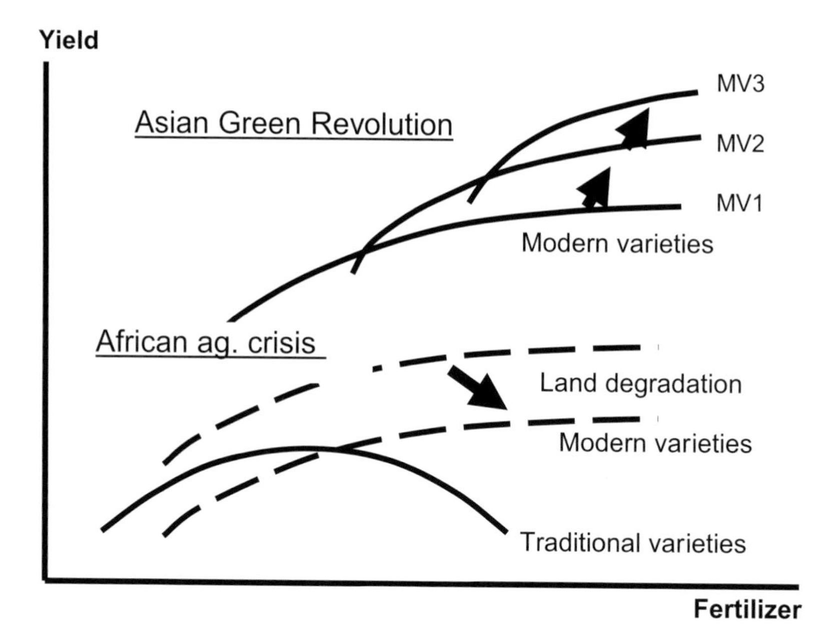

Fig. 13.3. Asian Green Revolution and African agricultural crisis.

released by IRRI starting in 1976, was far more significant than that of MV1. Thus, when MV1 dominated in the 1970s, the yield growth was less rapid than in the subsequent periods in Indonesia (see Fig. 13.1). Consequently, the actual yield curve of MV2 is located above that of MV1 in Fig. 13.3.

As will be discussed in the next section, the adoption area of MVs continued to expand, as national programmes developed location-specific MVs, termed MV3, suitable for diverse production environments, including drought-prone areas. MV3 uses MV1 and MV2 as parental materials and contribute to the improvement of production efficiency in agriculturally marginal areas of Asia (Hossain *et al.*, 2003; Evenson, 2004). In other words, MV1 and MV2 are characterized by regional public goods, whereas MV3 are closer to local public goods. Thus, the development of MV1 and MV2 by the international research institutions and that of MV3 by national research institutions can be justified from the theoretical point of view. Lastly, it is worth mentioning that the high yields in Asia were achieved owing importantly to the decreasing fertilizer price relative to the grain price, something which stimulated larger applications of fertilizer.

It is wrong to assume that no MVs are adopted in sub-Saharan Africa at present. On the contrary, improved maize and other varieties are widely adopted in various countries (see, e.g., Djurfeldt and Jirström, Chapter 4, and Seini and Nyanteng, Chapter 12, this volume; Ndjeunga and Bantilan, 2003; Sserunkuuma, 2003). The problem is that fertilizer is seldom applied and, hence, the impact of MVs on yield is low or negligible. There seem to be two explanations for this. First, the shift of the yield curve is not significant, as is illustrated in Fig. 13.3. Second, in addition to the small shift of the yield curve, the optimum amount of fertilizer application is close to nil due to high prices of chemical fertilizer. A recent paper by Jayne *et al.* (2003b) in a special issue of *Food Policy* on input markets in sub-Saharan Africa indicates that farm-gate prices of chemical fertilizer (such as DAP and Urea) are about twice the Cost, Insurance and Freight port prices. The lack of fertilizer application reduces soil fertility over time, thereby shifting the yield curve downwards (Fig. 13.3). The implication is that in order to raise grain yields in sub-Saharan Africa, we have not only to generate higher-yielding varieties but also to develop the system to apply nutrients to widely depleted soils.

Green Revolution in Asia

MVs were designed to have a significant yield advantage over TVs because of their capacity to respond favourably to high fertilizer application and to utilize solar energy effectively (Chandler, 1982). According to experimental data from IRRI, the first MV released in 1966, IR8, showed a higher yield response to different levels of nitrogen (N) application compared with TVs. The highest yields obtained from IR8 ranged from over 7.0 to 8.5 t/ha compared with 5.6 t/ha for a TV (IRRI, 1966:146). MVs, however, achieved high yields only under favourable production environments. It is also important to note that the MV technologies are scale-neutral so that small-scale farmers, who have advantages in monitoring labour activities, adopted the technologies rapidly.

The Indian case

The differential rates of MV adoption for rice among different regions in India and Bangladesh are shown in Fig. 13.4. MVs were first diffused rapidly in North India, including Punjab and Haryana, which are endowed with favourable conditions for pump irrigation, resulting in a sharp increase in its rice yield per hectare. The MV adoption in North India was followed by South India, which had the tradition of both tank and gravity irrigation. In contrast, MV adoption significantly lagged behind in other areas of India and Bangladesh, for which water control is difficult, and their average yields began to rise only as late as the 1980s.

It is noteworthy that the yield growth in North India slowed down significantly in the 1980s as MV adoption came close to saturation, while South India maintained a growth rate comparable to that of North India in the 1970s and 1980s. The rest of India and Bangladesh appear to have only very recently reached a situation similar to that prevailing in North India two decades earlier. These sequences were created by leads and lags in the exploitation and the consequent exhaustion of existing technology potential, across regions with different environmental conditions through technology transfer. This was not a simple transfer of a fixed technology, rather it involved adaptation of varieties and cultural practices (Evenson and Gollin, 2003). Also important was alteration of the environment itself. For example, MVs were adopted in East India and Bangladesh not so much in the traditional rice growing season under monsoon rain but were increasingly grown in flood-prone areas during the dry season with pump irrigation (Hossain *et al.*, 2003).[6] In this way, the rice production environment and cultural practices of the flood-prone deltaic regions were assimilated to North India. This process, by its nature, was similar to the

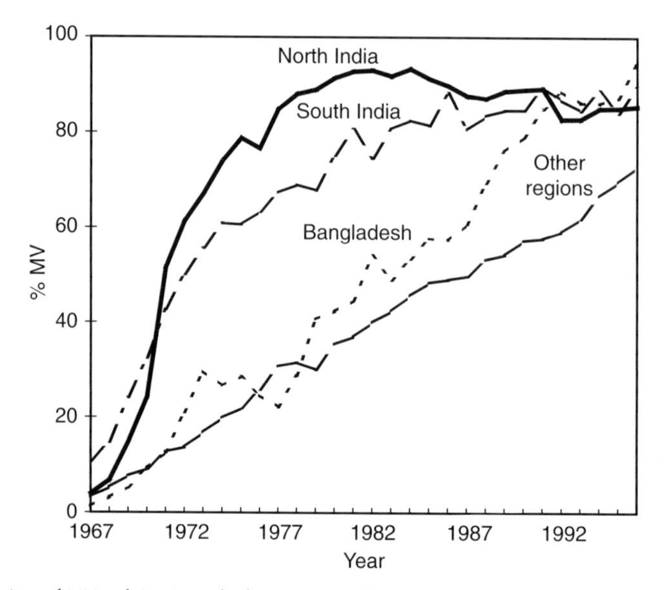

Fig. 13.4. Diffusion of MVs of rice in India by region and in Bangladesh in terms of percentage of rice harvested area. (Source: IRRI, 2003.)

technology transfer from the western to the eastern part of Japan during the Meiji period (Hayami and Ruttan, 1985).

In this regional technology transfer process, MVs developed for favourable northern and southern regions of India are used for the development of newer MVs, i.e. MV3, for less favourable production areas of East India and Bangladesh (Janaiah, 2002; Hossain *et al.*, 2003). In other words, yield growth in less favourable areas of South Asia would not have been possible without prior development of improved varieties in more favourable areas.

The Philippine case

As elsewhere in Asia, in the Philippines rice is cultivated in irrigated, rain-fed and upland ecosystems. The availability of irrigation is by far the most important physical factor affecting the adoption and productivity of MVs (David and Otsuka, 1994). In the Philippines, data on the adoption and yields of MVs by three representative ecosystems are available (see Fig. 13.5). The adoption of MVs was quick and widespread in a short period of time owing partly to the country's well-developed irrigation systems. Similarly, MV adoption rate in rain-fed ecosystems has increased consistently over time, indicating that MVs perform fairly well in rain-fed areas with sufficient rainfall. To the contrary, the MV adoption rate in the uplands has remained marginal even up to the mid-1980s. These contrasting MV adoption trends indicate that Green Revolution technology has spread most rapidly in irrigated ecosystems, to a lesser extent in rain-fed ecosystems, and not much at all in the uplands.[7]

Paddy yields of TVs and MVs in both irrigated and rain-fed ecosystems were close in the early 1970s. Yet yields of MVs, particularly in irrigated conditions, increased continually for the following two decades due importantly to the advent of MV2 and MV3. It is clear that both the adoption of MVs and the presence of irrigation are significant factors contributing to yield growth. Indeed, it has been established that socio-economic factors, such as land tenure and farm size, are not decisive factors affecting the adoption and yield performance of MVs (Ruttan, 1977; Hayami and Kikuchi 1982; Hayami and Otsuka, 1993; David and Otsuka, 1994). Contrary to a popular view, large-scale farmers

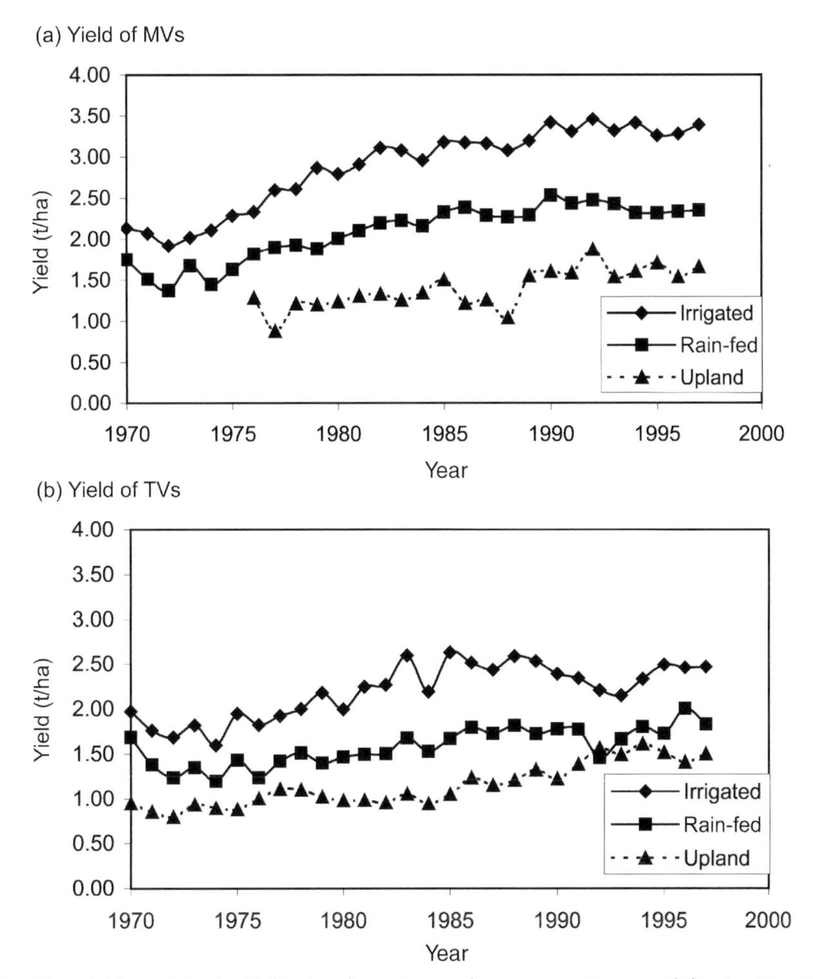

Fig. 13.5. Rice yield trends in the Philippines by variety and ecosystem. (Source: Philippine Rice Research Institute, 2000.)

are found to be less efficient than small-scale farmers in South Asia (Otsuka, 2004).

Regional Inequality in Asia

As we have seen, the regional gap in productivity arises between favourable and unfavourable areas because of the regionally differential impacts of the Green Revolution technology. It is therefore argued that agricultural research should focus on unfavourable areas to improve the welfare of poor people in such areas (Lipton and Longhurst, 1989). While this argument has some force, we must not overlook the fact that income of people in unfavourable areas tends to increase due to factor market adjustments, including interregional migration from such areas to favourable areas.

Favourable versus unfavourable areas in Asia

Data comparing yields of MVs and TVs, labour use and income across production environments are rare. We first examine here the community and household data collected by the IRRI in the late 1980s (David and Otsuka, 1994), which is followed by an examination of village data collected in the Philippines in 1985 and 1988. Table 13.1 compares MV adoption and yields of MVs and TVs across production environments. We have chosen only three countries because MVs completely dominated in many survey areas, so that the comparison of yields of MVs and TVs under similar conditions is not feasible. As would be expected, the adoption rates of MVs are higher in more favourable production environments, but MVs are also widely adopted in rain-fed areas.

It is important to observe that the yields of MVs are higher in more favourable areas, whereas the yields of TVs are less sensitive to differences in production environments. This observation implies that the yield gains associated with MV adoption are larger the more favourable the rice production environments. Thus, as Byerlee (1996:701) aptly points out, based on a more comprehensive literature review, 'the yield advantage of MVs is lower in marginal areas.'

The use of hired labour is closely associated with the production environments and the adoption rate of MVs. This hired-labour usage effects of MVs arise partly from the short maturity of MVs, which leads to sharp peak demands for labour, and partly from negative income effects of MV adoption on the supply of family labour of farm households (David and Otsuka, 1994; Otsuka *et al.*, 1994b). The increased demand for hired labour would have expanded the employment opportunities for the poor in rice production, since the main source of hired labour is the landless labourers, who belong to the poorest segment of the poor rural societies.

The greater labour demand, particularly for hired labour, from modern rice technology would increase wage rates in the favourable areas faster than in the unfavourable areas. If a labour market adjusts through interregional permanent and seasonal migration, however, wage rates will tend to equalize across production environments. In such a case, benefits from technical change in the favourable areas will be shared with people in the unfavourable areas, particularly with landless workers, who tend to be geographically more mobile than farmers. Those who remain in unfavourable areas, as well as migrant workers, benefit from MV adoption in the favourable areas because wage rates increase in unfavourable areas as a result of out-migration. Although village-level migration data were not available, the hypothesis that interregional migration occurs in response to differential MV adoption can be tested empirically by examining trends in village population growth rates and the proportion of landless households across production

Table 13.1. MV adoption and rice yields by variety across production environments in selected locations of Asia, 1985–1987[a]. (Source: David and Otsuka, 1994.)

		Yields (t/ha)	
Location/environment	MV adoption (%)	MVs	TVs
Central Luzon and Panay in the Philippines in 1985:			
Irrigated	97	3.6	2.4[b]
Favourable rain-fed	99	3.3	2.2[b]
Unfavourable rain-fed	40	2.6	2.0
Central Thailand in 1986:			
Irrigated	71	4.4	1.9
Rain-fed	11	3.3	2.1
Deep-water	1	1.8	1.9
Tamil Nadu in India in 1987:			
Canal irrigation	100	5.6	n.a.
Tank irrigation	72	4.3	2.6
Rain-fed	66	3.9	2.8

[a]Based on a survey of 50 villages in the Philippines, 33 villages in Thailand, and 30 villages in India.
[b]Yields when TVs were grown in the 1970s.
n.a. = Not available.

environments. The results of six-country comparative studies included in David and Otsuka (1994) indicate that rural labour markets in different production environments are closely integrated through interregional migration. In fact, agricultural wage rates tend to be equalized across production environments; even if there were regional wage differentials, the differences were not large. Such tendency for regional wage equalization cannot be understood without considering the contribution of interregional labour migration. Thus, David and Otsuka (1994:418) conclude: 'As far as the well-being of poor landless labourers is concerned, the impact of modern rice technology does not seem to be as inequitable as generally believed.'[8] Yet household income differed significantly across production environments and between farm and landless-worker households, as is demonstrated in Table 13.2, which shows per capita income in terms of US dollars. Average incomes of farm households are clearly much higher in irrigated and favourable rain-fed villages than in unfavourable rain-fed villages. The importance of rice production as a source of household income tends to be lower in less favourable areas, and the profitable opportunities for growing other crops and non-farm employment opportunities help reduce the income gap across production environments. For example, farmers in deep-water areas in Thailand, which are unfavourable for rice production, are not much worse off than those in irrigated areas because of the proximity to Bangkok.

According to the regression analyses of the determinants of farm household income by source, land income is positively associated with MV adoption and the availability of irrigation, but labour income from rice production and income from other sources are not clearly correlated with the technology and environmental factors. Thus, inter-village income differences of farm households shown in Table 13.2 can be explained largely by differences in returns to land associated with differential technology adoption between favourable and unfavourable areas.[9]

Landless households are generally poorer than farm households in the same production environment, mainly because of lack of access to land. In fact, in unfavourable rain-fed areas where the rate of return to land is relatively low, the difference in income between farm and landless households is also low. Income

Table 13.2. Average annual income per capita (US$) of farm and landless households by production environment in selected locations of Asia, 1985–1987[a]. (Source: David and Otsuka, 1994.)

	Irrigated	Favourable, rain-fed	Unfavourable, rain-fed
Farm households:			
Thailand	437	n.a.	198 (373)[b]
Bangladesh	163	135	121
Nepal	174	n.a.	149
India	266	262	131
Indonesia	150	119	n.a.
Philippines	228	201	86
Landless labourer households:			
Thailand	245	n.a.	115 (120)[b]
Bangladesh	114	115	96
Nepal	64	n.a.	54
India	120	92	100
Indonesia	n.a.	n.a.	n.a.
Philippines	126	121	75

[a]Based on household surveys in several selected villages in each country.
[b]Deep-water area.
n.a. = Not available.

disparities across production environments are much lower among the landless than among farm households, which is consistent with the findings of wage equalization across production environments through labour migration. It seems clear that regional productivity differentials associated with differential technology adoption did not significantly widen income inequality among the poor landless population across production environments (Renkow, 1993).

Possibility of a Green Revolution in East Africa

So far, we have focused on the first purpose of this chapter by describing the Green Revolution in Asia. In this section, we turn our attention to the second purpose: to propose a promising strategy to realize a new 'Green Revolution' in East African highlands which utilizes manure produced by cross-bred dairy cows.

Since the success of the Green Revolution in Asia, many African governments have made substantial efforts to promote high-yielding seed/fertilizer technologies (IFDC, 2001). Despite serious efforts made by newly independent African governments and a short-lived success, high-yielding seed/fertilizer technologies are currently adopted by only a small fraction of African farmers, as discussed in Holmén, Chapter 5 (this volume). Although many experts have already proposed various reasons for the failure and suggestions of what to do about it (Crawford *et al.*, 2003), we would like to make another attempt to propose a new farming system, called 'Organic Green Revolution' (OGR), that could dramatically increase crop productivity in favourable areas of Africa. Like the Asian Green Revolution, OGR requires improved varieties. However, OGR relies heavily, but not exclusively, on the use of manure and composts, which are essential under fragile conditions in sub-Saharan Africa.

Figure 13.6 shows a basic mechanism of Organic Green Revolution. The central component of this mechanism is an improved dairy production system which uses cross breeds of local and European cows (or goats). This production system is typically found in the Central and Western Highlands of Kenya and South-western regions of Uganda (Staal *et al.*, 2001; Waithaka *et al.*, 2002; Staal and Kaguongo, 2003).

It is much easier to collect manure from animals raised in stalls than from animals grazed on open fields. Moreover, with proper floor materials (such as cement) and management, farmers can gather manure with a minimum loss of organic matter and make highly productive compost by mixing it with refuse, dry grass, dry leaves, water, and/or domestic ashes. The compost in turn can be applied to crops whose yields are responsive to it.

In Uganda, a crop that is typically involved in this production system is banana, where farmers fertilize soil with compost and other organic matter before and after planting. In Table 13.3, we provide some empirical

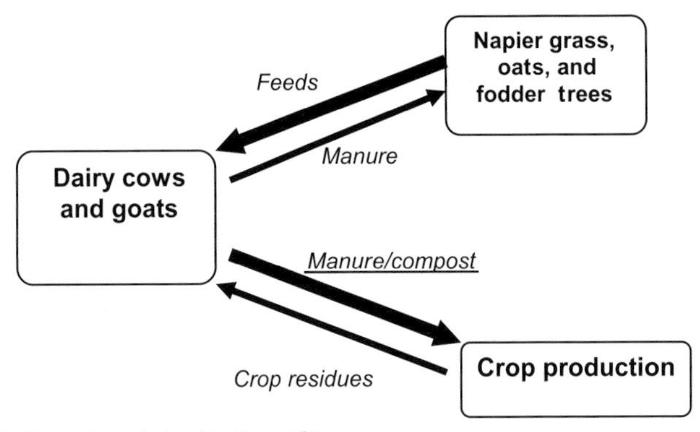

Fig. 13.6. 'Organic Green Revolution' in East Africa.

Table 13.3. Grain yields and Organic Green Revolution: evidence from Uganda and Ethiopia. (Source: FASID, 2003.)

Crop	Seed type	No. of communities	Yield without any fertilizer (kg/ha)	Yield with organic fertilizer only (kg/ha)	Yield with chemical and other fertilizer (kg/ha)
Uganda					
Banana	Local	42	3859	5921	n.a.
Maize	Local	37	1645	3380	n.a.
	Improved	28	1825	3334	n.a.
Ethiopia					
Maize	Local	26	903	1230	1707
	Improved	19	458	906	2169
	High yielding	8	843	1604	2267
Wheat	Local	18	507	775	897
	Improved	30	355	950	1621
Teff	Local	27	208	430	733

n.a. = Not available.

evidence, based on the 2003 FASID-REPEAT Community survey conducted jointly by FASID and Makerere University, to show the effectiveness of this practice. Out of the 42 communities, 15 (36%) did not apply any fertilizer to banana, while the remaining 27 communities (64%) applied some form of organic fertilizer. The difference in banana yields between the two groups is large. While the communities without any fertilizer obtained only 3859 kg of banana/ha on average, the communities with organic fertilizer produced 5921 kg of banana/ha.

In Ethiopia, nationwide fertilizer consumption dramatically increased in the 1990s largely due to a large-scale government-supported credit programme, the New Extension Intervention Program (NEIP) (Jayne *et al.*, 2003b). Community-level data from the 2003 FASID-REPEAT Community survey, jointly carried out by FASID and the International Livestock Research Institute (ILRI), are consistent with the macro-level expansion of fertilizer consumption (Table 13.3).

Table 13.3 shows average yields for three major crops in Ethiopia under different fertilizer application scenarios. Maize yield of local varieties increases from 903 kg/ha without any fertilizer application to 1707 kg/ha with chemical fertilizer (a 90% increase). The rate of increase for high-yielding varieties, however, is much higher: a 170% increase from 843 kg/ha without any fertilizer application to 2267 kg/ha with chemical fertilizer. We find a similar pattern for wheat.

Note that the OGR technology is based on improved dairy production and is highly labour intensive; hence, it is particularly appropriate for areas suitable for dairy production and land-scarce areas of Africa. The land scarcity, however, seems to be growing in many African countries with high population growth rates and slow economic transfer out of agriculture. Nationwide household surveys also suggest severe land shortages among smallholders in Africa (Jayne *et al.*, 2003a).[10] Thus, it seems that the demand for this type of technology is likely to increase in wide areas.

Concluding Remarks

Not only rice research but also research on maize and wheat has focused on favourable environments because of a high probability of scientific success (David and Otsuka, 1994; Byerlee, 1996). The homogeneous nature of irrigated and favourable rain-fed areas also implies wide adaptability of new technologies across country borders, ensuring significant effects on grain production. In this sense, such research is characterized by regional public goods and should be undertaken by the international agricultural research institutions. In contrast, it is scientifically much more difficult to develop new varieties for unfavourable production environments. Moreover, unfavourable environments are highly heterogeneous, suffering from drought, flooding, salinity and soil erosions, so that superior varieties, even if successfully developed, can be diffused only in limited areas. Because of such diversity, research for unfavourable areas is characterized by local public goods, whose research ought to be carried out by national agricultural research institutions. It must also be emphasized that such research is costly and unproductive unless research knowledge accumulated for favourable areas is available. In all likelihood, it is a mistake to attempt to generate new technologies from scratch for unfavourable areas in order to reduce poverty in such areas.

Since production environments of sub-Saharan Africa are much harsher than in Asia in general, it is much more difficult or costly to generate Green Revolution technology in this region. None the less, our analysis of ongoing Organic Green Revolution strongly suggests that the Green Revolution is possible in favourable areas of sub-Saharan Africa (see also Larsson, Chapter 7, this volume). Key features of Organic Green Revolution technology are similar to the Asian Green Revolution, i.e. combined use of improved seeds and fertilizer (inorganic in Asia and organic in Africa). In order to turn the possibility into reality, conscious and focused efforts in both international and national research programmes, as well as extension and dissemination programmes, will be required.

First of all, research on OGR is far from adequate: neither serious breeding research nor systematic research on the best farming system consisting of appropriate improved cereal varieties complemented with the management of soil, cows and trees, has been conducted. Obviously we need to strengthen international agricultural research programmes at this stage, because benefits of such research will spill over to wide areas. Second, in order to disseminate the OGR technology, not only extension programmes but also subsidy programmes for the adoption of dairy cows and goats are needed. Furthermore, the development of milk marketing systems must be facilitated to make OGR sufficiently profitable for farmers. Third, while agricultural research by the international organizations is expected to generate useful technologies for favourable areas, we need to strengthen national research programmes which can contribute to the generation of improved technologies for less favourable areas characterized by diverse and heterogeneous environments.

We have to recognize the fact that the Green Revolution is not a one-shot phenomenon but entails long processes involving a sequence of continuous research and development efforts with clear regional focuses in various stages of technology development. We are hopeful and confident that if the appropriate development strategies are pursued with the mobilization of sufficient human and financial resources, the Green Revolution in sub-Saharan Africa can be a reality.

Acknowledgements

We are indebted to useful comments from Kjell Havnevik, Magnus Jirström, Ruth Oniango and participants of the workshop *African Food Crisis: the Relevance of Asian Models* in Nairobi, January 2004.

Notes

1 This does not imply, however, that the state played only a limited role. See Chapters 3 and 4 (this volume) on the role of state in the Asian Green Revolution.

2 There are only Africa Rice Center (WARDA) and IITA, both in West Africa, which are much smaller than IRRI and CIMMYT.

3 At the time of writing this chapter, household-level surveys are under way in Kenya and Ethiopia. Please visit FASID's websites at www.fasid.or.jp for further information.

4 Since resistance of new varieties to pests and diseases declines over time, 'maintenance' research needs to continue to prevent yields from declining.

5 NERICA (new rice for Africa) varieties generated recently by WARDA have had no discernible impacts on the average yield performance in sub-Saharan Africa (Fig. 13.1).

[6] The diffusion of MVs induced investments in irrigation in subsequent periods not only in South Asia but also in the Philippines (Hayami and Kikuchi, 1982).

[7] The adoption rate of MVs in the uplands is noticeably low in 1996 and 1997 due to the El Niño effect.

[8] As Binswanger and Quizon (1989) demonstrate, a major way by which the Green Revolution improved the welfare of the poor was through reductions in grain prices.

[9] See also Quisumbing *et al.* (2004) on the issue of changing income sources among farm households.

[10] Although there is a growing concern about labour shortage caused by the AIDS epidemic in some African countries, empirical evidence of its impacts on farm production is still thin and mixed. For instance, by using panel data, Yamano and Jayne (2004) found a switch from high-value crops to food crops when a male member dies in a rural household in Kenya. This could be interpreted both as a result of a loss of labour and a result of a discrimination against women farmers in high-value crop production and marketing.

References

Barker, R. and Herdt, R. (1985) *The Rice Economy of Asia*. Resources for the Future, Washington, DC.

Byerlee, D. (1996) Modern varieties, productivity, and sustainability: recent experience and emerging challenges. *World Development* 24, 697–718.

Chandler, R.J. (1982) *An Adventure in Applied Science: a History of the International Rice Research Institute*. IRRI, International Rice Research Institute, Los Banos, Philippines.

Crawford, E., Kelly, V., Jayne, T.S. and Howard, J. (2003) Input use and market development in sub-Saharan Africa: an overview. *Food Policy* 28, 277–292.

David, C. and Otsuka, K. (1994) *Modern Rice Technology and Income Distribution in Asia*. Lynne Reinner, Boulder, Colorado.

Evenson, R. (2004) Food and population: D. Gale Johnson and the Green Revolution. *Economic Development and Cultural Change* 52(3), 543–569.

Evenson, R. and Gollin, D. (eds) (2003) *Crop Variety Improvement and its Effect on Productivity: the Impact of International Agricultural Research*. CAB International, Wallingford, UK.

FAOSTAT (2004) *Changes in Rice Yields in Selected Countries of Asia and Africa*. Food and Agriculture Organization, Rome.

FASID (2003) *Community Survey in Rural Uganda*. Foundation for Advanced Studies in International Development, Tokyo.

Hayami, Y. and Kikuchi, M. (1982) *Asian Village Economy at the Crossroads*. The Johns Hopkins University Press, Baltimore, Maryland.

Hayami, Y. and Otsuka, K. (1993) *The Economics of Contract Choice: an Agrarian Perspective*. Clarendon Press, Oxford, UK.

Hayami, Y. and Otsuka, K. (1994) Beyond the Green Revolution: agricultural development strategies into the new century. In: Anderson, J. (ed.) *Agricultural Technology: Policy Issue for International Community*. CAB International, Wallingford, UK.

Hayami, Y. and Ruttan, V. (1985) *Agricultural Development: an International Perspective*. The Johns Hopkins University Press, Baltimore, Maryland.

Hossain, M., Bose, M.L. and Mustafi, B.A.A. (2003) *Adoption and Productivity Impact of Modern Rice Varieties in Bangladesh*. FASID workshop on the Green Revolution in Asia and its applicability to sub-Saharan Africa, Durban. Foundation for Advanced Studies in International Development, Tokyo.

IFDC (2001) *A Strategic Framework for African Agricultural Input Supply System Development*. International Fertilizer Development Center, Mussel Shoals, Alabama.

IRRI (1966) *Annual Report*. International Rice Research Institute, Los Banos, Philippines.

IRRI (2003) *Diffusion of MVs of Rice in India by Region and in Bangladesh in terms of Percentage of Rice Harvested Area*. International Rice Research Institute, Los Banos, Philippines.

Janaiah, A. (2002) *Green Revolution in India*. FASID workshop on the Green Revolution in Asia and its applicability to sub-Saharan Africa, Durban. Foundation for Advanced Studies in International Development, Tokyo.

Jatileksono, T. and Otsuka, K. (1993) Impact of modern rice technology on land prices: the case of Lampung in Indonesia. *American Journal of Agricultural Economics* 75, 652–665.

Jayne, T.S., Yamano, T., Weber, M.T., Tschirley, D., Benfica, R., Chipoto, A. and Zulu, B. (2003a) Smallholder income and land distribution in Africa: implications for poverty reduction strategies. *Food Policy* 28, 253–275.

Jayne, T.S., Govereh, J., Wanzala, M. and Demeke, M. (2003b) Fertilizer market development: a comparative analysis of Ethiopia, Kenya, and Zambia. *Food Policy* 28, 293–316.

Lipton, M. and Longhurst, R. (1989) *New Seeds and Poor People*. Unwin Hyman, London.

Ndjeunga, J. and Bantilan, C. (2003) *Uptake of Improved Technologies in the Semi-arid Tropics of West Africa: Why is Agricultural Transformation Lagging behind*? FASID workshop on the Green Revolution in Asia and its applicability to sub-Saharan Africa, Durban. Foundation for Advanced Studies in International Development, Tokyo.

Otsuka, K. (2004) Efficiency and equity effects of land markets. In: Evenson, R. and Pingali, P. (eds) *A Handbook of Agricultural Economics*, Volume 3. Elsevier, Amsterdam.

Otsuka, K. and Place, F. (2001) *Land Tenure and Natural Resource Management: a Comparative Study of Agrarian Communities in Asia and Africa*. The Johns Hopkins University Press, Baltimore, Maryland.

Otsuka, K., Gascon, F. and Asano, S. (1994a) Second generation MVs and the evolution of the Green Revolution: the case of Central Luzon, 1966–1990. *Agricultural Economics* 10, 283–295.

Otsuka, K., Gascon, F. and Asano, S. (1994b) Green Revolution and labour demand in rice farming: the case of Central Luzon, 1966–90. *Journal of Development Studies* 31, 82–109.

Philippine Rice Research Institute (2000) *Rice Yield Trends in the Philippines by Variety and Ecosystem: Rice Statistics Handbook, 1970–97*. Philippine Rice Research Institute and Department of Agriculture of the Philippines.

Pingali, P., Hossain, M. and Gerpacio, R. (1997) *Asian Rice Bowls: the Returning Crisis?* CAB International, Wallingford, UK.

Quisumbing, A.R., Estudillo, J.P. and Otsuka, K. (2004) *Land and Schooling: Transferring Wealth across Generations*. The Johns Hopkins University Press, Baltimore, Maryland.

Quizon, J. and Binswanger, H. (1989) Modeling the Impacts of Agricultural Growth and Government Policy on Income Distribution in India. *World Bank Economic Review* 1, 103–148.

Renkow, M. (1993) Differential technology adoption and income distribution in Pakistan: implications for research resource allocation. *American Journal of Agricultural Economics* 75, 33–43.

Ruttan, V. (1977) The Green Revolution: Seven Generalizations. *International Development Review* 19, 16–23.

Sserunkuuma, D. (2003) *The Adoption and Impact of Improved Maize and Land Management Technologies in Uganda*. FASID workshop on the Green Revolution in Asia and its applicability to sub-Saharan Africa. Foundation for Advanced Studies in International Development Tokyo.

Staal, S. and Kaguongo, W. (2003) *The Uganda Dairy Sub-sector: Targeting Development Opportunities*. International Livestock Research Institute, Nairobi.

Staal, S., Owango, M., Muriuki, H., Kenyanjui, M., Lukuyu, B., Njoroge, L., Njubi, D., Baltenweck, I., Musembi, F., Bwana, O., Muriuki, K., Gichungu, G., Omore, A. and Thorpe, W. (2001) *Dairy Systems Characterisation of the Greater Nairobi Milk Shed*. SDP Research Report. MOARD Ministry of Agriculture and Rural Development, Nairobi, 73.

Waithaka, M.M., Nyangaga, J.N., Staal, S.J., Wokabi, A.W., Njubi, D., Muriuki, K.G., Njoroge, L.N. and Wanjoh, P.N. (2002) *Characterization of Dairy Systems in the Western Kenya Region*. SDP Collaborative Research Report. Small Dairy Project, Nairobi.

Yamano, T. and Jayne, T.S. (2004) Measuring the impacts of working-age adult mortality on small-scale farm households in Kenya. *World Development* 32, 99–119.

14 Conclusions and a Look Ahead

Tunji Akande,[1] Göran Djurfeldt,[2] Hans Holmén,[3] and Aida C. Isinika[4]
[1]*Agriculture and Rural Development Department, Nigerian Institute of Social and Economic Research (NISER), Ibadan, Nigeria;* [2]*Department of Sociology, Lund University, Lund, Sweden;* [3]*Department of Geography, Linköping University, Linköping, Sweden;* [4]*Institute of Continuing Education, Sokoine University of Agriculture, Morogoro, Tanzania*

Instead of a summary, this concluding chapter attempts to spell out the implications of the findings in this book. An African-Swedish team of authors will address a number of questions tending to recur in the debate on the relevance of the Asian experience for the Green Revolution in Africa. In the following, seven questions are posed and short answers are given, summarising our results and their implications for the debate.

What are the Important Lessons to be Learnt from the Asian Green Revolution (or Revolutions)?

A fundamental argument made in this book is that the Green Revolution was not merely a 'package of technology'. We have used a holistic model stressing that Asian Green Revolutions were *state-driven*, but that they provided for important roles for the private sector and, most importantly, included the smallholders in the process (Djurfeldt, Chapter 2, this volume).

Furthermore, we stress the geopolitical situation facing Asian governments from the mid-1960s and onwards. They did not opt for Green Revolutions in the 1960s out of enlightenment or altruism. All faced serious threats, which could have removed them from power and even threatened their very physical survival (conflict with China and Pakistan in India, simmering conflicts in Taiwan and Korea). The threat of famines loomed, as did the fear that food scarcity could lead to uprisings or communist revolutions. Ruling elites felt that, at the very least, they had to make sure that their constituencies had enough food to stay calm. These circumstances translated into far-reaching modernization programmes often propagated under a nationalistic rhetoric. Some governments used harsh methods to spread new yield-boosting technologies. Others won elections on promises to improve food crop agriculture. All had in common Green Revolution policies deliberately aimed at including the smallholders; if not, one could hardly have talked about revolutions. This inclusiveness gave governments widespread legitimacy and strengthened the various states.

The Asian Green Revolutions were concentrated on the major staple crops and moreover were first initiated in high-potential areas, where returns to investments were higher and, hence, made possible further investments elsewhere. From these core areas they spread spatially into other areas (and crops). Had the Asian governments at the time instead diverted investments, extension,

 The African Food Crisis (eds G. Djurfeldt, H. Holmén, M. Jirström and R. Larsson)

etc. to the most remote regions and the most place-specific crops, the Asian Green Revolution would never have come about.

What were the consequences of the Asian Green Revolutions for food security in the region? As brought out in Chapters 3–4, it is clear that, since the late 1950s in China and with the exception of North Korea more recently, Asia has avoided major famines. Countries that were food scarce then no longer are, and several have turned into net exporters. It is difficult to see how this could have been achieved without a Green Revolution. In the terms of this book, without state-driven efforts to promote intensification of food crop production, relying on smallholders and mediated by markets, famine would still have been the order of the day in Asia.

It is a widespread myth about the Asian Green Revolution that it worsened ecological crises and increased poverty and inequality (Shiva, 1991; deGrassi and Rosset, 2003). Although this issue has not been dealt with in this book, there is sufficient evidence that this myth does not stand the test of empirical evidence. There exists today a large body of literature which shows that on the whole the Asian Green Revolution has been scale neutral and, in fact, smallholders have tended to benefit rather than lose (Lipton, 1989; Hazell and Ramasamy, 1991; Jirström, 1996; Djurfeldt, 2001; Mosley, 2002). Moreover, as demonstrated by Otsuka and Yamano (Chapter 13, this volume), initial regional income inequalities due to uneven implementation were overcome by migration and interregional factor market adjustments. A growing number of researchers find that the Green Revolution can be an effective pro-poor development strategy.

Ecologically, the results have been mixed. This is natural, given that no strategy is entirely good or bad, and that there are trade-offs in all decision-making. Asian *riziculture*, like its counterparts in the West, faces problems of nutrient and pesticide leakage, and of salinization where irrigation systems are poorly managed or designed. Better management, continued research, including research on genetically engineered crops (GMOs) have the potential to reduce problems with leakage and pesticide residuals as more crops are made to develop pest-resistance.

Adversities notwithstanding, the Asian Green Revolution has also had major positive ecological effects. As pointed out by Borlaugh – the 'father of the green revolution' – 'the high yields of the Green Revolution . . . had a dramatic conservation effect: saving millions of acres of wild-lands all over the Third World from being cleared for more low-yield crops' (Borlaugh, 2002). Thus, 'if Asia's average cereal yields of 1961 (930 kg/ha) would have been maintained, the world would have needed nearly an additional 600 million hectares of the same quality to realize the total harvest of 1997' (Borlaugh, 2000). One might ask: 'Would there have been any forests left in Asia without the Green Revolution?' While such counter-factual questions have but speculative answers, they are sobering to reflect on!

Although monoculture is often presented as a particularly disastrous effect of the Green Revolution, it is a fact that in Asia monoculture of rice pre-dates the Green Revolution by centuries. Actually, Asian agriculture has been found to be more ecologically diverse after the Green Revolution than it was before it (Dawe, 2003, in deGregori, 2004).

It is for reasons like these that we found it urgent to consider the relevance of the Green Revolution also for sub-Saharan Africa.

Are the Agroecological, Demographic and Technological Obstacles to African Agricultural Development a Hindrance to Sub-continental Food Security?

Often arid and always diverse, African agroecologies are commonly seen as hindrances, but should rather be viewed as limitations. This is evidenced, e.g. by the recorded yield-gaps, which indicate a vast potential in African smallholder agriculture (Larsson, Chapter 7, this volume). Appropriate technology is largely available 'on the shelf'. With more support for crop research (e.g. through the Consultative Group on International Agricultural Research, CGIAR) more are 'in the pipeline'. During the last decades, Green

Revolution technologies have become more diverse, include more crops (and crop varieties) and are adapted to a wider range of agroecologies than was previously the case. Thus, technologies are now more Africa-friendly than they used to be (Haggblade, Chapter 8, this volume).

Although irrigation cannot be developed on a massive scale (as in parts of Asia), it can still make significant contributions to food security and agricultural development in sub-Saharan Africa. With irrigation it is possible to extend the cropping season, compensate unreliable rainfall patterns, control soil moisture and allow new crops to be grown both for consumption and for the market.

According to FAO statistics, only about 3% of the cultivated area is currently irrigated. There are reasons to believe that this is an under-estimate and in our sample we record around 7%. Our teams encountered numerous small-scale systems, built and managed by farmers, which are likely to be under-represented in official statistics. There is plenty of evidence that these types of systems continue to expand, driven by commercial opportunities in vegetables and off-season crops like green maize (Larsson, Chapter 7, this volume).

Technologies – high-yielding varieties, drought tolerant and pest resistant seeds, fertilizer, etc. – are available and peasants want them. Adoption rates are often high, not infrequently on a par with or even exceeding those in Asia in the early years of Asia's Green Revolution. A serious problem, however, is the price of fertilizer. African smallholders presently pay the highest prices in the world for inorganic fertilizers with serious consequences for the performance of other Green Revolution technologies, let alone food security (Holmén, Chapter 6 and Larsson, Chapter 7, this volume).

Demography is not the problem – poverty is, but poverty has other causes (e.g. lack of infrastructure and market integration). Sub-Saharan Africa has long been considered under-populated and it is only recently that densities are approaching those in Asia when the Asian Green Revolutions took off. Following Boserup (1965), it is in such a situation that intensification on a broad scale is likely to happen. 'Islands of intensification' (Widgren and Sutton, 1999) have been there all the time but only in more densely populated areas. Thus, whereas low density of population may have constituted an obstacle in the past, it is now rapidly being removed. Time is working *for* a Green Revolution also in Africa.

Can African Smallholders take the Subcontinent towards Food Security or are they so much Constrained by Poverty and by Engagements outside Agriculture that they are, like Classical Marxists used to Claim, a Species Threatened by Extinction?

Left to themselves, African smallholders are unlikely to achieve food security or to successfully engage in spontaneous, trial-and-error types of intensification. The average age of a Nigerian smallholder farmer is about 52 years (NISER, 2003), indicating that the farmer has little physical ability and motive energy left for farm activities. The more vigorous youths who are expected to replace the elderly farm operators are often unwilling to take to farming because of the drudgery and the poor returns from farming. We presently witness shrinking farm sizes and declining incomes among a large proportion of the farm households. Most small-scale farmers have no surplus to sell and quite a few are forced to sell due to distress. Many do not meet their subsistence requirements by their own production. With poor opportunities outside agriculture, they resort to soil-mining practices and often struggle to prevent yields from declining rather than increasing.

With adequate support it is quite possible that African smallholders will arrive at food security and food self-sufficiency from 'their own' effort. Even today, our investigation revealed major yield-gaps among the smallholders (Larsson, Chapter 7, this volume). State efforts to drive the intensification of agriculture in sub-Saharan Africa have periodically met with substantial success, giving another indication of the potential (Holmén, Chapter 6, this volume). These efforts have, however, only occasionally included the smallholders, Zimbabwe after independence and Ethiopia after the Derg being two of the exceptions. It is difficult to escape the conclusion that the next-to-universal exclusion of the African smallholders from the Green Revolution is due to two persistent myths: (i) the peasant's alleged hostility to change, and (ii) the superiority of large-scale production, and to the grip of these myths on the urban élites, whose influence on agricultural policies is easy to demonstrate (Holmén, Chapter 6, this volume).

Food security, for smallholders as well as for the population at large, requires deeper integration of farmers into the markets, especially for staple foods. This again presupposes demand for their products, as well as extension of the technologies (seed, fertilizer, etc.) that they need in order to produce for sale. Markets, however, do not pop up from nowhere and they take time to become established, as the experience of Structural Adjustment Programmes testifies to. For markets to evolve, it is essential that governments – as they did in Asia a few decades ago – not only oversee the operation of markets but also encourage actors to be active on them. It is not the smallness of the African peasants which constrains them, but the deep economic and political crisis afflicting African agriculture and discouraging smallholders from realizing their potentials.

Will Increased Market Integration of Small-scale Agriculture Increase the Economic and Ecological Crisis in Sub-Saharan Africa?

To judge from the Asian experience summarized above, increased market integration of smallholders and ecological sustainability are not antithetical. It is a green myth that market development and ecology are necessarily in conflict with one another. In fact, much of the ecological crisis in contemporary sub-Saharan Africa is a consequence of the economic crisis, rather than the other way round.

The discussion in preceding chapters clearly shows the strong link between the level of market integration and the total volume of food crop sales as well as yield at the farm level. While there is evidence of significant gains from purely organic farming systems for the production of local staple crops, the *Afrint* data set demonstrates that it is necessary to supplement with inorganic fertilizers to gain sustainably in yields of main staple cereals (Larsson, Chapter 7, this volume).

The *Afrint* studies have also shown that where market opportunities exist for farmers to realize reasonable returns from their labour and capital investments, and the necessary inputs are accessible, they are likely to respond not only by intensifying food crop production in order to meet their own subsistence needs as well as market demand, but also by investing in maintaining or improving land fertility. Where such opportunities are weak or non-existent – i.e. without market integration – much smallholder agriculture is unsustainable, both socially and ecologically. With few incentives and no money to invest, no livestock to produce manure and with fertilizer prices prohibitively high, these peasants are forced into mining the soil of nutrients. Hence, instead of being damaged by being integrated into the market, smallholders suffer when excluded from it.

Can Sub-Saharan Governments Handle Agricultural Development?

Another myth, especially popular among Western academia, politicians and aid agencies, is that African governments are incapable of driving development, let alone maintaining law and order in their territories. Here it should be remembered that exactly the same arguments were made about Asian governments, especially loudly at the very same time that they were initiating their Green Revolution (Djurfeldt and Jirström, Chapters 3 and 4, this volume).

African governments are not incapable of driving agricultural development, but their capacity obviously varies, both currently and in recent history. What regularly appears to be forgotten is that African political leaders have had priorities other than their critics.

Much recent critique of the African state is ill-founded in the sense that it presupposes one 'good policy' which is believed to be applicable everywhere and under all circumstances. It does not consider the various situations in which states (or peasants, for that matter) find themselves. States have different and shifting imperatives as well as constraints. For decades post-colonial governments in sub-Saharan Africa did not experience strong pressures to take the lead in agricultural development, especially not in the inclusive way that a Green Revolution demands.

In the first decades after independence, food was not a great problem and for smallholders extensification was still an option. Thus there was room for other priorities on the part of the state, as well as of the peasants. National leaders, to some extent, could live a life quite detached from their 'constituencies'. This was made all the easier during the Cold War when governments received 'development aid' irrespective of whether they had any developmental ambitions or not. This external life-line helped in keeping governments in power and allowed them quite selectively to disseminate spoils (credit, subsidies, strategic inputs, etc.) among a minority of supporters.

The situation is different today. In large parts of sub-Saharan Africa the land frontier has been reached or is about to be reached. Intensification thus has to substitute for extensification. At the same time, the external lifeline has been weakened. SAP and declining aid have deprived governments of resources previously used in co-optation policies. To an increasing degree African governments, if they want to remain in power, have to make serious efforts to develop the internal resources of their countries. With smallholders making up an overwhelmingly large part of the population, implementation of Green Revolution policies seems a natural option. It is therefore no surprise that we have documented signs that governments in some sub-Saharan countries 'obstruct' donors' demands of a reduced role for governments and instead resume a more active role in promoting food crop agriculture.

Can Sub-Saharan Africa Export Itself out of its Agricultural Crisis?

The stance of donors in general, and the IMF and the World Bank in particular, following the implementation of Structural Adjustments in many African countries, is to emphasize food security rather than food self-sufficiency. Food security implies that resources should be directed at developing those sectors where countries have a comparative advantage and obtain their food supplies from the global market. The implication is that sub-Saharan Africa should refrain from developing national or regional self-sufficiency in staples, unlike what Asia did, and unlike what the industrialized countries currently are doing, if not preaching.

Under present conditions, it seems highly unlikely that a 'food security' policy would actually lead to African food *security*. This is so for several reasons. First, due to European and US subsidies to their own farmers, the world market for cereals is saturated and world market prices are artificially low. As long as such unfair policies prevail, grain exports are obviously no option for sub-Saharan Africa. Second, rich-country protectionism annually deprives poor-country producers of billions of dollars of would-be export earnings (e.g. cotton, sugar). Third, for some non-food export crops limited demand (in rich countries) and rapidly increasing supply outside Africa dramatically reduce profitability (coffee prices have been reduced by 70% in only a few years since Vietnam became a major exporter) and put former exporters out of business. In January 2004, for example, Kenyan coffee growers threatened to cut down their coffee trees because coffee production is no longer profitable for them (*Daily Nation*, 30 January, 2004).

Much of the imports into the EU and the US is controlled by big supermarket chains. Since the private sector in many African countries is weak and infrastructure often is inadequate, African would-be exporters have difficulty in competing in world markets. They are at a disadvantage when it comes to meeting prescriptions about quality, packaging, timeliness etc. This effectively puts most African peasants out of the export option. At the same time, rich-country dumping of their own subsidized surpluses in poor countries deprives peasants in sub-Saharan Africa of most incentives to invest in and modernize agriculture. This further reduces their export potential. Policy prescriptions by donor countries, the World Bank and the IMF apparently have not enhanced Africa's export potential. UNCTAD (2003:7) found that most sub-Saharan African countries 'have been unable to diversify into manufactured exports'. Hence, 'on the whole, Africa's share in world exports fell from about 6% in 1980 [pre-SAP] to 2% in 2002 [post-SAP]' (UNCTAD, 2003:1).

Whereas many constraints to African agricultural exports remain internal to the continent, the answer to the question,'Can Africa export itself out of its agricultural crisis?' to a large extent depends, not on the African actors or leadership, but upon measures taken by political institutions outside Africa. Are they willing to create markets for African exporters?

In the meantime, African governments have reasons to overhaul their import policies and look closer at the room that the WTO gives them for protecting their own domestic production of staples.

What is the Relevance of the Asian Model for Sub-Saharan Africa?

Too many attempts have been made in Africa to copy the Asian Green Revolution. It cannot be copied. As is often pointed out, conditions are entirely different, both in terms of agroecology and in terms of economic, political and global circumstances. The African Green Revolution is and must be different, as the 'limping' development in Africa amply shows (Holmén, Chapter 6, this volume).

That it cannot be copied does not mean that the Asian Green Revolution is irrelevant. We have tried to show that prevalent definitions of the Asian Green Revolution are too narrow and too much focused on technology. This narrow focus prevents one from discovering the true relevance of the Asian experience.

The research reported upon in this volume corroborates the relevance of the Asian Green Revolution. As there was in Asia, there is scope in Africa for a state-driven, market-mediated and small farmer based Green Revolution. For it to progress from a one-legged limp to a two-legged stride requires African governments to get up walking.

Currently, too many governments do not own their agricultural policies. To be effective, governments have to invest in building the infrastructure and the institutions needed to better integrate smallholders into markets. They further have to invest in 'farmer friendly' technologies. Although present geopolitical conditions are perhaps more constraining than those facing an earlier generation of Asian leaders, there is still room for action. It can be made bigger by creative interventions on the part of the donors.

References

Borlaugh, N. (2000) Biotechnology will be the salvation of the poorest. *International Herald Tribune* 15 March, 2000.

Borlaugh, N. (2002) We can feed the world. Here's how. *The Wall Street Journal* http://www.ifpri.org/media/innews/2002/051302.htm (13 May, 2003).

Boserup, E. (1965) *Conditions of Agricultural Growth*. George Allen & Unwin, London.

deGrassi, A. and Rosset, P. (2003) A new Green Revolution for Africa? Myths and realities of agriculture, technology and development. Open Computing Facility (23 September, 2003).

deGregori, T.R. (2004) Green Myth vs. the Green Revolution. *The Daily Nation* http://www.butterfliesandwheels.com/printer_friendly.php?num=50 (7 March, 2004).

Djurfeldt, G. (2001) *Mera Mat: Att brödföda en Växande Befolkning*. Arkiv förlag, Lund, Sweden.

Hazell, P.B. and Ramasamy, C. (1991) *The Green Revolution Reconsidered: the Impact of High-yielding Varieties in South India*. The Johns Hopkins University Press, Baltimore, Maryland.

Jirström, M. (1996) *In the Wake of the Green Revolution. Environmental and Socio-economic Consequences of Intensive Rice Agriculture – the problems of Weeds in Muda, Malaysia*. Lund University Press, Lund, Sweden.

Lipton, M. (1989) *New Seeds and Poor People*. The Johns Hopkins University Press, Baltimore, Maryland.

Mosley, P. (2002) The African green revolution as a pro-poor policy instrument. *Journal of International Development* 14, 695–724.

NISER (2003) *NISER Annual Survey of Crop Production Conditions in Nigeria*. Nigerian Institute for Social and Economic Research, Ibadan.

Shiva, V. (1991) *The Violence of the Green Revolution: Third World Agriculture, Ecology and Politics*. Zed Books, London; Atlantic Highlands, New Jersey, USA.

UNCTAD (2003) *Economic Development in Africa – Trade Performance and Commodity Dependence*. United Nations, New York and Geneva.

Widgren, M. and Sutton, J.E.G. (1999) 'Islands' of intensive agriculture in the East African Rift and highlands: a 500-year perspective. African environments past and present. Department of Human Geography, Department of Physical Geography, Stockholm University, Stockholm.

Index